Euphemia Vale Blake

Arctic experiences

Containing Capt. George E. Tyson's wonderful drift on the ice-floe

Euphemia Vale Blake

Arctic experiences
Containing Capt. George E. Tyson's wonderful drift on the ice-floe

ISBN/EAN: 9783337111007

Printed in Europe, USA, Canada, Australia, Japan

Cover: Foto ©Andreas Hilbeck / pixelio.de

More available books at **www.hansebooks.com**

ARCTIC EXPERIENCES:

CONTAINING

CAPT. GEORGE E. TYSON'S

WONDERFUL DRIFT ON THE ICE-FLOE,

A HISTORY OF THE POLARIS EXPEDITION,

THE

CRUISE OF THE TIGRESS,

AND RESCUE OF THE POLARIS SURVIVORS.

TO WHICH IS ADDED

A GENERAL ARCTIC CHRONOLOGY.

Edited by E. VALE BLAKE.

NEW YORK:
HARPER & BROTHERS, PUBLISHERS,
FRANKLIN SQUARE.
1874.

Entered according to Act of Congress, in the year 1874, by
HARPER & BROTHERS,
In the Office of the Librarian of Congress, at Washington.

PREFACE.

A LEADING object of this work is to present, in a popular form, the entire history of the Polaris Exploring Expedition, not only giving the valuable results accomplished by it, but going deep enough into causes to trace out the weak points in the organization of the party—indicating, without fear or favor, those elements of disintegration which were at work from the outset, calculated to impair, though it could not wholly destroy, its efficiency.

The truth must be told, in order that succeeding expeditions may avoid the errors which prevented the development of that *esprit de corps* which is essential to the highest success of exploring parties.

Notwithstanding its unfortunate features, the Polaris Expedition was not a failure, but a grand success; for though far more and better might have been accomplished with a united and harmonious company, we can proudly point to the record of the voyage, in its geographical achievements, as unrivaled; nor do the scientific results interest the world the less because of any cloud resting upon any member of the expedition.

Whoever reads this book to the end will naturally be led to ask, If so much could be accomplished by a divided and disaffected party, what might not be done by a united and properly disciplined body equally well equipped?

In regard to the personal experiences of Captain Tyson, the natural reticence and modesty of that officer has compelled the editor to underestimate and suppress much that is fairly due to him; and the reader is reminded that far more emphasis might fairly be given to his share, in whatever of success was achieved by the Polaris Expedition, but for this peculiarity of the Assistant Navigator,

> "Who, like a statue solid set,
> And moulded in colossal calm,"

appears quite unaware that he has done any thing extraordinary, or more than what any right-minded, honest man would have done under the same circumstances.

All the original data possessed by Captain Tyson (except his journal written on the *Polaris*, which was left on board at the time of the separation) was placed in the hands of the editor, with every necessary verbal explanation, before the former sailed in the *Tigress*. During the interval between his rescue and his return to the Arctic regions in search of the *Polaris* survivors, he recompiled from memory and a few brief notes his lost journal, and we are thus enabled to give it, with but slight verbal variations, from the original diary.

Captain Tyson's "Early Experience" will show that amateur Arctic explorers, physically fitted for the work, may be found in every whaling ship that sails.

In addition to the narrative portion, the introductory chapter contains a general résumé of Arctic experiences; and in the chronology will be found epitomized all the principal events of interest relating to previous and contemporary Polar expeditions, adding greatly to the value of the work as a book of reference.

In accordance with the popular character designed, scientific terms and mere details of work performed have been avoided. These will be published in other forms, for the special benefit of students and scientists.

In conclusion, we have only to express our thanks to those who have kindly assisted us by supplying original documents, official or other information, and facilitating our work by gratefully remembered courtesies.

Among those to whom we are greatly indebted are Hon. George M. Robeson, Secretary of the Navy; Hon. John G. Schumacher, of Brooklyn; Prof. Spencer F. Baird, of Washington; Dr. I. I. Hayes, of New York; Mr. Archibald, British Consul at New York; Col. Jas. Lupton, Washington, D. C.; Messrs. J. Carson Brevoort, and S. B. Noyes, of Brooklyn, N. Y.; Messrs. H. R. Bond, T. W. Perkins, E. W. White, and Mr. Barnes, of New London, Ct.

Others might be named but for whose kindly offices our labors would have been greatly embarrassed; and, if their names do not illumine this page, they have none the less shed a bright and cheering light on our progress from the earliest inception of this work to its end, and their many courtesies will ever dwell in our grateful remembrance. E. V. B.

Brooklyn, January 1, 1874.

CONTENTS.

INTRODUCTORY CHAPTER.

The Northern Sphinx.—Arctic Nomenclature.—Geographical Mistakes.—The Hyperboreans.—The Pre-Columbian Era.—Frobisher's Gold.—Gilbert and Others.—Henry Hudson. — Russian Explorers. — Government Rewards. — Early American Enterprise.—The Whaler Scoresby.—Remarkable Land Journeys.—Combined Sea and Land Explorations.—The Era of Modern Discoveries.—Parry's Drift.—Steam first used in the Arctic Seas. — The Magnetic Pole fixed. — Back's Discoveries.—Dease and Simpson.—Rae on Boothia.—Sir John Franklin's Last Expedition.—Relief Parties.—A glorious Spectacle.—First Grinnell Expedition.—Ten Exploring Vessels meet at Beechey Island.—Dr. Kane.—Rumors of Cannibalism.—The Problem of the North-west Passage solved.—Bellot.—Obtuseness of the British Naval Board.—Providential Mental Coercion.—The Forlorn Hope.—Dr. Hayes.—Profit and Loss.—What is the Use of Arctic Explorations?—Remote Advantages. — Ancient Gradgrinds. — Arctic Failures and Successes. — Unexplored Area. — Modern Chivalry.—A pure Ambition..Page 19

CHAPTER II.
CAPTAIN TYSON'S EARLY ARCTIC EXPERIENCE.

Captain Tyson's Reflections on the Ice-floe.—Nativity.—Early Life.—Ships as Whaler.—Death of Shipmate.—Arrives at the Greenland Seas.—The "Middle Ice."—The "North Water."—First Sight of Esquimaux.—The Danes in Greenland.—The Devil's Thumb.—Meets De Haven.—Whales and their Haunts.—A prolonged Struggle with a Whale.—Sailors' Tricks.—Cheating the Mollimokes.—Young Tyson volunteers to winter ashore at Cumberland Gulf.—The Pet Seal.—Life Ashore.—Relieved by the *True Love.*—Is taken to England.—Returns to the Arctic Regions.—Sights the abandoned British Ship *Resolute.*—With three Companions boards the *Resolute.*—Finds Wine in the Glasses.—All have a good Time.—Don the Officers' Uniforms.—Returns to his Ship.—Ships as Second Mate in the *George Henry.*—As First Officer.—As Captain of the Brig *Georgiana.*—Meets Captain Charles F. Hall.—Witnesses and tries to prevent the Loss of the *Rescue.*—Sails as Master of the *Orray Taft,* of New Bedford.—Of the *Antelope.* —Sails to Repulse Bay, and takes the first Whale captured in those Waters.—Again meets Captain Hall, and supplies him with a Boat.—Peculiar Electrical Phenomena at Repulse Bay.—Sails in the Top-sail Schooner *Era.*—Meets Captain Hall, then "in training" with the Esquimaux.—Log-book Records.—Winters ashore at Niountelik Harbor.—Removes from New London to Brooklyn.—Sails in the *Polaris* as Assistant Navigator.. 75

CHAPTER III.
THE POLARIS EXPEDITION.

The North Polar Expedition authorized by Congress.—Captain Hall's Commission.—The *Periwinkle*, afterward *Polaris*, selected.—Letter of Captain Hall's.—Description of the Steamer *Polaris*.—Liberal Supplies.—A patent Canvas Boat.—Books presented by J. Carson Brevoort.—A characteristic Letter of Captain Hall's.—An Invitation to visit him at the North Pole..........................Page 100

CHAPTER IV.

The *Polaris* put into Commission.—Official Instructions to the Commander.—Scientific Directions.—Letter of Captain Hall's.—List of the Officers and Crew..... 107

CHAPTER V.
BIOGRAPHICAL SKETCH OF CAPTAIN HALL.

Nativity and early Life of Charles Francis Hall.—Leaves his native State of New Hampshire and settles in Ohio.—Takes to Journalism.—Attracted by Arctic Literature.—Unsuccessful Effort to join M'Clintock.—Sails for the Arctic Regions in the *George Henry*, of New London.—The Tender *Rescue* and the Expedition Boat lost in a Storm.—He explores Frobisher Bay and Countess of Warwick Sound.—Collects Relics of Franklin's Expedition.—Returns to the United States.—His Theories regarding the Franklin Expedition.—Sails for the North, 1864, in the Bark *Monticello*.—His Discoveries.—Skeletons of Franklin's Men scattered over King William Land.—Annual Reports.—His Life with the Esquimaux.—Return to the United States.—Physical Appearance.—Mental Traits.—In the Innuit Land he did as the Innuits do.—Persevering Efforts to organize the North Polar Expedition.—President Grant personally interested.—"That Historical Flag."—How he would know when he got to the Pole.—His Premonitions.—His last Dispatch... 113

CHAPTER VI.

Dr. Emil Bessel.—Sergeant Frederick Meyers.—Mr. R. W. D. Bryan.—Sidney O. Buddington.—Hubbard C. Chester.—Emil Schuman.—William Morton.—Letter of Captain Hall's.—The *Polaris* sails.—Disaffection on Board.—Meets the Swedish Exploring Expedition.—Favorable condition of the Ice.—United States Ship *Congress* arrives at Disco with Supplies for the *Polaris*.—Insubordination on Board.—Captain Hall's Idiosyncrasy.—He "bids Adieu to the Civilized World".... 129

CHAPTER VII.
NOTES BY CAPTAIN TYSON ON BOARD THE POLARIS.

Captain Tyson's Soliloquy on leaving Harbor.—A Thunder-storm.—Arrive at St. Johns.—Icebergs in Sight.—Religious Services on board the *Polaris* by Dr. Newman, of Washington.—Prayer at Sea.—Esquimau Hans, with Wife, Children, and "Vermin," taken on board.—Firing at Walrus.—The Sailing-master wants to stop at Port Foulk.—The *Polaris* passes Kane's Winter-quarters.—An impassable Barrier of Ice.—Misleading Charts.—The open Polar Sea recedes from Sight.—

Afraid of "Symme's Hole."—*Polaris* enters Robeson Channel.—Surrounded by Icefields.—Council of Officers.—Puerile Fears.—Sir Edward Belcher.—The American Flag raised on "Hall Land."—Seeking a Harbor.—Repulse Harbor.—Thank God Harbor.—Providence Berg.—Housing the Ship for Winter-quarters..Page 141

CHAPTER VIII.

A Hunting-party.—A cold Survey.—Description of Coast-hills.—A Musk-ox shot.—Landing Provisions.—Arctic Foxes.—Captain Hall prepares for a Sledge-journey.—Conversation with Captain Tyson.—Off at last.—Captain Hall "forgets something."—Twenty "somethings."—The Sun disappears.—Banking the Ship.... 152

CHAPTER IX.

Putting Provisions Ashore.—Return of Captain Hall.—"Prayer on leaving the Ships."—Captain Hall taken Sick.—What was seen on his Sledge-journey.—Apoplexy?—M'Clintock's Engineer—Death of Captain Hall.—A strange Remark.—Preparing the Grave.—The Funeral.—"I walk on with my Lantern."—Thus end his ambitious Projects.. 159

CHAPTER X.

Captain Buddington passes to the Command.—Scientific Observations.—The first Aurora of the Season.—Sunday Prayers discontinued.—Dr. Bessel Storm-bound in the Observatory.—Meyers to the Rescue.—An Arctic Hurricane.—Fast to the Iceberg.—Sawing through the Ice.—Electric Clouds.—Pressure of Floe-ice.—The Iceberg splits in two.—The *Polaris* on her Beam-ends.—Hannah, Hans's Wife, and the Children put Ashore.. 166

CHAPTER XI.

Thanksgiving.—A Paraselene.—Dr. Bessel's bad Luck.—"It is very dark now."—Oppressive Silence of the Arctic Night.—The Voracity of Shrimps.—"In Hall's Time it was Heaven to this."—A natural Gentleman.—No Service on Christmas.—The *Polaris* rises and falls with the Tide.—Futile Blasting.—The New Year.—Atmospheric Phenomena.—The Twilight brightens.—Trip to Cape Lupton.—Height of the Tides at Thank God Harbor.................................. 169

CHAPTER XII.

An impressive Discussion.—Daylight gains on the Night.—Barometer drops like a Cannon-ball.—Four mock Moons.—Day begins to look like Day.—The Fox-traps.—The Sun re-appears after an Absence of one hundred and thirty-five Days.—Mock Suns.—Spring coming.—An Exploring-party in Search of Cape Constitution.—A Bear-fight with Dogs.—New light on Cartography.—Tired of canned Meat. 174

CHAPTER XIII.

Sledge *vs.* Boat.—What Chester would do when he got Home.—Photographing a Failure.—Off on a Sledge-journey with Mr. Meyers, Joe, and Hans.—Habits of the Musk-cattle.—Peculiar strategic Position.—Encounter a Herd.—How the Young are concealed.—Dull Sport.—Newman Bay.—Preparing for Boat-journeys.—What does he mean?—Climatic Changes.—Glaciers.—Wonderful Sportsmen.—The Ice thick and hummocky.—A dangerous Leak .... Page 179

CHAPTER XIV.

Two Boat-parties arranged.—A Disaster.—Chester's Boat crushed in the Ice.—The "Historical Flag" lost.—Chester takes the patent Canvas Boat.—Captain Tyson's Boat-party.—Reach Newman Bay.—Dr. Bessel's Snow-blindness.—Drift-wood. —Extinct Glaciers.—Unfavorable Condition of the Ice.—A Proposal rejected.— Return to the Ship............................ 185

CHAPTER XV.

Engineer's Report.—A new Inscription.—A gentle Awakening.—Providence Berg disrupted.—Having "enough of it."—Lost Opportunities.—The Advent of little Esquimau "Charlie Polaris."—Beset near Cape Frazier.—Alcohol Master.— Interruption of his morning "Nip."—Drifting with the Floe.—Pack-ice in Smith Sound.—The Oil-boiler.—The bearded Seal.—Preparations for spending another Winter in the North.—A south-westerly Gale... 190

CHAPTER XVI.

JOURNAL OF GEORGE E. TYSON, ASSISTANT NAVIGATOR ON UNITED STATES STEAMER POLARIS, KEPT ON THE ICE-FLOE.

Adrift.—The fatal Ice Pressure.—"Heave every thing Overboard!"—The Ship breaks away in the Darkness.—Children in the Ox-skins.—First Night adrift.— Snowed under.—Roll-call on the Ice-floe.—Efforts to regain the Ship.—The *Polaris* coming!—A terrible Disappointment.—The overladen Boat.—Three Oars, and no Rudder.—The Ice breaks beneath us.—Drifting to the South-west.—Regain the large Floe.—Hope of regaining the *Polaris* abandoned.—Building Huts. —Native Igloos.—Estimating Provisions.—Locality of the Separation.—Meyers's and Tyson's Opinion.—Two Meals a Day.—Mice in the Chocolate.—Too cold for a Watch.—Too weak to stand firmly.—Hans kills and eats two Dogs.—Natives improvident.—Lose Sight of the Sun.—The Dogs follow the Food............... 197

CHAPTER XVII.

A vain Hunt for Seal.—Pemmican.—The Dogs starving.—Blow-holes of the Seal.— Mode of Capture.—Sight Cary Island.—Hans mistaken for a Bear.—Down with Rheumatism.—One Boat used for Fuel.—The Children crying with Hunger.—Joe the best Man.—The Bread walks off.—One square Meal.—Bear and Fox-tracks.— Effects of lax Discipline.—Joe and Hannah.—Our Thanksgiving-dinner........ 215

CHAPTER XVIII.

Can see the Land.—Hans's Hut.—Nearly dark: two Hours of Twilight.—Economizing Paper.—Northern Lights.—Lying still to save Food.—"All Hair and Tail." —Weighing out Rations by Ounces.—Heavy Ice goes with the Current.—The Esquimaux afraid of Cannibalism.—Fox-trap.—Set a Seal-net—Great Responsibility, but little Authority.—All well, but hungry.—The fear of Death starved and frozen out of me.—The shortest and darkest Day.—Christmas..............Page 225

CHAPTER XIX.

Taking account of Stock.—Hope lies to the South.—Eating Seal-skin.—Find it very tough.—How to divide a Seal *a la* Esquimau.—Give the Baby the Eyes.—Different Species of Seal.—New-year's Day, 1873.—Economizing our Lives away.—Just see the Western Shore.—"Plenty at Disco."—Thirty-six below Zero.—Clothing disappears.—A glorious Sound.—"Kyack! Kyack!"—Starvation postponed.—Thoroughly frightened.—Little Tobias sick.—Oh, for a sound-headed Man!—Four ounces for a Meal.—The Sun re-appears after an Absence of eighty-three Days.. 234

CHAPTER XX.

Belated Joe.—Wrong Calculations.—Drift past Disco.—Beauty of the Northern Constellations.—Hans unreliable.—"Where Rum and Tobacco grow."—Forty below Zero.—An impolite Visitor.—One hundred and third Day on the Ice.—Perseverance of the Natives in hunting.—Hans loses a good Dog.—Beautiful Aurora.—The Mercury freezes.—Too cold for the Natives to hunt.—A little Blubber left.—Trust in Providence.—Effects of Refraction.—Relieving Parties on the Ice.—Our Lunch, Seal-skin with the Hair on.—A natural Death.—One hundred and seven Days without seeing printed Words................................... 246

CHAPTER XXI.

A solemn Entry made in the Journal, in View of Death.—More Security on the Ice-floe than on board the *Polaris*.—Eating the Offal of better Days.—Tobias very low.—Anticipations of a Break-up.—Hope.—Joe, Hannah, and Puney.—"I am *so* hungry."—An interior View of Hans's Hut; his Family.—Talk about reaching the Land.—Inexperience of the Men misleads their Judgment.............. 259

CHAPTER XXII.

Dreary, yet beautiful.—The Formation of Icebergs.—Where and how they grow.—Variety of Form and History.—"The Land of Desolation."—Strength failing.—Travel and Rations.—Unhealthy Influence of mistaken Views.—Managing a Kyack on young Ice.—Secures the Seal.—"Clubbing their Loneliness."—Poor little Puney's Amusement.—Any Thing good to eat that don't poison.—Narwhals, or Sea-unicorns.—A royal Seat.—Hans criticised.—Cleaning House.—"Pounding-day."—Our Carpet.—Lunching by the Yard on Seal's Entrails.—"Oh! give me my Harpoon."—No Clothing fit to hunt in.—Inventory of Wardrobe.—Narwhals useful in carrying off Ball and Ammunition.—Pleasant Sensations in Retrospect.—The Skin of the Nose.—Castles in the Air.—Violent Gale and Snow-storm.—Digging out.—Three Feet square for Exercise.—Dante's Ice-hell..................... 266

CHAPTER XXIII.

Patching up Clothes.—Captain Hall's Rifle.—Cutting Fresh-water Ice for Drink.—Salt-water Ice to season Soup.—Four months' Dirt.—Sun Revelations.—"You are nothing but Bone."—That chronic Snow-drift.—Seal-flipper for Lunch.—Watching a Seal-hole.—Eating his "Jacket."—Dovekies.—The Solace of a Smoke.—Native Mode of cleansing Cooking Utensils.—The West Coast in Sight.—Joe's

Valuation of Seals.—Prospects dark and gloomy.—Bill falls Overboard.—Death to the Front.—Evidences of Weakness.—The Natives alarmed.—Washington's Birthday.—A novel Sledge.—The "right Way of the Hair."—Discussions about reaching Shore ....Page 279

CHAPTER XXIV.

Decide to make the Attempt.—Foiled by successive Snow-storms.—Down to one scant Meal a Day.—Land thirty-five Miles off.—God alone can help us.—Canary-bird Rations.—Bear-tracks.—A Bird Supper.—A Monster Oogjook.—Six or seven hundred Pounds of fresh Meat! Thirty Gallons of Oil!—Oogjook Sausage.—Our Huts resemble Slaughter-houses.—Hands and Faces smeared with Blood.—Content restored.—Taking Observations.—Out of the Weed.—A Present from Joe.—Heat of Esquimaux Huts.—Desponding Thoughts.—"So I sit and dream of Plans for Release."—Terrific Noises portend the breaking up of the Floe.—An unbroken Sea of Ice.—Hans Astray again.—That "Oogjook Liver."—The Steward convinced.—An Ice-quake in the Night.—The Floe breaks twenty yards from the Hut.—Floe shattered into hundreds of Pieces.—Sixty Hours of Ice, Turmoil, and utter Darkness.—The "Floes" become a "Pack."—Storm abates.—Quietly Drifting.—A Choice for Bradford.—Our Domain wearing away.—Twenty Paces only to the Water.—Whistling to charm an Oogjook.—A Relapse into Barbarism 288

CHAPTER XXV.

A Bear prospecting for a Meal.—The Ice in an Uproar.—Seven Seals in one Day.—Spring by Date.—The "Bladder-noses" appear.—Off Hudson Strait.—A Bear comes too close.—A lucky Shot in the Dark.—Description of *Ursus maritimus*.—Milk in the young Seal.—Fools of Fortune.—We take to the Boat.—Rig Washboards.—A desperate Struggle to keep Afloat.—Alternate between Boat and Floe.—Striving to gain the west Shore.—Dead-weights.—Ice splits.—Joe's Hut carried off.—Rebuild it.—Ice splits again, and destroys Joe's new Hut.—Standing ready for a Jump.—Our Breakfast goes down into the Sea.—No Blubber for our Lamps.—The Ice splits once more, separating Mr. Meyers from the Party.—We stand helpless, looking at each other.—Meyers unable to manage the Boat.—Joe and Hans go to his Relief.—All of us but two follow.—Springing from Piece to Piece of the Ice.—Meyers rescued.—He is badly frozen.—Mishaps in the Water.—High Sea running.—Washed out of our Tent by the Sea.—Women and Children stowed in the Boat.—Not a dry Place to stand on.—Ice recloses.—Sea subsides.—Land Birds appear.—No Seal.—Very Hungry.. 305

CHAPTER XXVI.

Easter-Sunday.—Flashes of Divinity.—Meyers's Suffering from want of Food.—Men very Weak.—Fearful Thoughts.—A timely Relief.—Land once more in Sight.—Flocks of Ducks.—Grotesque Misery.—A Statue of Famine.—A desolating Wave.—A Foretaste of worse.—Manning the Boat in a new Fashion.—A Battery of Ice-blocks.—All Night "standing by" the Boat.—A fearful Struggle for Life.—Worse off than St. Paul.—Daylight at last.—Launched once more.—Watch and Watch.—The Sport and Jest of the Elements.—Lack of Food.—Half drowned,

CONTENTS.   13

cold, and hungry.—Eat dried Skin saved for Clothing.—A Bear! a Bear!—Anxious Moments.—Poor Polar! God has sent us Food.—Recuperating on Bear-meat.—A crippled, overloaded Boat.—A Battle of the Bergs.—Shooting young Bladder-noses.—Hoping for Relief.. Page 317

CHAPTER XXVII.

A joyful Sight!—A Steamer in View.—Lost again.—She disappears.—Once more we seek Rest upon a small Piece of Ice.—The Hope of Rescue keeps us awake.—Another Steamer.—We hoist our Colors, muster our Fire-arms, fire, and shout.—She does not see us.—She falls off.—Re-appears.—Gone again.—Still another Steamer.—Deliverance can not be far off.—Another Night on the Ice.—Hans catches a Baby Seal.—"There's a Steamer!"—Very Foggy, and we fear to lose her.—Hans goes for her in his Kyack.—She approaches.—We are saved!—All safe on board the *Tigress*.—Amusing Questions.—A good Smoke and a glorious Breakfast.—Once more able "to wash and be clean."—Boarded by Captain De Lane, of the *Walrus*.—Meyers slowly recovering.—A severe Gale.—Six hundred Seals killed.—Captain Bartlett heading for St. Johns.—The Esquimaux Children the "Lions."—Awaiting the Tailor.—Going Home in the United States Steamship *Frolic* 326

CHAPTER XXVIII.

THE SEARCH FOR THE POLARIS AND THE SURVIVORS OF THE EXPEDITION.

The News of the Rescue.—Captain Tyson and Party arrive at Washington.—Board of Inquiry organized.—Testimony given as to lax Discipline.—The *Juniata*, Commander Braine, dispatched, with Coal and Stores, to Disco.—Captain James Buddington, Ice-pilot.—Captain Braine's Interview with Inspector Karrup Smith, of North Greenland.—*Juniata* at Upernavik.—Small Steam-launch *Little Juniata* essays to cross Melville Bay.—Repelled by the Ice.—President Grant in Council with Members of the National Academy of Sciences.—Purchase of the *Tigress*.—Description of the Vessel.—Necessary Alterations.—List of Officers.—Captain Tyson Acting Lieutenant and Ice-pilot.—A Reporter to the *New York Herald* ships as ordinary Seaman.—Esquimau Joe ships as Interpreter.—Several Seamen belonging to the Ice-floe Company ship in the *Tigress*.—Extra Equipments...... 340

CHAPTER XXIX.

The *Tigress*, Commander James A. Greer, sets sail.—Enthusiasm at her Departure.—Hans and Family as Passengers. — "Knowledge is Power."—Arrive at Tessuisak.—Governor Jansen.—*Tigress* proceeds North.—Approach Northumberland Island.—Not the place of Separation.—Make Littleton Island.—Excitement on Board on hearing Human Voices.—Encampment of the *Polaris* Survivors found.—Commander Greer's Success.—Esquimaux in Possession of the deserted House.—Captain Tyson's Advice to seek the Whalers 350

CHAPTER XXX.

CAPTAIN TYSON'S CRUISE IN THE TIGRESS.

Captain Tyson's Journal on board the *Tigress*. — "Too late." — Fire training on board.—*Mal de mer*.—A tall Story.—Angling for Porpoises with Pork.—A nautical Joke.—Beware of the *Tigress*.—Fog at Sea.—Naïve Comments on Icebergs.—

Tender Hearts among the Blue-jackets.—Illusions.—Aurora.—Whistling to frighten the Bergs.—Splendid Northern Lights.—Heavy Gales.—The Doctor's Clerk.—Two old Whalers.—We leave Night behind us.—Poor Hans's Affliction.—Family returned to Greenland.—The *Tigress* pitching and rolling.—The Fog-blanket.—Cheese for Bait.—An Iceberg turns a Somersault.—A beautiful Display.—A slight Accident.—Meet the Steam-launch.—Official Correspondence with Commander Greer.—Ashore at Littleton Island and Life-boat Cove.—Sounding for the foundered *Polaris*.—Abundance of Food abandoned by the *Polaris* Survivors..Page 356

CHAPTER XXXI.

Homeward-bound.—Fire! Fire!—An honored Custom.—Contrast of the Sailor's Life.—A Set-off to the Midnight Sun.—Heavy Gale.—All want to shoot a Bear.—Executive Officer White the "killing" Man.—A narrow Escape.—Thoughts of Home.—At Upernavik for Repairs.—The Danish and half-breed Girls.—Dress.—Dancing.—A startling Record.—At Goodhavn Harbor.—Captain Tyson visits the *Juniata*.—Continued bad Weather.—Sight Cape Mercy.—The Sea sweeps the Galley.—The Cook disgusted.—Effects of the Gale in the Wardroom.—"At home" in Niountelik Harbor, Cumberland Gulf... 368

CHAPTER XXXII.

A Change for the better.—Repairing Damages.—Company in the Gulf.—Looking for Scotch Whalers.—The Natives bring Deer-meat to the Ship.—Arctic Birds flying South.—Captain Hall's old Protégés.—Demoralization of the Natives of the west Coast.—Collecting "Specimens."—Bad Case of "Stone Fever."—"Time and Tide wait for no Man."—Billy's Curiosities.—Captain Tyson meets his late Rescuer, Captain Bartlett.—Mica Speculation.—Short of Coal.—How we lost our Dinner.—A saltatory Dining-table.—Sight a Scotch Whaler.—Arrival at Ivgitut, South Greenland.—Meet the *Fox*, of Arctic Fame.—Kryolite, Coal, Fish, and another Gale.—Friend Schnider, the fat Dane.—Canaries, Pigeons, etc., domesticated here.—The Crew overworked.—A Hurricane.—Antics of the Furniture.—Force of Sea-waves.. 376

CHAPTER XXXIII.

The Gale abates.—Consultation as to Course.—Useless Cruising.—Start for Home.—More bad Weather.—Land-birds blown out to Sea.—Reminiscences of the Ice-floe Drift.—A narrow Escape.—A black Fog.—Interviewing a Hawk at the Masthead.—Arrive at St. Johns.—News of the *Polaris* Party.—Return to Brooklyn.—What the *Tigress* accomplished.—Lessons in Arctic Navigation.—Bravery of the Officers.—A stormy but agreeable Cruise... 383

CHAPTER XXXIV.

THEORY OF NORTH POLAR CURRENTS.

The Hydrography of Smith Sound.—The Currents forbid the Theory of an "Open Polar Sea."—Movements of the Ice.—A northern Archipelago a reasonable Supposition.—Velocity of Current along the east and west Coasts.—No Current in the Middle.—Experience of the *Polaris*.—Absence of large Bergs in Smith Sound.—Open nearly all Winter.—Radiant Heat preserved by Cloud Strata.—Deflection of the Current at Cape York.—Robeson Channel described.—Land seen from the

Mast-head both east and west.—Coast-line beyond Cape Union.—Two Headlands to the east-north-east of Repulse Harbor.—Absence of Snow on Coast of North Greenland above Humboldt Glacier.—Elevated Plateaus in the Interior.—The Land around Polaris Bay.—Clam-shells at an Elevation of two thousand Feet.—Variegated but odorless Flora.—Animal Life.—Insects.—Skeletons of Musk-cattle...Page 388

CHAPTER XXXV.

How to reach the North Pole.—Smith Sound the true Gate-way.—This course offers the Alternative of Land Travel.—Plenty of Game in Summer.—April and May the Months for Sledging.—Proper Model of Vessel's Hull.—Twenty-five Men enough.—A Tender necessary.—A Dépôt at Port Foulke with a detail of Men.—Ice at Rensselaer Harbor.—Avoid Pack-ice in Smith Sound.—Go direct for west Coast.—Form *Caches* at intervals of fifty Miles.—Deposit Reserve Boats.—Style of Traveling-sledge.—Native preferred.—Selecting Dogs.—Keep them well fed.—Keep Stores on Deck.—Winter as far north as the Ship can get.—How to get out of a Trap.—Provision a Floe, and trust to the Current.—Take your Boats along.—Replenish at *Caches*.—Two Months from a high Latitude sufficient.—It will yet be done... 393

CHAPTER XXXVI.
THE FATE OF THE POLARIS.

The *Polaris* Survivors.—Ship driven to the North-east.—Her Position on the Night of October 15.—Darkness and Confusion.—Anchors and Boats gone.—The Leak gains.—Steam up.—Roll-call on Board.—Lookout for the Floe Party.—Storm abated.—Inspection of Stores.—The *Polaris* fast to grounded Hummocks.—"Let her fill!"—Life-boat Cove.—The *Polaris* left a Legacy to an Esquimau Chief.—She founders in his Sight... 398

CHAPTER XXXVII.
THE FORTUNES OF THE POLARIS SURVIVORS.

Life on Shore.—A House built.—Visitors.—Womanly Assistance.—Scientific Observations.—Amusements.—Old Myoney.—Hunting.—Boat built.—Starting for Home.—A Summer-trip.—Sight a Vessel.—Rescue by Captain Allen, of the *Ravenscraig*.—Romance of the *Polaris* Expedition.—Safe Arrival of all the Survivors at New York.—Consul Molloy... 402

CHAPTER XXXVIII.
SCIENTIFIC NOTES.

The Pacific Tidal Wave.—Meteorological and Magnetic Records.—Glaciers.—Fauna.—Entomology.—Flora.. 410

APPENDIX.. 423

INDEX.. 481

ILLUSTRATIONS.

	PAGE
A FEARFUL STRUGGLE FOR LIFE	*Frontispiece.*
TRACK OF THE POLARIS, AND VELOCITY OF CURRENTS	*Map.*
SEBASTIAN CABOT	26
HALL DISCOVERING FROBISHER RELICS	28
BARENTZ'S WINTER-QUARTERS	29
HENRY HUDSON	31
BARON VON WRANGEL	33
WILLIAM SCORESBY	36
CAPTAIN PARRY	39
SIR JOHN ROSS	46
SIR JOHN FRANKLIN	52
ADVANCE AND RESCUE	57
ARCTIC DISCOVERY SHIPS	58
FINDING REMAINS OF SKELETONS IN A BOAT	65
DR. KANE	68
DR. HAYES	70
ESQUIMAU WOMAN'S KNIFE	74
CAPTAIN TYSON	76
CAPTURING THE SEAL	79
CONGRESS AND POLARIS AT GOODHAVN	82
THE "DEVIL'S THUMB"	84
EIDER-DUCKS	88
"EVERY THING PRESENTED A MOULDY APPEARANCE"	94
THE POLARIS	101
A SNOW-SQUALL	106
CHARLES FRANCIS HALL	114
JOE, HANNAH, AND CHILD	118
RELICS OF FRANKLIN'S EXPEDITION	120
FAC-SIMILE OF CAPTAIN HALL'S WRITING	128
DR. EMIL BESSEL	129
SIDNEY O. BUDDINGTON	130
HUBBARD C. CHESTER	131
EMIL SCHUMAN	132
WILLIAM MORTON	133
UPERNAVIK	138
THE FISCANAES PILOT	143
ICE BREAKING UP	151
POLARIS AT CAPE LUPTON—WINTER-QUARTERS, 1871–'72	152
CAPTAIN HALL'S SLEDGE-JOURNEY	156

ILLUSTRATIONS.

	PAGE
GOTHIC ICEBERG	158
BURIAL OF CAPTAIN HALL	164
HEAD AND ANTLERS OF THE ARCTIC REINDEER	173
SEALS	178
MUSK-OX	180
ESQUIMAU DOG	184
ARCTIC WOLVES	189
GRAVE OF CAPTAIN HALL	190
THE LUMME OF THE NORTH	196
"THE SHIP BROKE AWAY IN THE DARKNESS, AND WE LOST SIGHT OF HER IN A MOMENT"	199
NATIVE LAMP	210
THE GREAT AUK	214
JOE WATCHING SEAL-HOLE	216
A PERILOUS SITUATION	224
AN AURORA	227
PLACING STORES ON THE ICE	233
CAPTAIN TYSON IN HIS ARCTIC COSTUME	239
GOING THROUGH AN ICEBERG	245
HANS, WIFE, AUGUSTINA, AND TOBIAS	249
ARCTIC HOSPITALITY	258
HANNAH AND JOE PLAYING CHECKERS	262
SURROUNDED BY ICEBERGS	265
HANS GOING FOR A SEAL ON YOUNG ICE	270
NARWHAL	273
BREAKING UP OF ICE-RAFT	300
AN ESQUIMAU PILOT	304
OOMIAK, OR WOMAN'S BOAT	316
ICE-DRIFT OF THE TYSON PARTY	327
THE RESCUE	329
THE COMPANY WHO WERE ON THE ICE-DRIFT WITH CAPTAIN TYSON	338
THE JUNIATA	342
THE TIGRESS	345
GOVERNOR JANSEN AND FAMILY	352
POLARIS CAMP, 1872-'73	354
SCENE IN SOUTHERN GREENLAND	367
ENCAMPMENT NEAR IVGITUT	375
KRYOLITE MINE	380
A SUMMER ENCAMPMENT	401
THE LATEST STYLE	409

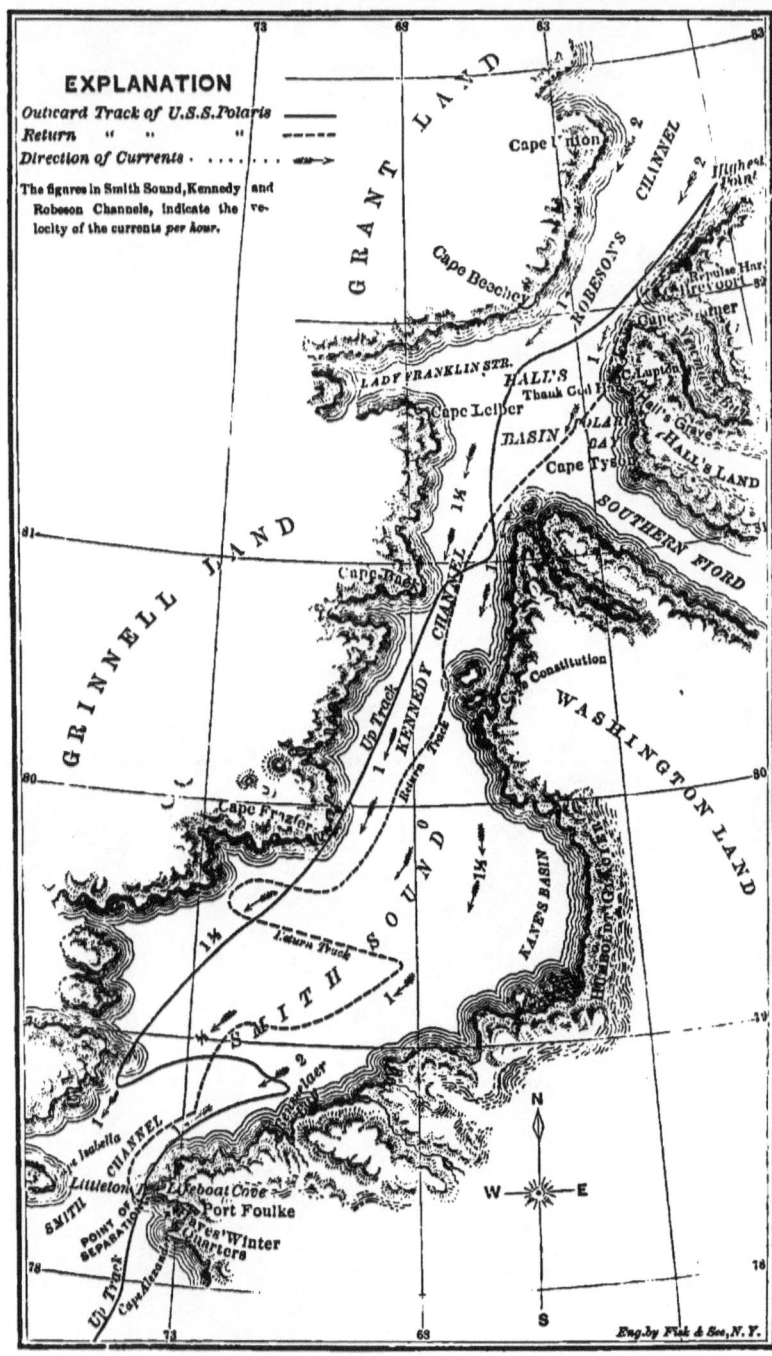

MAP SHOWING THE TRACK OF UNITED STATES STEAMER "POLARIS," AND VELOCITY OF CURRENTS.

ARCTIC EXPERIENCES

BY

LAND AND SEA.

INTRODUCTORY CHAPTER.

The Northern Sphinx.—Arctic Nomenclature.—Geographical Mistakes.—The Hyperboreans.—The Pre-Columbian Era.—Frobisher's Gold.—Gilbert and Others.—Henry Hudson. — Russian Explorers. — Government Rewards. — Early American Enterprise.—The Whaler Scoresby.—Remarkable Land Journeys.—Combined Sea and Land Explorations.—The Era of Modern Discoveries.—Parry's Drift.—Steam first used in the Arctic Seas. — The Magnetic Pole fixed. — Back's Discoveries.—Dease and Simpson.—Rae on Boothia.—Sir John Franklin's Last Expedition.—Relief Parties.—A glorious Spectacle.—First Grinnell Expedition.—Ten Exploring Vessels meet at Beechey Island.—Dr. Kane.—Rumors of Cannibalism.—The Problem of the North-west Passage solved.—Bellot.—Obtuseness of the British Naval Board.—Providential Mental Coercion.—The Forlorn Hope.—Dr. Hayes.—Profit and Loss.—What is the Use of Arctic Explorations?—Remote Advantages. — Ancient Gradgrinds. — Arctic Failures and Successes. — Unexplored Area. — Modern Chivalry.—A pure Ambition.

THE invisible Sphinx of the uttermost North still protects with jealous vigilance the arcana of her ice-bound mystery. Her fingers still clutch with tenacious grasp the clue which leads to her coveted secret; ages have come and gone; generations of heroic men have striven and failed, wrestling with Hope on the one side and Death on the other; philosophers have hypothesized, sometimes truly, but often with misleading theories: she still clasps, in solemn silence, the riddle in her icy palm — remaining a fascination and a hope, while persistently baffling the reason, the skill, and the courage of man.

Skirmishers have entered at the outer portals, and anon retreated, bearing back with them trophies of varying value. Whole divisions, as of a grand army, have approached her domains with all the paraphernalia of a regular siege, and the area of attack been proportionably widened; important breaches have been effected, the varied fortunes of war befalling the assailants; some retaining possession of the fields they have won; some falling back with but small gain; others, with appalling loss and death, have vainly sought escape and safety from her fatal toils. Nor has the citadel been won. "UNDISCOVERED" is still written over the face of the geographical pole.

Yet as brave men as ever trod the earth or sailed the wide salt seas have time and again returned to the encounter, defying this Polar tyrant, who hurls from her mysterious abode the vengeful storms of wind and hail and snow; smiting some with ice-blindness, and others with the dread consuming scurvy; while others still she decoys into the perils of a frozen solitude whence there is no return, and the terrors of starvation meet them; for still others she spreads the treacherous crevasse, or sets upon them the cruel, unpitying savage; while the rotting ribs of noble vessels lie scattered through all her borders. Worst fate of all, some noble souls have been sent empty-handed back, to die of disappointed hopes, and grand ambitions quenched! Hitherto repulsing all—victor over all—save the indomitable will; but that, enduring, man shall yet overcome even the terrific elemental forces with which she defends her domain.

For, strange as it may seem, while she defies, she tempts; while baffling effort, she encourages hope; while foiling the bravest, she holds out inducement to renewed attack. As with one hand uplifted, she swears, "Hitherto shalt thou come, and no further," with the other she beckons delusively to the next aspirant. So that each brave enthusiast says to himself, "I shall conquer—she has betrayed all others; I shall win;" and thus the hope of final success never has been and never will be quenched, until full fruition satisfies the questionings of science and the longings of adventure.

That we may be the better prepared to judge what will be the future of Arctic exploration, we will take a retrospective view of what the ancient mariners of other centuries have accomplished, and what the scientists of our own age have endured, in the hopes of solving the Polar mystery. And we may be assured that terrors which could not repel the little shallops of the early adventurers, will not dismay the better-equipped explorers of the present and the future.

ARCTIC NOMENCLATURE.

But it is well to premise that, unless the reader is familiar with the details of Arctic explorations, he is very apt to get bewildered with the mixed nomenclature which he encounters, with each successive publication; and this is no fault of the authors, but the result of peculiar circumstances and conflicting vanities, added to the fact that the Arctic region is unlike every other portion of the earth, except its southern antipodes, in the fact that much of its surface, both land and water, has no aboriginal names, being destitute of inhabitants; while those places which have received names from successive explorers have, in many instances, been given titles unknown to the old geographers.

This has sometimes arisen from the fact that what has been named as an island turns out, on more accurate survey, to be a peninsula, or a portion of the main-land, and, of course, the reverse experience is liable to occur. What some early voyager has called a strait or a channel, a later explorer determines to be a bay, and then that gets a new name. But what complicates the Polar geography and hydrography much more than these simple reversals of contour or superior accuracy, results from the practice—especially with modest travelers, of naming their discoveries for friends and patrons—often obscure in every thing but wealth; and then, later in history, the explorer's own name is considered more suitable, and influential admirers bring it to the front and affix it, like the writing on an ancient palimpsest, over those which he selected—the patron's name giving way, with various prefixes or suffixes, to that of the discoverer.

Thus one needs to be familiar with each successive addition to Arctic literature; indeed, to be able to carry in the mind's eye the contour of headlands, islands, shore lines, gulfs, bays, and rivers, in order to be enabled to trace the minuter history and daily movements of any particular party. To exemplify. In a map published in a work on Arctic affairs, just previous to Parry's first voyage, Baffin Bay was treated as a "phantom," and found no place, though it had been accurately described by the discoverer. In the chart furnished to Sir John Franklin, in 1845, the name of Barrow Strait is given to all the water-course extending from Lancaster Sound to Banks Land. In a map drawn from official documents, published by J. Arrowsmith, of London, in

1857, we have this same water subdivided into Barrow Strait, Parry Sound, and M'Clure Strait; while in "Monteith's Physical Atlas," dated 1866, we find Melville Strait substituted for Parry Sound; and, instead of M'Clure Strait, Banks Strait and M'Clintock Channel; while in "Guyot's Atlas" Baffin Land supplements and obliterates Cockburn Land. What were formerly called the Parry Islands are now termed the Arctic Archipelago; and the new edition of "Appleton's Cyclopedia" has changed the well-known Pond Bay of the whalers to Eclipse Sound. In many English maps Grinnell Land is called Albert Land, it being so named by Captain Penny, who did not know that De Haven had been before him; and the error, though sufficiently exposed, has been persisted in. Thus the whole Arctic regions have been subjected to a continued change of nomenclature, and, of course, where hundreds of names are concerned, a familiar knowledge of events, and great care in transcription is requisite to a clear understanding of the position of a given party at a definite period. Without such circumspection, writers on Arctic affairs are apt to sadly confuse their narratives and bewilder their readers.

Then, too, voyagers themselves make mistakes of this description which mislead the chart-makers. Though this is embarrassing, it is not surprising, when we consider the difficulties under which surveys are often made in that intensely cold climate; and the fact, too, is considered, that very often the whole of the land visible, as well as the ice-closed waters, are all of one nearly uniform whiteness, so that it is exceedingly difficult to distinguish at any great distance the one from the other.

GEOGRAPHICAL MISTAKES.

In illustration of these possible mistakes we will only refer to a few of those which are well known, and have been made by usually careful and experienced travelers. In 1819, Buchan and Lieutenant (afterward Sir John) Franklin sailed for a considerable distance through Lancaster Sound, and then concluding it to be a bay, "seeing land at the end," they turned back. Captain John Ross also made the same mistake.

The famous Captain Kellet reported a mythical land off the Herald Islands. Speaking of Wrangel Sea, or what we should call the Polar Sea, he wrote, so late as November 15, 1851: "We have certain proof of there being land in this sea (Wrangel's), for on August 17, 1849, I landed on an island in lat. 71° 19′ N., long. 175° W.; it is almost inaccessible, and literally alive with birds. From the neighborhood of this island I saw, as far as a man can be positive of his sight, in those seas to the westward an extensive land, very high and rugged, distant from my position I conjecture fifty or sixty miles. I could not approach it with my ship, but might possibly have done so with a steamer."

Three years later the United States steamer *Vincennes*, Commodore Rogers, visited Herald Island, and sailed around in all directions, as well as to the westward, looking for the "extensive" land described by Captain Kellet, of the Royal Navy, but found none; neither that above described, nor some other land reported in the Arctic Parliamentary papers of 1849–51. Well might Captain Kellet say, as he did in his report: "It becomes a nervous thing to report a discovery of land in these regions without actually landing on it; but, as far as a man can be certain who has one hundred and thirty pairs of eyes to assist him, and all agreeing, I am certain I have discovered an extensive land. I think it is also more than probable that those peaks we saw are a continuation of a range of mountains seen by the natives off Cape Jaken, and mentioned by Baron Wrangel."—*Par. Papers*, 107.

And yet he was mistaken—there was no land there!

Again, Captain Kennedy, of the *Prince Albert*, in his report to Lady Franklin, in October, 1852, describes how he and the young French officer, Réné Bellot, *walked over the land* which Sir J. C. Ross, the great Antarctic as well as Arctic traveler, had reported to be a sea. This place was between 72° and 73° N. lat., and about 100° W. long. Réné Bellot, with the instinctive politeness of his nation, wrote in his journal: "Hitherto I had hoped Sir James Ross was right in his conjectures, but there can be no doubt now that he was mistaken, for we have walked over the land."

And then this same careful Kennedy, at Cape Walker, himself walks over a cairn erected by Captain Austin, and mistakes it for a natural production of the cliff.

Among the more modern explorers, Dr. Kane frequently refers to the mistakes of his predecessors. He says (Appendix, page 303): "The island named Louis Napoleon by Captain Inglefield *does not exist;* the resemblance of ice to land will readily explain the mistake."

Again he says: "There is no correspondence between my own and the Admiralty charts north of 78° 18' N. Not only do I remove the general coast-line some 2° of long. to the east, but its trend is altered 60° of angular measurement. There are no landmarks of my predecessor recognizable."

These mistakes he attributes in part to the "sluggishness of the compass, and in part to the eccentricities of refraction."

Dr. Kane's successor—Dr. Hayes—corrected the western coast-line of his friend, saying also of the opposite coast: "He was much tempted to switch it off twenty miles to the eastward." While the *Polaris* has sailed into what he and others thought to be the Polar Sea, north of Kennedy Channel, finding a strait and bays, obliterating the Polar Ocean in the latitude where it was supposed to exist, but confirming the idea that it will yet be found, only farther to the north than any human eye has yet penetrated.

But, though many mistakes have been made, much more of tangible fact has been revealed. Certain lands and waters, once as mythical as the "Hyperborean" of the ancients, are now as familiar to the geographer and Arctic mariner as the coasts of Europe, or our own Atlantic sea-board.

There is also an additional perplexity arising from the peculiar refracting power of the atmosphere, which at times throws up the lowlands into plateaus, and slight elevations into precipitous capes and headlands, so that the most careful observers have been deceived by a phenomenon not suspected to exist. In view of all these embarrassments likely to affect the accuracy of the Arctic explorer, we heartily concur in the wisdom of that energetic and successful navigator, Captain Kennedy, when he declared that he "would never report any thing as land which he had not walked over, nor any thing as water which he had not sailed through."

THE HYPERBOREANS.

What has been really discovered, instead of only imagined, we shall now briefly note.

Without going into the details of the old Norwegian colonization of Greenland, and the exploration of the American coast by the Norsemen of the tenth and eleventh centuries, *via* Iceland, which are matter of separate record, and have no direct bearing on the history of modern Arctic exploration, we will only briefly advert to the fact of such communication with the Old and New World having taken place, showing that in those comparatively early ages, while the rich southern plains of Europe and Asia were but sparsely populated, and millions of square miles lay open to the natural pre-emption of the first comer, there were still always to be found whole nationalities who preferred the cold and rugged districts of the North wherein to build their homes, to what would seem to us the more attractive regions of the temperate zone; but as the white whale and the Polar bear would perish in a warmer clime, so there have ever been races of men who have courted the Polar cold, and avoided, as a stifling furnace, the genial breezes of the luxurious South.

THE PRE-COLUMBIAN ERA.

Approaching the era of the modern discovery of America, but preceding it by little over a century, we find that the north-west passage to India was attempted by two Venetian brothers named Zeni, who were but the precursors of a long list of mercantile adventurers who essayed the same course; for at first it was not scientific enthusiasm or even a morbid curiosity which sent so many ships and expeditions vainly beating out their strength against the north-western barrier. Gain was the motive power which mainly ruled all these efforts for more than two centuries.

RIVAL EXPLORERS.

The English, Dutch, Danes, and Russians were, with reason, anxiously jealous of the rapid strides which Spain, in the sixteenth century, was making toward universal dominion; and to offset her power and gains in Mexico, Peru, and elsewhere, the English in particular made desperate efforts to find a shorter and easier way to the East Indies than that which the tedious sail round the Cape of Good Hope afforded; and what the English attempted by the north-west, Russia, somewhat later, tried to secure, both by land and sea, following a north-east course.

And even after this fanciful idea, based on geographical ignorance, was finally exploded, mercantile enterprise mingled with the pride of national acquisition in stimulating Arctic explorations. For though in all, or nearly all of the more modern attempts, scientific results were recognized as subordinate subjects of interest, it was not until the time of Franklin and Parry that any expedition was fitted out for the sole purpose of geographical and scientific inquiry.

SEBASTIAN CABOT.

Sebastian Cabot made his first voyage to the north-west coast of America under letters patent from Henry VIII., empowering the elder Cabot (John) and his three sons "to discover and conquer unknown lands," they being the first (of the Columbian era) who ever saw the main-land of North America, and on these north-western voyages he was the first to note the variations of the needle; but the subject of trade and commerce was always a prominent object with himself and royal patron. Later he projected a voyage to the North Pole; but though he penetrated the Arctic circle he succeeded in getting only to 67° 30′, sailing through Davis Strait; but neither he nor John Cabot had divested themselves of the idea that the ancient Cathay might be thus reached.

After the Cabots came the Cortereal brothers, who, from 1500–03, made three voyages, disastrous in loss of life, and not attaining any higher latitude than 60° N.

The results of these voyages were not particularly encouraging, and the thoughts of kings and the merchant princes of those times began to dwell on other means and routes to the spice lands of the Orient; and in consonance with this change in the tide of public opinion, an expedition was prepared by the Muscovy Company of London, under the leadership of the ill-fated Sir Hugh Willoughby, with instructions to find a north-east passage to Cathay and India. He succeeded in reaching Nova Zembla; there he encountered the formidable ice-fields of the Arctic Ocean, was forced back in a south-westerly direction to the coast of Lapland, where he and his whole ship's company were found frozen to death!

Richard Chancellor, who was the real navigator of this expedition, and sailed in one of the three vessels composing it, reached the north coast of Russia, landed and made his way to the presence of the Czar, from whom he obtained the mercantile privileges which resulted in founding the famous "Muscovy Company" of London.

FROBISHER'S GOLD.

The next movement of importance were the voyages made in 1576–78 by the renowned Frobisher. He was an early and zealous advocate of the north-west route, and spent many years in fruitless attempts to get his mercantile friends to invest in the project of a voyage of exploration, which he believed would be successful under his leadership; but so many of this class had suffered pecuniary losses in previous expeditions that he was unable to procure a ship.

Failing with the "mercenarie men of trade," he next turned to the Court, and finally succeeded in enlisting the sympathy and aid of Elizabeth's ministers. On his first voyage he collected, from the shores of what he called a strait, but what Charles Francis Hall discovered to be a bay, a quantity of black ore, thinking that it contained gold, and with this treasure returned to England.

To those who have read Captain Hall's work, narrating his explorations in that vicinity, the whole subject of "Frobisher's gold" must be familiar. Some of the metallurgists of London appear to have been either deceived themselves, or connived at deceiving others into the belief that mining could be profitably conducted in the country north of what was then called Frobisher Strait; and for a while Sir Martin Frobisher and the riches of

HALL DISCOVERING FROBISHER RELICS.

the new Cathay was the latest sensation of the Court circle. He received the encouragement and patronage of Elizabeth herself on two succeeding voyages; but neither his own private fortune nor the royal coffers appear to have been replenished by the "witches gold." It is proper to add, that scientific observations, as understood in those days, were not neglected.

GILBERT AND OTHERS.

The chivalrous and courtly Sir Humphrey Gilbert was another of the Elizabethan courtiers who was persuaded of the practicability of a north-west passage to China, if not India. He was a navigator of great skill and experience, and made two voyages of discovery to the north coast of America; and, on his second, he took formal possession of the island of Newfoundland in the name of the British Queen. But he was not permitted to participate in the honors which awaited him in his own country. His ship foundered at sea, and all on board perished, thus experiencing, as the poet sings of him in the ballad,

> "It was as near to heaven
> By water as by land."

John Davis, the discoverer of the strait which bears his name, also surveyed a considerable part of the coast of Greenland as far north as the seventy-third degree.

During all this time the Dutch, the French, and the Danes were not idle; but they went principally to the north-east. Barentz made three voyages, 1594-96. He started under great disadvantages, being inexperienced and far from properly furnished; but he was brave and persevering, and what man could do under such circumstances he did; on his third voyage he had to abandon his

BARENTZ'S WINTER-QUARTERS.

ship, and with his crew take to the boats, but unfortunately perished from exposure and exhaustion when near Icy Cape, a headland of Russian America, in the Arctic Ocean. His house, which he built on land for winter-quarters, was discovered by a Norwegian whaler, named Carlsen, in 1871, on an island E.S.E. of Nova Zembla.

Many others, whom we have not space to mention, fill out the long list of bold and hardy adventurers whom neither continued disaster nor threatened death could turn from their purpose; and no doubt some nameless heroes, who did not happen to rank high enough to catch the "sounding trump of fame," might, if we knew their humble history, their faithful courage and endurance, outshine in merit all the rest.

But, regardless of individual virtue, history inexorably fixes her pivotal points upon those men and events which form a necessary connecting link with the times past and the time coming. In accordance with this mode of selection, the name of Henry Hudson starts to the front as a prominent standard-bearer in the work of Arctic exploration. His first voyage was made under the direction of the old Muscovy Company, in 1607. Considering the previous history and the many failures of preceding explorers, he received the somewhat astonishing order "to go direct to the North Pole!" He did what he could to obey orders, and reached 81° 30', steering due north along Spitzbergen, until he proved that course to be impossible. The next year he started out again, with the intent, we presume, to accomplish indirectly what he had failed to do directly; at least, on this voyage he stood to the north-east, but got only to 75° N. Once more, in the succeeding year, he tried the same course, but meeting with heavy ice, he turned about and sailed toward the west, and, reaching the American coast, began anew the search for a north-west passage. He did not find that, but he found something better; he discovered New York Bay and the Hudson River, and then, needing to be reprovisioned, sailed for home.

Returning in 1610—his fourth voyage—he directed his course farther north, struck the straits, and sailed through to the magnificent bay, both of which waters bear his name. On the great bay he sailed several hundred miles, farther to the west than any one had yet penetrated, and wintered on an island in its mouth—Southampton Island; and then tried again, in the spring, to find

HENRY HUDSON.

the long-sought passage to the Pacific. But the long cold winter, with insufficient food, had told on the moral as well as physical condition of the men, the hardier portion of whom were completely demoralized, and finally mutinied against any further detention in these Western waters. The end of this noble man was sad indeed: with his son and several sick sailors he was turned adrift in an open boat, while the mutinous crew took possession of the vessel and stores. One noble-hearted, faithful man, John King, the ship's carpenter, voluntarily accompanied him, and shared his fate. The ringleader of the mutinous crew, with five others, was killed by the natives: several others died, some of starvation; and the rest managed to get the ship back to England; but Henry Hudson, with his seven companions, was never heard of more.

As the sad story finally leaked out, there arose, mingled with pity for Hudson's fate, and indignation against the mutineers, a buoyant feeling of expectancy over the great discoveries which had been made. It was now confidently believed that the passage was absolutely found, that it was only necessary to sail on and on through the water which we now know to be a bay, to reach the China Seas. In consequence of this impression, the next few years saw several other voyagers sailing for "Hudson's great sea," in the pursuit of which several minor discoveries were achieved. Fox Channel, Sir Thomas Rowe's Welcome, and other waters were partially explored; the excitement was kept up to an exceptionally high tone; and this prolific period culminated in the discovery of the great bay to the north of Davis Strait by William Baffin in 1616. He explored the western coast of this water to the mouth of Lancaster Sound, and none went farther than he to the north-west for another half-century.

The hopes and expectations which the discovery of Hudson's Bay had excited finally faded, until anticipation was extinguished by the ever-recurring fact that all the discoverers eventually came back to England, and, whatever else they found, they did not find a practicable passage to the Indies. In addition to these reasons, enterprise was now in a measure directed to the colonization of the Atlantic coast, now within the limits of the United States; and though voyages continued to be made, both to the north-east and the north-west, and in the former direction many sledge expeditions were planned, yet no important discovery for many years again aroused the enthusiasm of the English nation.

RUSSIAN EXPLORERS.

During this time the Russians were particularly active in their scientific experiments upon the variation of the magnetic needle, and in the examination of other phenomena in such portions of the Arctic regions as lay accessible to them. The most enduring results obtained by the Russians in the early part of the eighteenth century was achieved by Vitus Behring, a captain in the Russian Navy, who, for his tried courage and skillful seamanship, was appointed by Peter the Great to the command of a voyage of discovery. In 1728 he explored the northern coasts of Kamtschatka as far north as 67° 18′, thus making the discovery of the straits which separate Asia from America, previous to which the impression prevailed that the continents were there united. But it was still uncertain whether the land to the east of the straits was a part of the main-land, or only islands scattered along the coast. To determine this, in 1741 he sailed from Okhotsk, intending to explore the American coast; he twice made the land, but was driven back by violent storms, and at last he was cast upon a desolate ice-covered island, since named for him, where he died. The crew managed to subsist with the aid derived from the wrecked vessel, out of which, in the spring, they built a small sailing craft, and in August reached the coast of Kamtschatka;

BARON VON WRANGEL.

but the gallant Behring lives only in the straits and island which preserve his name. Other Russian expeditions followed, among which was that of Shalaeloff in 1760, who died of starvation, and some others, which accomplished little, concluding this series with the important sledge journey of Baron Von Wrangel and Anjou in 1820-23, which had a marked influence upon the opinions and subsequent course adopted by nearly all of the succeeding British explorers. These intelligent and persevering Russians attained to lat. 70° 51' N., long. 155° 25' W., then met the open sea, for which they were not prepared. Thus, in all the expeditions so far sent out in ships, the way had been barred at different points by impenetrable ice, while those who had essayed the trans-glacial plan had been met with interposing arms of the sea which as effectively stayed their progress.

GOVERNMENT REWARDS OFFERED.

As early as 1743 the British Parliament had offered £20,000 for the discovery of a passage by the Hudson Bay route, which stimulated once more the flagging enthusiasm, and several voyagers sailed; some through Behring Strait to the east, hoping thus to reach Hudson Bay by the imaginary ocean, which then existed in the brains of nearly all Arctic explorers.

Between 1769 and 1772, Hearne made three land trips, on the last of which he discovered the Coppermine River, which he traced to its source. The next year Captain Phipps, afterward Lord Mulgrave, was sent out by the Admiralty, with orders to make for the North Pole—this object to take precedence of all others; meteorological, magnetical, and other scientific observations were also to be made objects of investigation; and thereafter geographical science became a successful rival to the mercantile spirit, which had hitherto dictated the instructions given in previous expeditions. Phipps went the Spitzbergen route, but reached only 80° 48'—not as far north as Hudson attained sixty-six years before.

Undiscouraged, the British Parliament again took up the subject, and, though now involved in the preliminary quarrel which resulted in the loss of her American colonies, her ministry had still eyes, ears, and thoughts for discoveries in the far North. In 1776 the British Government offered, in addition to the standing reward of £20,000 for the actual discovery of the pole, the same sum for any through route, and £5000 to any one who should reach to within one degree of the pole.

In the mean time, the famous Captain Cook was ordered to the search for the pole. He went through Behring Strait, and got only to 70° 45'. A vessel had gone out to Baffin Bay in the hope of meeting him, but, as is well known, his voyage terminated fatally to himself, and unsuccessfully as regarded the object in view.

The next important discovery was that of the Mackenzie River in 1789.

EARLY AMERICAN ENTERPRISE.

In the American colonies, too, emulation was ripe, though the means of fitting out large expeditions did not exist; but as early as 1754 we find that private enterprise was directed to the same point of attraction. In the *Gentleman's Magazine* of that year is an account of the voyages of the *Argo*, of Philadelphia. Captain Charles Swayne had made two voyages in search of a north-west passage, obtaining valuable information of the coast of Labrador and Hudson Bay, but failing to get north of lat. 65°.

In 1772 some gentlemen in Virginia, moved by the same desire which had actuated the enterprise of the civilized world for centuries, fitted out the brig *Diligence*, under the command of Captain Wilder, who also made Hudson Bay, and sailed about its broad waters north and west, thinking to find a passage, and believing there was one; but, repelled by the ice, he retreated, and afterward made the latitude of 69° 11' in Davis Strait.

THE WHALER SCORESBY.

The name of William Scoresby may justly be considered as the connecting link between the old explorers—the adventures made almost solely in the interest of commerce, and those more liberal modern enterprises, conducted in the spirit of the newly-dawning scientific era.

And yet Scoresby's name scarcely figures, even incidentally, in any general record of Arctic heroes, for the simple reason that the British Government, though availing itself of his knowledge and experience, was unwilling to confer its honors on any except those of the Royal Navy.

William Scoresby, though an eminently learned and scientific man, was for many years known only as a successful and enterprising whaler. It was on one of these voyages, in the year 1806, while lying-to for whales in what is known as the "Greenland Seas," on the east side of Greenland, in lat. 78° 46' N., that he thought that he would venture to deviate from the usual whaleman's track, and penetrate, if possible, to the "Polar Sea," in which he fully believed. Spreading his sails, and with a good wind, he soon left the whaling fleet behind him, and shortly after encountered the heavy ice which he knew he must penetrate to reach the open water beyond. With consummate skill, tact, and boldness he bored his way through the pack-ice, and, undismayed at the novelty of his position, separated from his companion vessels; with the great ice barrier between him and civilization, he bravely pushed on toward the north, where his hopes were gratified, and his opinions confirmed, by finding a "great openness or sea of water." He reached the high latitude of 81° 30' N., 19° E. long., seas never before visited by whalemen, and never previously attained in either hemisphere except by Hudson. Parry after-

WILLIAM SCORESBY.

ward went higher in his sledge journey, but not in a sailing vessel. But Scoresby was something more than a whaler. On each voyage he added something to accurate geographical knowledge by surveying the coast and islands which he visited, and by him a large portion of the eastern coast of Greenland was first accurately traced, and prominent points named. He corrected the thermometrical statements and other incorrect so-called scientific information of his day; he experimented on the temperature of deep-sea water, on terrestrial magnetism, and other natural phenomena, and published many interesting papers relating to the meteorology and zoology of the Arctic regions.

Ross and Franklin had both dilated upon the curious phenomenon of red snow observed in their Arctic voyages; and in 1828 Scoresby analyzed a portion of the colored snow of Greenland, and found that the coloring matter consisted of exceedingly minute marine infusoria.

As early as 1814 he had published a paper on the "Polar Ice," including a "Project for reaching the North Pole." He made fifteen voyages, in which he touched 80° N., the results of which were made public in a book entitled the "Arctic Regions," in 1816. At this time he was considered by all the intelligent friends of Arctic exploration as an authority upon all matters connected with the Polar region.

It was out of a correspondence which he held with Sir Joseph Banks in 1817, that was evolved the combination of events which led to the equipment of those mixed land and water explorations commanded by Parry, Ross, and Franklin.

The eminent French savant, M. de la Roquette, in his memoirs of the latter, addressed to the Geographical Society of France, says: "In spite of previous discoveries, the subject of Arctic explorations was again almost forgotten, when an English whaler, an intelligent and intrepid sailor, who had for many years navigated the Greenland seas, demonstrated the possibility of effecting a per-glacial voyage across to the Pacific. In a letter written by him to Sir Joseph Banks, *this whaler*, Scoresby the younger, narrated a remarkable circumstance which he had witnessed during his last voyage in 1817." (This statement referred to a great disruption or removal of the usual ice barrier, which occurred in 1816–17, in the parallel of the island of Jan Mayen, and near the eastern coast of Greenland.)

"This information, a similar condition of the ice occurring also in 1806, awakened in England the long-dormant projects for attaining the North Pole, and for opening up the north-west passage."

In 1835 Sir John Ross made the same admissions in the preface to a work on his own voyages, observing, "that a sort of *renaissance* of public interest in Arctic affairs had followed upon the publication of Scoresby's views, as given to Sir Joseph Banks." From one of these letters we extract the following:

"Scoresby says: 'I mentioned the fact of a large body of the usual ices having disappeared out of the Greenland Sea, and the consequent openness of the navigation toward the west, whereby

I was enabled to penetrate, within sight of the east coast of Greenland, to a meridian which had been usually considered quite inaccessible. After some account of the state and configuration of the ice, and our progress among it, I proceeded to remark on the facilities which on this occasion were presented for making researches in these interesting regions, * * * *toward deciding whether or not a navigation into the Pacific, either by a north-east or north-west passage, existed.* I also expressed a wish to be employed in such researches through a series of voyages, that the most favorable seasons might be improved to the best advantage, and that the most complete investigation might be accomplished; and, by the way of avoiding unnecessary expense, I proposed to combine the object of the whale-fishery with that of discovery, on every occasion when the situation of the ice was unfavorable for scientific research. Since no one can possibly state, from observation of the ice in any one season, what opportunity may occur on a subsequent occasion, it would be well to have this reserve (whaling) for the reduction of the expenditure, in the event of the opportunity for discovery failing.'" This was evidently too sensible an idea to penetrate the brains of the British Naval Office.

Seven weeks after this letter was written, a notice appeared in the public prints of the day, "that, owing to the statements of the *Greenland captains* respecting the diminution of the Polar ice, the Royal Society had applied to ministers to send out vessels in the Polar Seas."

It was reasonably expected by Scoresby and his friends that he would have been appointed to the command, if an expedition was planned; but red tape prevailed: the Admiralty were fixed in their opinion that none but officers of the Royal Navy were capable of commanding an exploring expedition. Scoresby was offered a subordinate position; but this he naturally refused to accept.

In August the British expedition entered Lancaster Sound, and sailed up it for sixty miles, when they thought they saw land at the end, and thence concluded it to be a bay. The weather was bad, which prevented their examining its contour more closely, and they put about, exploring the sound to the south and east, and then returned to England in October of the same year.

Captain Ross, who also visited the sound, likewise thought it a bay, but some of his officers, including Parry, were of a different opinion, and, on the return of the expedition to England, the question of "sound" or "bay" was the topic of much interested and not a little angry discussion. The English public were dissatisfied, and Parry's followers being the more energetic party, aided him in preparing a private expedition to go back, and, by actual survey, to settle the point.

He sailed in May, 1819, in the *Hecla*, with a consort, the *Griper*, under command of Lieutenant Lyon; these vessels carried a combined crew of ninety-four men, and were furnished with provisions

CAPTAIN PARRY.

for two years. On their way up Baffin Bay, they encountered ice on the 18th of June, and were temporarily "beset" on the 25th; but a lead opening, they reached Lancaster Sound on the 30th of July, but not without trouble, though they were fortunate enough, early in August, to find the sound free, and a channel, which they followed to the mouth of Barrow Strait, thus finally exploding the idea of its being a bay. The strait Parry entered and sailed through as far as Prince Regent Inlet, which, with many other capes, points, bays, headlands, and so forth, he named. As he approached the magnetic pole, he found his compasses of but little use, so great was the dip of the needle. The hopes of officers and crew were greatly excited, and when, after encountering immense difficulties, he, on September 4, crossed the one hundred and thirteenth degree of west longitude, he told the men that the *Hecla* had earned the reward of £5000 offered by the Government, the enthusiasm knew no bounds. Two weeks later he was beset; but the crew cut a passage through the ice till a lead was reached, and the party attained Melville Island in safety. Here Parry wintered, using every opportunity to explore the country in different directions, and adding largely to the topographical and hydrographical knowledge of the day respecting that region of country. In June of that year (1820) it was yet very cold; but a thaw set in early in July, and on the 2d of August the ice broke up and set them at liberty. Two weeks later they were again beset for a time; but getting clear with great exertions, they started for home, where they were received with hearty welcomes; and on a report of the discoveries made being published, the utmost satisfaction was expressed both by the Government and the public press.

The successes of Parry had, however, but whetted the public appetite, and the next year he sailed again, with instructions to go to Repulse Bay by the way of Hudson Strait, with the hope that thus the dangerous encounters with the "middle ice" might be avoided. On this occasion he again sailed in the *Hecla*, with the *Fury* as consort, of which Captain Lyon was in command. They reached the terminus of Hudson Strait in August, 1821, and from there sailed north to Fox Channel, and thence to Repulse Bay, in hopes of finding an outlet to the north or west, and for that purpose made careful and extensive explorations; but were early beset in the ice, and in September cut a dock for the vessels in a heavy floe, from which they were not released until the next July. During the winter they occupied the time in sledge journeys of exploration, and in recording the results of their scientific experiments. They went carefully over the course, including Lyon Inlet, then through Fox Channel to the strait uniting the latter with Boothia Gulf, naming the strait Fury and Hecla. They reached the middle of these straits in September, 1822. Here they wintered, remaining until August, 1823, when they returned to England.

EXPLORATIONS CONDUCTED ON FOOT.

During the period in which Parry had made two voyages, the other expedition (overland), which had started in September, 1819, from York Factory, on the west side of Hudson Bay, and which was expected to explore the coast from the Coppermine River east, was undergoing a fearful experience. The leaders were Sir John Franklin (then lieutenant), and Dr. Richardson. There were also two midshipmen, Messrs. Hood and Back (afterward Sir George), and a seaman named Hepburn. It had been arranged in England that if Parry made the coast on his first voyage, he was to co-operate with this small but energetic land party.

The latter, leaving York Factory in September, after almost unparalleled sufferings — with cold beyond measurement, for their thermometer was frozen — finally reached Chipewyan, a dépôt of the Hudson Bay Company, after a foot journey of eight hundred and fifty-six miles! Resting here for a while in July, 1820, they traveled to Fort Enterprise, where was a small hut containing stores, making five hundred miles more. Here they wintered, while Mr. Back returned to Fort Chipewyan to hurry on supplies for the next season. It was during the absence of Mr. Back that an Iroquois hunter, in the employment of the party, shot Midshipman Hood, with the intention, as Franklin and Richardson supposed, of eating him; whereupon Dr. Richardson took the responsibility, and deliberately shot the Indian through the head.

The hardships which they had endured had reduced their strength of body and mind almost to inanity; and Mr. Back also suffered great hardships on his journey, but his indomitable will and great physical endurance brought him through, and he reached Fort Enterprise, with supplies of provisions, on the 17th of March, 1821. He traveled eleven hundred miles on this journey, sometimes for two or three days without food, and at night having for covering but one blanket and a deer-skin, the thermometer much of the time registering from 47° to 57° below zero.

Mr. Back having rejoined his party with supplies from Fort Chipewyan, they started again from Fort Franklin, where they had halted, dragging their provisions and canoes to the Coppermine River, eighty miles distant. Embarking in these frail boats, they sailed seaward, and reached the coast of what they supposed to be the sea about the middle of July. They then turned to the east, sailing and paddling alternately, as circumstances required, for five hundred and fifty miles—all the time thinking they were going toward the Arctic Ocean: at the end of that time they found they had only been navigating an immense bay. Convinced at last of this, on reaching Dense Strait they called the headland Cape Turnagain, and sadly prepared to retrace their course. A more disappointed party could scarcely be imagined. To add to their perplexity, they found they had only food for a few days, and no signs of animal life which promised them a substitute. However, they manfully set to work and built two canoes, with which they entered Hood River a short distance west of Point Turnagain. Food failing them, they were reduced to the utmost extremity, and became so weak in consequence that they abandoned the canoes they had constructed, being unable to drag them around certain rapids which they encountered. Some days they managed to gather a little rock-tripe or moss, and finally ate their old shoes and scraps of leather attached to other articles. Two of their number died of exhaustion; but at last, when all were nearly at the point of death from starvation, their eyes were cheered by the sight of York Factory, from which they had started out *three years before,* having in their absence traveled over fifty-five hundred miles—notable specimens of what the human frame, when controlled by an intelligent will, is capable of enduring. They brought up at this haven of rest in July, 1823, and soon after returned to England.

COMBINED SEA AND LAND EXPLORATIONS.

A few months only elapsed before another expedition was proposed, on a larger scale than any which had yet been projected. This consisted of four divisions.

One vessel, under Parry, was destined for Prince Regent Inlet, which it was thought opened at the south. The second party, under Franklin, was ordered to go down the Mackenzie River to the sea, and then divide, part to travel to the eastward, and the others with Franklin to the westward until they struck Behring Strait. Captain Beechey was ordered to sail round Cape Horn to Behring Strait, and thence to make Kotzebue Sound, and wait there for Franklin. The fourth party, under Captain Lyon, in the *Griper*, was to go to the south of Southampton Island, up Rowe's Welcome to Repulse Bay, then cross Melville Isthmus to Point Turnagain. The object of the whole expedition being to secure, if possible, a thorough exploration of the space between the eastern and western shores of the North American continent, and the correct configuration of its northern boundary, the expedition, therefore, contemplated and was prepared for both land and sea travel.

Captain Lyon's part was soon finished. His vessel was twice nearly wrecked, and he abandoned the further pursuit eighty miles from Repulse Bay.

Parry sailed in May, 1824, in the *Fury*, with the *Hecla* as consort, and reached Lancaster Sound; but was there caught in the ice and had to winter at Port Bowen. The *Fury* was afterward wrecked, and Parry took both crews back to England in the *Hecla*.

Franklin's party had a more extended service. With him was Dr. Richardson, Lieutenant Back, and Messrs. Kendall and Drummond, the latter a naturalist of reputation. They got to Fort Chipewyan in July, 1825, and from there went to the Great Bear Lake to winter. From thence, in pursuance of orders, Franklin undertook the descent of the Mackenzie River, which he accomplished, reaching the sea at lat. 69° 14' N., long. 135° 57' W., a distance of one thousand and forty-five miles.

On the 28th of June, 1826, the whole remaining portion of Franklin's party also went down the river to its mouth, and there separated, Franklin going to the west, and Dr. Richardson to the east. The former skirted the coast, which trended to the north-north-west till he reached lat. 70° 24', and long. 149° 39' W. Here his further progress was barred, and he named the place Return Reef. The weather was excessively bad, and, as usual, provisions were short. He was also unaware of the fact that Captain Beechey was waiting for him only one hundred and forty-six miles farther west; for Beechey, in the *Blossom*, had passed Behring Strait, had gone to Chamisso Island, in Kotzebue Sound, where, getting no information of Franklin, he went north-north-east to Point Barrow, and from there, forwarding boat parties, he awaited their return until it became dangerously late in the season, when he put off for winter-quarters in Petropaulovski. One of his boat parties returned in time to accompany him; the other proceeded to the south-east (overland) to the posts of the Hudson Bay Company.

In the mean time Franklin returned to the Mackenzie, having explored the whole coast for three hundred and seventy-four miles to the north-north-west, which in its intricacies involved, in coming and going, over two thousand miles.

Dr. Richardson had during the time made an extended journey to the east, but without developing any special points of interest.

The whole expedition once more met and wintered at Great Bear Lake, where they established a series of valuable observations on terrestrial magnetism. And it was a curious incident that Parry's quarters, at only an interval of one year apart, were situated at the opposite side of the magnetic pole, just eight hundred and fifty-five miles distant, both parties making the same observations. And thus, while the needle at Port Bowen was regularly increasing its western direction, that at Fort Franklin, pointing directly toward it, was increasing its easterly—a beautiful and conclusive proof of solar influence upon the daily variation. Captain Beechey returned to his appointed rendezvous the succeeding year; but he and Franklin never again met.

THE ERA OF MODERN DISCOVERIES.

In 1818 commenced what may be called the modern era of Arctic exploration, primarily induced, as we have shown, by the writings and influence of Scoresby, and aided to the last by Sir John Barrow, the faithful advocate of Arctic explorations. In this year two expeditions were fitted out by the British Government, the one under Captain Ross and Lieutenant Parry, the other under Captain Buchan and Lieutenant (afterward Sir John) Franklin, the last being more particularly devoted to scientific investigations.

The orders of the scientific party were to go, between Spitzbergen and Greenland, as far north as possible. Here they found the temperature far milder than they expected, and attained the highest latitude yet reached; but it was not without great danger—the ice floes surrounded them on all sides, and one ship, the *Dorothea*, was completely shattered. Nevertheless the philosophical experiments, on the elliptical figure of the earth especially, were conducted with very interesting results; also experiments in refraction and magnetic phenomena. In April they started to return, and were beset with ice not far from Waggat Island, but cleared themselves, and made for the coast of Greenland.

PARRY'S DRIFT.

The year 1827 saw Captain Parry at the head of another expedition destined for the north shore of Spitzbergen, supplied with two well-built covered boats, so arranged that they could be put on runners, and thus dragged as a sledge where they could not be floated. Arrived at Spitzbergen, he started on the ice, provided with food estimated for seventy-one days; but the journey was not to prove so easy in reality as it did in the instructions of the Naval Office. First, they were impeded with thin ice, through which the boat could not sail, and which was not strong enough to travel over; next, it was rough ice, which threatened continually to rack the sledge-runners to pieces — and worse, snow-blindness attacked nearly the whole party. This evil they endeavored to circumvent by abandoning day travel entirely, and moving forward only at night — a night, however, which was by no means dark in that latitude in summer.

Considering the outlay of exertion, the gains appeared insignificant. The first five days they had made only ten miles. They had hoped this time surely to reach the Pole; but appreciating the difficulty with every step, the leading officers agreed with Parry that they would be content could they make the eighty-third parallel; but in their problem was an unknown quantity which they had not taken into the account. Unperceived by them for a while, and still longer unaccounted for, was the strange fact that, no matter how many miles they traveled toward the north, at each observation they found themselves steadily moving south. *The ice was moving beneath them*, carrying them south with every hour. This was an obstacle which no human ingenuity could remove. At 82° 45' they gave up the contest, finding that, though they had traveled nearly three hundred miles over ice and through water, they were yet but one hundred and seventy-two miles from the *Hecla*. Burying their great hopes in a sad but blameless failure, they got back to the ship on the 21st of August, and returned to England.

It was no wonder that the zeal of the Government officers began to flag under such repeated disappointments, and that in consequence we find that the next serious effort was made under the auspices and with the means of a private enthusiast.

STEAM FIRST USED IN THE ARCTIC SEAS.

Sir Felix Booth, an ardent friend of Arctic exploration, fitted out the *Victory,* putting her under the command of Captain John Ross, who was accompanied by his nephew, Sir James Ross. With

SIR JOHN ROSS.

the *Victory* a new element appears, hitherto a stranger to Arctic waters—STEAM. The *Victory* was fitted with a steam boiler, to be "used in calm weather." The expectation still was that a north-west passage could be made through Prince Regent Inlet.

The *Victory* sailed in May, 1829, and reached the inlet on the 9th of August, and came up with the wreck of the *Fury* on the 12th; on the 15th they got to "Parry's farthest;" here they encountered serious difficulty with ice, but, persevering, managed to work along three hundred miles on a coast-line not hitherto explored, reaching to within two hundred miles of the extreme point reached by Franklin on his last expedition.

Here the shore trended to the west, and though now closed by ice, Ross thought that these two hundred miles would be navigable at some time of the year, and he would await his opportunity; but the present season was now over. October had overtaken them, and on the 7th inst. they went into winter-quarters at what is now known as Felix Harbor. There ice fetters held them fast for eleven months. Not until September, 1830, did they get under way, and then only made *three* miles, when they were again beset, and obliged to winter until August, 1831, when they made *four* miles more; and on the 27th of September they were once more fast for the season. *Seven miles in two years!*

ROSS REACHES THE MAGNETIC POLE.

Ross could stand that rate no longer. In April of 1832 his nephew, James C. Ross, made a sledge excursion to the west, and reached and FIXED THE MAGNETIC POLE in lat. 70° 5' 17" N., long. 96° 46' 45" W.

But two Arctic winters had told upon the health of the crew; the scurvy broke out, and with it the despondency which usually accompanies, and is often the precursor of, that disease. The ship was obliged to be abandoned, and the whole company started east, taking their boats on sledges. Their first objective point was the wreck of the *Fury*, where at least some shelter could be obtained, and also material, and possibly stores; they endured terrible hardships on the way, but managed to get there on the 1st of July, but were too much reduced to go farther at that time, and before they were recuperated sufficiently, winter weather was upon them, and at Fury beach they were compelled to remain through the season of 1832-33. The suffering among all parties, especially the crew, was intense; many were fearfully sick, and several died. With the opening of early summer they made desperate efforts, and on July 8 they reached the open sea. Here they launched their boats, which they had dragged much of the way, trusting that they might be seen and relieved by some whaler. This fortunately happened; they were picked up on the

26th of August, 1833, by Captain Humphreys, of the *Isabella*. Though he willingly received them on board, he did not recognize the identity of Ross, nor at first believe their story: he "thought Captain Ross's party had all been dead for two years," but was finally convinced of his error, and in September landed them at the Orkneys, whence they might get conveyance to England. They had been absent four years, from 1829 to 1833.

DISCOVERY OF GREAT FISH RIVER.

Seven months previous to their rescue and return, Lieutenant Back, accompanied by Dr. King, naturalist, had left England in search of Ross and his party, and reached Fort Resolution, on Great Bear Lake, in August, and from there went on to Musk Ox Lake, to the north-east. Finding nothing of the parties, they returned to winter at Fort Reliance, suffering much from lack of sufficient food and the intensity of the cold. In April they had just planned a movement to the sea-coast, when they learned of Ross's safety. In June they started on a boat excursion down the *Thlew-ee-choh* River, which he called the Great Fish River, since named Back, which they hoped would lead them to the Polar Sea; and on July 29, after a hard and fatiguing journey of five hundred and thirty miles, they reached the open water, at lat. 67° 11′ N., long. 94° 30′ W.; but it was not the Polar Ocean. The river they had followed led through a most desolate country; neither trees nor vegetation were visible, except of diminutive lichens and mosses. But as game was sometimes observed, it is quite possible that in the sheltered valleys, which the rocky bluffs of the river hid from their sight, the aspect of the country was less forbidding.

After exploring the shores for some distance, meeting with many obstacles, they turned back, and after four months of continuous travel reached again Fort Reliance, on Great Bear Lake, in September, and from there returned home in the autumn of 1835.

DEASE AND SIMPSON'S EXPLORATIONS.

Two years later the Hudson Bay Company sent out two men, Dease and Simpson, with orders to go down the Mackenzie River to the sea, and then to move along the coast to the west until they reached the point where Beechey had waited for Franklin;

this they considered would complete the survey of the north shore of the North American coast. They reached Return Reef (Franklin's farthest west) in July, 1826, and up to this time no one had been beyond this point from the east. They pushed on, and finally got to Point Barrow (Beechey's extreme east), and thus their task was happily completed. On their way they discovered two large rivers, which they named respectively the Colville and Garry; then returned to winter-quarters at Great Bear Lake.

In June of 1838 Messrs. Dease and Simpson, with others, started again for the north coast, by way of the Coppermine River, intending this time to travel toward the east, but found themselves stopped by accumulations of ice. On this the party divided: some took sledges and proceeded overland. They passed Point Turnagain, the farthest point reached on the west on that route; found Dease Strait filled with ice, and at the eastern termination of the straits a large, bold headland, and to the north an extensive tract of land, new to explorers, which they named Victoria. There they clambered with great difficulty to the top of a high, ice-bound cape, from which they were surprised to see a broad sea beyond free from ice. They explored forty miles of Victoria Land to the east-north-east, concluding their survey in the summer of 1838.

The next year they sailed through Dease Strait, and settled the exact coast-line up to the point which Back reached in 1834, and beyond found that the estuary of Back's, in which they sailed, separates Boothia on the west from the American continent. They almost joined their discoveries to Ross's, and came within ninety miles of the point he had fixed upon as the magnetic pole. These excursions of Dease and Simpson were among the most useful which had been made, yet their names and labor have been almost completely overshadowed by some who had led much more expensive and sensational expeditions, but who really added little to the general fund of geographical information. Indeed, all the American coast north was now explored, except that portion lying between Dease and Simpson's extreme point west of Boothia and Ross's winter-quarters, on the east of the same land, and that tract between Ross's quarters and the extreme point reached by Parry in 1822, at the entrance of Fury and Hecla straits.

IS BOOTHIA A PENINSULA OR AN ISLAND?

The question now to be settled was this: Could ships pass betwen Boothia and the main-land? On this parties were formed, and while many book-geographers and parlor-sailors were quite sure that they could, and vehemently argued that "nothing was easier than to sail through Prince Regent Inlet and round the southern coast of Boothia through to the north-west," others, and these mostly Arctic travelers, held the matter in much doubt; some affirming that Boothia was a peninsula, and not an island. To settle the point, the Hudson Bay Company, which had now become a steady patron of Arctic explorers, sent out Dr. John Rae in the summer of 1846, he and his party reaching Chesterfield Inlet in July of that year. This expedition was prolific of stupid opinions put forth by the British Board of Admiralty, as will be seen by the context.

From Chesterfield Inlet Dr. Rae passed on to Repulse Bay, then conveyed his boats to the Gulf of Akole. Being unable to get farther that season, he returned to winter in Repulse Bay; but in April of 1847 he reached the inlet which Ross had found on a land trip while wintering on the coast of Boothia, and on which occasion the latter had proved the continuity of the coast to that point; and thus, between them, it was proved beyond a doubt that Boothia is joined to the main-land.

We said without a doubt, but it was not exactly so. Dr. Rae, in a letter to Charles Dickens, then editor of *Household Words*, and published in vol. x., No. 19, says: "The Esquimaux tracing, or delineation of coast, was entered in the Admiralty charts in dotted lines [indicating doubt], until my survey of eighteen hundred and forty-seven, which showed that in all material points the accounts given by the natives were perfectly correct. When Sir John Ross wintered three years in Prince Regent Inlet, the natives drew charts of the coast-line to the southward of his position, and informed him that in that direction *there was no water communication* leading to the western sea. Sir John Ross's statements, founded on those of the natives, were not believed at the Admiralty, *nor were my own*, in eighteen hundred and forty-seven, although *I saw the land all the way*, and in which I was supported by the Esquimaux information. The authorities at the Admiralty would still have Boothia an island. Last spring

I proved beyond the possibility of a doubt the correctness of my former report, * * * *for where parties of high standing at home would insist on having nothing but salt-water, I traveled over a neck of land, or isthmus, sixty miles broad.*" This was in accordance with the spirit which treated Baffin Bay as a myth when its existence was first announced by the original discoverer, and which excluded it from the maps until rediscovered by Ross.

It was on the 29th of May, 1814, that Dr. Rae saw, during a break in the clouds, which seemed to lift during a fearful storm for the very purpose, a headland, which he named Cape Ellice, in lat. 69° 42' N., and 85° 8' W., and which was within ten miles of Fury and Hecla straits; and this made the entire survey complete, with the exception of the straits, and they were partially known by the accounts of Dease and Simpson as well as Parry.

SIR JOHN FRANKLIN'S LAST EXPEDITION.

We now approach the most exciting era of Arctic research. In May of 1845 the far-famed expedition of Sir John Franklin, in the *Erebus*, with Captain Richard Crozier, in the *Terror*, started from England—that swarming place of Arctic adventurers. Accompanying them was a tender, which, however, after unloading her provisions, was sent back from Davis Strait. The *Erebus* and *Terror* were supplied with every thing which the ample means of the Government could at that time provide. A long detention was anticipated, and they were fully provisioned for three years.

On the 26th of July, 1845, these ships were last seen by civilized man. The master of the whaler *Prince of Wales*, Captain Dannet, met them in lat. 74° 48' N., long. 66° 13' W. They were then moored to an iceberg, apparently waiting for an opportunity to get into Lancaster Sound. This was the final glance of recognition between those brave explorers and the representative sailor of the race to which they belonged—the last intelligence which reached the civilized world for years respecting them. Where they went, how they reached Cape Riley and Beechey Island, and what became of the two noble vessels, is yet a mystery only partially solved.

The general instructions of the Admiralty to Franklin directed him to go to Baffin Bay, then through Lancaster Sound, on

SIR JOHN FRANKLIN.

through Barrow Strait, which the Admiralty "thought would be free;" to Cape Walker, about 98° W. long.; and then, turning to the south and west, to use his best judgment in getting through to Behring Strait. This course would have led him through Melville, then Parry Sound.

Much was expected from this expedition. The previous experience of the commander, and of many of those who accompanied him, was such as to inspire the greatest confidence in the results; but when two years had elapsed, and no tidings were received of even the whereabouts of the travelers, anxiety for their safety began to be excited; and as the winter of 1848 wore away without intelligence, the painful silence became intolerable. Inaction was no longer endurable, and with a bound of enthusiasm a noble expedition was planned to go for their relief, scarcely any doubt at this time being entertained but that they could be found, and the survivors rescued.

FIRST RELIEF EXPEDITION.

Early in the spring of 1848, the *Plover*, Commander Thomas Moore, and the *Herald*, Captain Kellet, started to go by the Behring Strait route to Chamisso Island, in Kotzebue Sound, with orders, if Sir John had not arrived there, to go thence to the eastward as far as they could in their ships, and then to forward parties in boats, in the hope of meeting him. This expedition was accompanied by an amateur Arctic explorer, Mr. Robert Sheddon, in his pleasure-yacht, the *Nancy Dawson*.

Mr. Sheddon was an active, energetic man, who rendered very acceptable aid to the expedition. They reached Chamisso Island on the 14th of July, 1849, and with their boats got as far east as Icy Point, and then sent a party forward to try and reach the Mackenzie River. The vessels got north to lat. 72° 51′, and to long. 163° 48′ W., and were then beset in the ice; however, they sent out land expeditions, discovered new lands and islands— one large tract of land about lat. 71° 30′, long. 175° W. On the 24th of August a portion of the boat expedition returned, reporting no signs of the lost; but two whale-boats had still gone on up the Mackenzie River, not intending to return to the ships, but to work homeward by the way of Fort Hope and York Factory.

Those who came back had been as far as Dease Inlet. The expedition remained until the summer of 1850, exploring in different directions, and the next season Captain Kellet, with the *Plover*, wintered in Grantley Harbor, and the *Herald* returned home.

While these parties had been operating from the Behring Strait side of the continent and toward the north-east, Sir John Richardson, the friend and late fellow-traveler of Franklin, had gone with a land searching expedition down the Mackenzie River toward the Polar Sea, which he reached August 4, 1847, leaving dépôts of provisions at intervals all along the route, so that if any of the lost party should stray that way they might at least find some food to sustain them. They traveled eight hundred miles on this journey. In the summer of 1849 Dr. Sir John Richardson returned to England. At or nearly the same time Dr. Rae was exploring the shores of Wollaston Sound, and repeated the exploration in 1850.

SIR JAMES ROSS.

The third expedition, under Commander Sir James Ross, left England on the 12th of May, 1848, for the express purpose of searching the south side of Lancaster Sound to Cape York, and then, if nothing was found, to cross the mouth of Prince Regent Inlet, which they did, wintering at Leopold Harbor. The *North Star*, a transport ship, also followed in the summer of 1849, with stores for Sir James Ross, and wintered at the head of Wostenholm Sound, in lat. 76° 33', the farthest north in that direction any English vessel had then ventured, returning to England the next September.

In the spring of 1849 Sir James Ross extended his search to the shores of North Somerset, lat. 72° 38' N., long. 95° 40' W.; and though he found not Sir John Franklin, he discovered the fact that North Somerset and Boothia were united by a narrow isthmus; he also explored part of the shore north of Barrow Strait, and both sides of Prince Regent Inlet; but all the evidence collected was negative as to those parts having been visited by Franklin. He, too, returned to England in November, 1849.

As the various searching parties reached home, all with the same tale of ill success, the heart of the British public was chilled with the growing certainty that serious disaster must have overtaken the unfortunate Franklin. And in this the interest and sympathy of all intelligent persons in the United States, and we may add the civilized world, was deeply aroused.

SPECULATIONS ON FRANKLIN'S COURSE.

Speculation was rife conjecturing what possible course he could have taken to thus elude the search of so many indefatigable seekers; and, finally, the opinion worked uppermost, at least in England, that he was probably ice-bound among some of the many small islands west of Melville Island.

REWARD OFFERED BY THE BRITISH GOVERNMENT.

In March of 1849 the British Government offered a reward of £20,000 to any private exploring party, belonging to any nation, which should render efficient aid to the lost wanderers. Lady Franklin at the same time sent a large supply of coals and food, which were placed at Cape Hay, on the south side of Lancaster Sound, with the merest chance that some survivor might wander to that vicinity.

A GLORIOUS SPECTACLE.

Had some intelligent inhabitant from another sphere approached our globe in the direction of the North Pole in the year 1850, a sublime spectacle would have met his vision, and one which would have given the strange visitant an impression that the human race was endowed with the keenest sympathies and the noblest sentiments. No less than twelve vessels, besides sledge and boat parties, led by such men as Ross, Rae, Collinson, M'Clure, Osborne, Austin, Ommany, Penny, Forsyth, and De Haven, with many as noble companions, might have been seen all wending their way over the land and ice and snow, and through the waters of the Arctic regions, in search of a lost brother! Surely, if there was ever an exemplification of the humanizing effects of scientific pursuits, we have it in this evidence of chivalrous self-sacrifice. Volunteers had sprung to the rescue on the first intimation from the Government that relieving parties would be organized, and in the year named Great Britain had eight different expeditions abroad in search of Sir John Franklin and his companions.

Dr. Rae was instructed to go farther north than he had yet been; to get, if he could, to Banks's Island, and also to Cape Walker, on the north side of Victoria Land; and at the same time two small parties were to follow the main-land west to Point Barrow—one by the Mackenzie River, and the other by the Colville.

Then there was the Behring Strait expedition, consisting of the *Enterprise*, under Captain Collinson, and the *Investigator*, with M'Clure. These were expected to keep together, sailing to the eastward as far as they could, with special instructions to make friends of the natives, with the view of learning from the Esquimaux if they had any knowledge of the missing party. They were also ordered to place *cachés* in all suitable places, and particularly warned against allowing their vessels to get beset in the ice.

The *Plover*, it will be remembered, was still on the Pacific side, and all three were well provisioned, and officered by capable and determined men.

On the other side of the continent, bound for Baffin Bay, was Captain Austin with the *Resolute*, Captain Ommany in the *Assistance*, Lieutenants Cator and Osborne with the *Pioneer* and *Intrepid* —the two latter being screw-propellers. These all sailed in the spring of 1850.

All thus far named were Government vessels. Then there was raised and fitted out by public subscription the schooner *Felix*, with a small tender, the *Mary*, under Captain (Sir John) Ross. He started in April, and was provisioned for eighteen months, expecting to take, as he did, an entirely different course from the rest—going by Cape Hotham, on the west-side entrance of Wellington Channel, and intending to search all the headlands west to Banks's Land; then, unless he was happily successful, he was to send back the tender, and go on himself in the *Felix*, and to winter as far west as he could get.

LADY FRANKLIN.

Lady Franklin also fitted out a vessel, bearing her own name and title, officered by Captain Penny, having the brig *Sophia* in company. Captain Penny had no positive orders as to his course, which was left entirely to his own judgment. Lady Franklin also bore a considerable portion of the expense of fitting out the *Prince Albert*, commanded by Captain Charles Forsyth, who was accompanied by Mr. W. P. Snow, of New York, who went to England for that purpose—he as well as the commander being a volunteer in the humane work. They went to the shores of Prince Regent Inlet and the Gulf of Boothia. They also sent out overland parties to explore the west side of Boothia to Dease and Simpson straits. This expedition sailed in June of 1850.

THE FIRST AMERICAN EXPEDITION.

The first American expedition recognized by the Government was chiefly indebted, both for vessels and equipment, to the liberality of Mr. Henry Grinnell, a merchant of New York—the

United States Naval Department furnishing an able commander for the two vessels, *Advance* and *Rescue*, in the person of Lieutenant De Haven. This expedition sailed from New York on the 24th of May, 1850. De Haven's plan was to reach, if possible, Banks's Land and Melville Island, and then to use his discretion, as events indicated, as to going west or north.

DISCOVERIES AT CAPE RILEY.

Of all these parties, Captain Ommany was the first to find any evidences of the missing party. At Cape Riley, August 23, 1850, he came upon the site of what had evidently been an encampment, namely, the stone flooring of a tent, a quantity of birds'

ADVANCE AND RESCUE.

bones, and the indubitable proofs of civilization in a number of empty meat-canisters, with other small relics; and this occurred in the fifth year since Sir John had been absent and unreported.

DISCOVERIES AT BEECHEY ISLAND.

At Beechey Island, three miles west of Cape Riley, at the entrance of Wellington Channel, Lieutenant Osborne found the first winter-quarters of the missing expedition. First, there was an embankment for a house, with carpenters' and armorers' workshops, the inevitable empty meat-cans, and, most conclusive of all, the graves of three men belonging to the *Erebus* and *Terror*.

58 ARCTIC EXPERIENCES.

These bore the date of the winter of 1845–46. Farther inland on the island were found some articles of wearing apparel.

Lieutenant De Haven arrived at Beechey Island on August 25, just two days after Lieutenant Osborne, and continued the search, as did also Captain Penny and the officers of the *Prince Albert*.

TEN EXPLORING VESSELS MEET.

On the 27th of August, 1850, there met at Beechey Island, as if drawn by an irresistible instinct, ten of the searching vessels. Besides those already named, were Sir John Ross, Austin, and

ARCTIC DISCOVERY SHIPS.

M'Clintock. But though the *débris* of Sir John Franklin's party was unmistakable, the searchers looked in vain for any record or document of any description. Nothing was found to indicate which way they had taken when breaking out of their winter-quarters. It was consequently inferred that their departure had been sudden—hastened, perhaps, by some unexpected movement of the ice—though the greater probability is that there was nothing special to record, or time would certainly have been found to deposit some writing.

But it is not impossible that some writing, really existing and concealed by the snows and *débris* of five winters, remained undiscovered; for we know that on another occasion, when a party landed from the *Prince Albert*, an experienced officer walked over a cairn, much more recently constructed, and mistook it for a part of the cliff.

Several vessels lay by Beechey Island during the winter, so as to be ready in the spring of 1851 to renew the search, organizing land expeditions in the mean while to explore the shores of Wellington Channel, the coasts of Banks's Land, and the waters leading from Barrow Strait to Melville Island. Different routes were selected, and six hundred and seventy-five miles of new coastland was discovered and examined. But Franklin was not found.

Of all the explorers entering through Baffin Bay, M'Clintock got the farthest west of these expeditions—namely, to 114° 20' W. long., and to 74° 83' N. lat. In this region the animals were so tame that it was quite evident they were unused to the presence of man, and hence the inference that the natives did not hunt in that direction.

The search in Wellington Channel having developed no signs of Franklin's presence, it was now thought by the most experienced that he had probably moved toward the south-west (rather late to come to the conclusion that he had gone where he was ordered). Captain Penny had gone to the northern limits of Wellington Channel until he found another stretch of water, which he called Victoria Channel.

DR. KANE'S DISCOVERY.

Dr. Kane, who at this time accompanied Lieutenant De Haven as surgeon, discovered what he thought to be traces of heavily-laden sledges, and judged from their direction that Franklin had certainly gone north from Cape Riley with his ships on the breaking up of the ice in 1846, and that through Wellington Channel he had reached the Polar basin, and in this direction the *Advance* sailed as far as it was possible to proceed. Dr. Kane's daring on this occasion earned for him among the British officers the sobriquet of the "mad Yankee." But no more relics were discovered at this time, though the record afterward found at Point Victory proved that Kane was right, and that Franklin did attempt that course, but was turned back.

The *Advance* returned to New York September 30, and the *Rescue*, under Lieutenant Griffith, October 2, 1851.

When the discoveries on Beechey Island were first made, the *Prince Albert* had been sent home to carry the news to England, but was almost immediately dispatched back again to search on the shores of Prince Regent Inlet and the neighborhood of Fury Beach and Cape Walker.

In the mean while Rae's search, in 1851, had been to the south of the others, and as his search had been very thorough, upon his report, it was concluded that Franklin had at least not been south of the American main-land coast or its connecting peninsulas.

RUMORS OF MURDER AND CANNIBALISM.

Sir James C. Ross reported a rumor to be current among the Esquimaux to the effect that Franklin's party had been murdered by the natives in Wostenholm Sound, and that cannibalism had been resorted to; and to verify or dispose of this story, Lady Franklin dispatched the *Isabel*, Commander Inglefield, to the section indicated. He found nothing to confirm the report, and subsequently went up Smith Sound to lat. 78° 28' 21"—one hundred and forty miles farther than any one had yet been in that direction.

He brought back the encouraging statement that in the highest latitude he had attained he had found the climate more genial, and that the winds from the north were less cold than those from the south. He established the hitherto only suspected fact of the existence of Kennedy Channel.

Dr. Kane's theory in regard to the Wellington Channel route appears to have made a deep impression in England, for we next learn that in April, 1852, Sir Edward Belcher, with five vessels, namely, the *Assistance, Resolute, North Star, Pioneer*, and *Intrepid*, with several tenders, were sent out; the *Assistance* and *Pioneer* were especially detailed to proceed to the extreme limits of Wellington Channel.

In the spring of 1853 Messrs. Henry Grinnell, of New York, and George Peabody, of London (the latter furnished $10,000), with other private parties, fitted out another expedition under the command of Dr. E. K. Kane—he who had been "surgeon, naturalist, and historian" of the first Grinnell expedition under De Haven.

Lady Franklin, in that year, also sent the *Rattlesnake* and *Isabel* to Behring Strait to assist Captains Collinson and M'Clure; and Dr. Rae again went to Boothia; and, lastly, the *Lady Franklin* and *Phœnix*, under Captain Inglefield, was sent to Barrow Strait to aid Sir Edward Belcher.

THE NORTH-WEST PROBLEM SOLVED.

This year witnessed the actual accomplishment of the passage from the west by M'Clure, and from the east by Captain Kellet, one of whose officers, Lieutenant Pim, met M'Clure on the ice between the latter's ship and Dealy Island. Twenty days later Captain Collinson came up in his ship; found the north-west passage solved, and turned to the south-east, completing the passage in another direction. M'Clure, having wintered in 1850 near where the connecting waters could be traced, had, by observation, established the passage as early as October 31 of that year.

After reaching England, Captain M'Clure was knighted by the Queen; but Collinson, who equally deserved the credit, received but an honorary medal.

BELLOT.

Out of the mass of records of suffering, danger, and death which had attended many of the expeditions, the imagination and sentiment of nearly all the Arctic historians have singled out for special sympathy and commiseration the fate of a gallant young Frenchman, named Réné Bellot, who accompanied Captain Kennedy, and afterward Captain Inglefield, as a volunteer explorer; and who, during a violent gale of wind, was blown from a piece of floating ice and drowned, August 18, 1853. On September 4, a boat, containing his chart, journal, and other personal effects, was floated down into the hands of Sir Edward Belcher, by whose party they were picked up and preserved.

OBTUSENESS OF THE BRITISH NAVAL BOARD.

Of all the inexplicable occurrences with which the history of Arctic expeditions has at times astonished the world, none appears to us so utterly unintelligible as the course of the English Admiralty in its instructions to the rescue parties sent out during the first six or seven years in search of Sir John Franklin. The amazing fact confronts us that not one of these exploring parties, ostensibly sent for his relief, were directed to that section of country where he was most likely to have been found.

The Admiralty instructions by which Franklin was to be guided, directed him to go (*vide* Sec. 5 of the "Instructions") "through Lancaster Sound and Barrow Strait, *without stopping to examine any openings* to the northward or southward of the latter; but to push on to the westward without loss of time on the parallel of about $74\frac{1}{4}°$ to Cape Walker. From that point we desire that every effort be used to penetrate to the *southward* and *westward* in a course as direct to Behring Strait as the position and extent of the ice, or of land at present unknown, may admit."

Yet, in the face of these instructions, not one of all the searching expeditions, whether fitted out by the Government, the Hudson Bay Company, or by private generosity, either from England or America, were directed to Melville Sound the only spot where these instructions could, if followed, have carried him. Almost every other accessible part of the Arctic regions was faithfully scoured from Baffin Bay to Behring Strait, but not the place to which he was sent!

PROVIDENTIAL MENTAL COERCION.

The extraordinary obtuseness on this point existing among so many men of intelligence, and many also of practical Arctic experience, accustomed to obey Admiralty orders, really looks like one of those Providential coercions by which the minds of men are controlled for purposes not perceived until the time has long passed, with all its exciting discussions and prejudiced interests.

We may now possibly perceive a utility in this abnormal condition of the reasoning powers which led the British Admiralty continually away from their own instructions to Sir John Franklin.

Let us suppose that the first rescue party had been sent to where he was—south of Melville Sound, and the whole mystery had been at once cleared up. It is quite possible that, there and then, Arctic explorations would have received their quietus for many years, and the splendid series of discoveries which have since followed by Ross, Parry, Rae, Back, Penny, De Haven, Kane, Hayes, Hall, and others—nearly all looking in the wrong direction—would have remained in silent obscurity, their thrilling stories all untold. On the theory of an ulterior Providential intent can we alone explain the singular conduct of the Naval Board.

It was not until five years after the question of Franklin's safety was mooted that Dr. Rae penetrated to Cape Walker; and beyond that there seemed a fatality, brooding over all the explorers which tabooed the only true and proper course to the south and west of Melville Sound. Every place to which he was not sent was thoroughly ransacked; whither he was sent not a single ship or man was ordered by the British Admiralty.

The region referred to lies between 103° and 115° W. long., and between the 73° and 74° parallels of N. lat.

THE FORLORN HOPE.

After there was none, or the faintest possible hope, that any survivors of the party remained, Lady Franklin succeeded in getting the little steam-yacht *Fox* fitted up for a final conclusive search. This vessel sailed from Aberdeen on the 1st of July, 1857, under the command of Lady Franklin's devoted friend, the experienced Arctic explorer, Captain M'Clintock. He met with a most unparalleled and provoking delay during his first season, by getting entangled in the pack-ice off Melville Bay, in which he was inclosed, and finally drifted to the vicinity of Disco, and southward.

Reaching free water on the 24th of April, 1848, after a drift of two hundred and forty-two days, and, as he estimated, eleven hundred and ninety-four geographical miles — the longest and most extraordinary on record until we come to that of Captain Tyson's, which was fifteen hundred miles, and under the greatest contrast of circumstances. The drift of De Haven had approached it in length, and that of the abandoned ship *Resolute* exceeded it in romantic interest; while the ice-floe drift of the captain and crew of the German exploring ship *Hansa* alone affords any sort of parallel to the ice-borne waifs of the *Polaris*.

Refitting at Disco, Captain M'Clintock started to recover his lost ground as early as the ice would permit. He had taken from Goodhavn the tombstone commemorative of Sir John Franklin, which was prepared in New York under Lady Franklin's orders, and which had been originally put in charge of Lieutenant Hartstene when he went to the relief of Dr. Kane, and which had been left by him in Greenland.

Proceeding direct to Beechey Island, Captain M'Clintock erected the monument in a suitable and conspicuous spot, and then proceeded, *via* Prince Regent Inlet, Bellot Strait, and Franklin Channel, to King William Land, which he searched as thoroughly as time permitted. At Erebus Bay he found a boat containing two skeletons, which were identified as belonging to the missing explorers; and while he was thus encouraged to hope that the solution of the mystery was close at hand, his lieutenant, Mr. Hobson, had actually found at Victory Point the record which told of the death of the gallant Sir John, for whom two nations had been in search for ten successive years. The date of Sir John Franklin's death was given as occurring on the 11th of July, 1847—two years from the time of his leaving England. But what became of the one hundred and five men living at that time remained, to a great extent, as mysterious as ever, until Captain C. F. Hall's explorations of 1864–69. The Esquimaux story of the death of the greater part by exhaustion and starvation, and probable cannibalism, as reported by Ross and Dr. Rae, might be true, but up to the time of Hall's researches they had not been positively proved, or that all who composed the expedition of the *Erebus* and *Terror* were dead; and of the ships no certain knowledge was obtained till Hall's visit. After collecting ample supplies of relics belonging to the lost party, Captain M'Clintock returned to Aberdeen, and thenceforward the search was considered ended by the English Government and people.

Not so in this country. There was one man, at least, in the United States, away to the westward, in the noble State of Ohio, who was pondering by day and dreaming by night of the possible fate of some poor soul yet surviving among the Esquimaux; that man was Charles Francis Hall. What he did toward clearing up the mystery will be found narrated in the chapter entitled "Biographical Sketch of Captain Hall."

During 1853–55 Dr. Kane was pursuing the search in a northerly direction, *via* Smith Sound, making extensive discoveries on the west coast of Greenland—once attempting to reach Beechey Island on a sledge-journey, but driven back by an impenetrable barrier of hummocky ice. But his own account is too recent and familiar to need extended notice here, except to say, in explanation of his taking what appears, at the first thought, so unlikely a route as Smith Sound to find Franklin, who was known

FINDING REMAINS OF SKELETONS IN A BOAT.

to have gone to the west; that having discovered, as our readers will remember, the sledge-tracks at the mouth of Wellington Channel, he believed the missing party had sailed north through those waters until they had reached the Polar Sea, and that they were there detained; and that, consequently, the most direct way of reaching and aiding them was to get to the same place by the most feasible route, which he believed was Smith Sound and connecting waters.

Dr. Hayes's journeys and subsequent publications have familiarized the reading public, not only with the history and resources of Greenland, but also with the contour of the western shore of Smith Sound and Grinnell Land, and beyond up to Cape Union, the most northern point observed up to that period, or until the observations made from the *Polaris*.

While Hall was searching Frobisher Bay, Hayes was heading due north along the shore of Grinnell Land with a perseverance and courage which, since the death of Captain Hall, leaves him with scarcely a rival in Arctic research.

PROFIT AND LOSS.

Among those general readers who have not made a specialty of Arctic literature, but are familiar only with the widely-bruited failures of certain unfortunate explorers, it naturally enough appears that the expense, suffering, and loss of life far outweigh any possible benefits to be derived from continued explorations. But the degree of publicity given to Arctic expeditions has usually followed the reverse rule which prevails in other more or less speculative enterprises. In most of these, the successes, like the drawing of a grand prize in a lottery, is heralded throughout the land, while the failures are quietly kept in obscurity. The very opposite course has habitually occurred in the matter of Arctic explorations—the disasters, the losses, the deaths have all been made the most of, published with every setting and surrounding, to make them appear the main results of an expedition, when, in fact, they have been but the accidents, the unavoidable incidents, of traveling over new and unknown lands or seas. All the disasters which have befallen Polar voyagers are not to be attributed to the climate. The latitude has been innocent of many misfortunes attributed to it. Very often has an expedition been marred by bringing together, without sufficient discrimination,

DR. KANE.

ill-assorted companions, whose incongruity has only become the more apparent the longer they were kept together; and still more frequently has their efficiency been greatly impeded by the continued use of disease-inducing food, and other reasons which may now be regarded as belonging to the past, and in no way necessary of repetition in organizing future expeditions.

WHAT IS THE USE OF ARCTIC EXPLORATIONS?

Notwithstanding the immense additions which Arctic explorations have made to our scientific knowledge, there are still people left who persist in asking, "What is the use of Arctic explorations?" and though we might attempt to answer them, we conceive that, to minds capable of propounding the above proposition, whatever the answer, it would prove unsatisfactory to these querists; for that sort of intellectual apathy which can regard any kind of knowledge as useless makes hopeless the discussion of its value.

If all study, inquiry, and experiment were suspended, except such as promised immediate and profitable return, a large proportion of the brain-force eventually prolific of the most interesting and valuable results would be incontinently obliterated. Had the scientists of past ages been obliged to show a present paying application of the crucial tests with which they endeavored to force open the sacred arcana of nature, the nineteenth century would scarcely have reached to the mental condition of the Dark Ages; or, rather, the nations would never have emerged from the sensuous bonds of barbarism—a showy and luxurious barbarism, perhaps, but certainly resting in a condition devoid of any high culture or worthy mental effort.

But to those, on the other hand, imbued with a genuine love of knowledge for its own sake, no expense, toil, or suffering seems too great to purchase even the slightest addition to our sum total of facts; still more valued is the elucidation of those suggestive ideas which lead to the discovery of governing laws and fixed principles in art, science, morals, or philosophy; and thus the answer to the query, "What is the use of Arctic explorations?" will be as different as the life aim of those who discuss it.

REMOTE ADVANTAGES.

Some brains appear organically incapable of comprehending or grasping remote and contingent uses; they must see the finished product turned out by the complicated machinery, or they can perceive no virtue in belts and wheels, cams and eccentrics; and to such as these the fact that a north-west passage through the Polar regions is obviously impracticable for commercial purposes, settles the question that all geographical and scientific research in that direction is a mere senseless battering of human endurance against the elemental forces of cold and ice and sunless atmospheres.

ANCIENT GRADGRINDS.

This sort of people, had they existed, as no doubt many did, in the B.C. eras of the world, would have stood mocking at the simple gnomon, and the apparently demented individuals who, hour by hour, day by day, month by month, and year by year, continued, with enthusiastic and tireless vigil, to watch, measure, mark, and record the exact length of the shadow which it cast. "What is the use of knowing the length of that shadow?—what a fool to spend life in that way!" cried the old Gradgrinds of China, Arabia, and Egypt. Yet out of this seeming frivolous

DR. HAYES.

proceeding grew up the noble science of astronomy, without a knowledge of which our mariners would still be creeping around the shores of continents, not daring to venture out of sight of land.

PENDULUM EXPERIMENTS.

Early in the present century various stations in the Old and New Worlds, ranging from the equator to Nova Zembla, were visited by a veteran traveler, who was looked upon by the ignorant as "crazy." He always carried with him various clocks and detached pendulums; and when he landed at any point, his first object was to secure a solid foundation whereon to place his clocks; his next, to find or construct, partly underground, a small building, within which to suspend his pendulums, and this inclosure he always insisted must be so contrived as to secure a perfectly equal temperature during the whole period of his observations.

He came and went in a vessel of the British navy, and much the seamen wondered; and the natives of various half-civilized countries looked on in astonishment, asking, "What could be the use of that old man burying his clocks and pendulums for weeks at a time, when no one but himself and assistant could see them?" But the scientific world has long known that the "crazy man" was the renowned Sabine, and that through his pendulum experiments was obtained the exact difference of clock rates at the equator and successive parallels of latitude approaching the poles, and thus was secured the necessary data for calculating the oblateness of the earth, and sequently accounting for the precession of the equinoxes.

SCIENTIFIC DEVOTEES.

The ultimate uses of scientific research, under which head may now be classed all Polar explorations, is scarcely ever realized in its fullness even by those most devoted to the pursuit; but it may safely be asserted that no kind or degree of positive knowledge has ever been obtained without yielding valuable fruit—yes, a hundred-fold as compared with the toil of obtaining it; not often, however, does the reward fall to the immediate discoverer, but more frequently on the world at large is the eventual benefit conferred. Thus it has been in mathematics, and notably in chemistry, in geographical research, and in every branch of science. The way is strewn with willing martyrs, whose enthusiasm has been inexplicable to those of cooler temperaments, greater caution, and lacking that insight which enables the scientific devotee to count a perilous or toilsome life-long effort as nothing, if the truth be learned, the experiment succeed, the end be gained, the victory won, and at last the doubters and mockers be overwhelmed with the practical benefits evolved, making the uses of science intelligible even to them.

ARCTIC FAILURES AND SUCCESSES.

In regard to Arctic expeditions, though all have so far failed of reaching the geographical pole, yet none have been wholly failures; from each something has been learned, by which succeeding ones have profited. And though to the inconsiderate, who draw their deductions rather from the great hopes of Captain Hall than from the actual results, the *Polaris* expedition has been called a great failure, we think it will be found by all who peruse these pages that, instead of being a failure in any scientific sense, it has been a great success, not only in the fact of Captain Hall having reached with his ship a higher latitude than any ever yet attained in that direction, but also in the added stock of scientific observations made on board, especially those which show entirely unsuspected conditions of magnetic polarity, as evidenced by the dip, and the amount of variation in the needle in the high latitudes reached by the *Polaris*, with many other facts of value secured by the Scientific Corps.

MODERN FACILITIES.

The probability is now becoming every year more apparent that, with the increased resources of modern art, difficulties to which the earlier navigators succumbed will be effectually surmounted. The art of preserving, by hermetically sealing, so many varieties of food, has already reduced the dangers of sickness in uninhabited regions to its minimum, while each year adds something to the mechanical contrivances which makes an extended residence in the Polar regions less and less hazardous; and there appears no reason why success in the now limited object of search should not, within a comparatively brief period, perch upon the banner of some succeeding Arctic explorer.

The famous navigator Captain Cook thought that no one would ever get nearer to the South Pole than he had done; yet Sir James Ross and our own Wilkes have made his discoveries insignificant by their greater daring. Pigafetta, the companion of Magellan, when he had circumnavigated the globe, thought he had performed a feat which "would never be repeated by mortal man!" while the larger-brained and larger-hearted Columbus, when he first touched the shore of San Domingo, exclaimed, "El mundo es poco!"—*the world is little!*—as if in his soul he longed for greater dangers to overcome, and wider seas over which to sail.

This was the true spirit of progress, he the true enthusiast; while the men who are forever asking "What is the use?" are the drag-anchors of society, who, if listened to, which, happily, they are not, would keep the world forever in its swaddling-clothes.

UNEXPLORED AREA.

Of the land surface of the earth, 11,600,000 square miles, or one-seventeenth of the whole, is grouped within the Polar regions—a tract, as the French *savant*, Réclus, points out, sixty times the area of France. Now, should it not convince the most obtuse that, while scientists remain practically unacquainted with so much of the land surface of the planet, that there must necessarily remain many problems, not geographical only, but meteorological, tidal, and especially electrical and magnetic, unanswered? How give a true solution of any complex scientific subject, while so many of its component parts remain unknown quantities?

That the north Polar region, with its grand focal problem, will yet be conquered by the courage of our navigators and the self-sacrificing spirit of science, during this or a near succeeding generation, we have not the shadow of a doubt.

MODERN CHIVALRY.

There is one other part of this subject which we can not overlook, namely, the development and encouragement of a noble, chivalrous sentiment, which in these latter days has but few opportunities for exercise. The giants and the dragons are all dead, and the chivalry of the Mediæval Ages is no longer needed by the gentle dames, who are learning so rapidly how to fight for themselves. Livingstone and Baker are bound to exhaust Africa, and what is there left? The Arctic regions alone remain a *terra incognita*, so attractive to the knights-errant of science. Where, then, shall the Mr. Greathearts disport themselves, if not in the land of the Aurora? Away with your calculating financiers, who count the cost of every thing to the uttermost farthing; and give place to the royal enthusiasts who are ready once more to try again—ready to attack and demolish the only geographical mystery left to this book-whelmed generation.

Do you doubt the courage? do you doubt the chivalry? Hunt up your books of travel, bring out your biographies, and see if you can find a parallel to the courage, skill, endurance, tact, self-control, and Christian trust in an ever-guiding Providence, which enabled the chief officer on the ice-floe, Captain Tyson, to maintain, without any positive exhibition of authority, a tranquil, firm, and careful oversight of the eighteen persons providentially thrown upon his direction. "He could have made his own way back to the ship, but he would not desert the women and children," said one. "There was not much commanding done on the ice; but if we went contrary to what he advised, it always turned out wrong," said another; and this for six months, on a voyage of over fifteen hundred miles on broken and shifting ice! "For eighty days the sun did not show itself above the horizon; and when it did, only for a few hours at a time:" denying himself needful food, that others might not lack; encouraging and supporting the desponding, and with his great physical strength, acquired by long acclimatization, holding the weaker to their places while the winds and waves contended for their hunger-smitten bodies.

A PURE AMBITION.

No; if it is recorded within the archives of fate that Arctic explorations are to be forever baffled of their great purpose, let us take no hand in crushing out the spirit which inspires them. It is a pure and healthful ambition to add to the world's knowledge; to carry the flag of our country where human foot has never trod; to unravel the mystery of ages, and to close up the hopes and efforts of the centuries with a successful invasion of those hidden realms which kings and princes have desired to see unveiled. How much nobler and purer is such an ambition than is the unceasing strife for gain, or the petty jealousies of office-hunting, or the belittling pursuit of ease and pleasure, as the end and aim of life? Instead of discouraging and repressing the spirit of adventure and research, it becomes every Government, and brings honor to every people, who systematically cherish and foster it; and that life is *not lost* which is sacrificed in such pursuits.

ESQUIMAU WOMAN'S KNIFE.

CHAPTER II.

CAPTAIN TYSON'S EARLY ARCTIC EXPERIENCE.

Captain Tyson's Reflections on the Ice-floe.—Nativity.—Early Life.—Ships as a Whaler.—Death of Shipmate.—Arrives at the Greenland Seas.—The "Middle Ice."—The "North Water."—First Sight of Esquimaux.—The Danes in Greenland.—The Devil's Thumb.—Meets De Haven.—Whales and their Haunts.—A prolonged Struggle with a Whale.—Sailors' Tricks.—Cheating the Mollimokes.—Young Tyson volunteers to winter ashore at Cumberland Gulf.—The Pet Seal.—Life Ashore.—Relieved by the *True Love*.—Is taken to England.—Returns to the Arctic Regions.—Sights the abandoned British Ship *Resolute*.—With three Companions boards the *Resolute*.—Finds Wine in the Glasses.—All have a good Time.—Don the Officers' Uniforms.—Returns to his Ship.—Ships as Second Mate in the *George Henry*.—As First Officer.—As Captain of the Brig *Georgiana*.—Meets Captain Charles F. Hall.—Witnesses and tries to prevent the Loss of the *Rescue*.—Sails as Master of the *Orray Taft*, of New Bedford.—Of the *Antelope*.—Sails to Repulse Bay, and takes the first Whale captured in those Waters.—Again meets Captain Hall, and supplies him with a Boat.—Peculiar Electrical Phenomena at Repulse Bay.—Sails in the Top-sail Schooner *Era*.—Meets Captain Hall, then "in training" with the Esquimaux.—Log-book Records.—Winters ashore at Niountelik Harbor.—Removes from New London to Brooklyn.—Sails in the *Polaris* as Assistant Navigator.

"WHILE floating down on the ice-floe, in the midst of dirt and darkness, hungry and cold, I often thought of friends at home, and wondered how many of them would have been able to endure the exposure to which our whole party was subjected; and, most of all, I wondered at myself that I could have learned, in a few short months, to have eaten such things, and submitted to such practices, as but few civilized persons have ever been called to endure. In regard to the physical strength and vital resistance of my system, that is no doubt to be accounted for, in great measure, by my previous life.

"I had been for twenty-three years sailing the Northern and Arctic seas; I had not seen a Fourth of July within the United States for twenty years, and but a few days of summer weather in this latitude for the same period; a part of the month of June, 1871, just before the *Polaris* sailed, was all the summer I had experienced in the temperate zone for two decades. I had plenty

CAPTAIN TYSON.

of time to think on the ice, but, unfortunately, most of the time, in consequence of being on short allowance, and much of the time feeling a grinding, tearing hunger, one could think of nothing but about 'something to eat.'

"But occasionally, after we had captured a good fat seal, and we had all been able to satisfy our hunger for a few hours at least, the stomach *could* be forgotten; and so at intervals I jotted down a short sketch of my previous voyages; and as they all show more or less of a contest with the ice-fiend, and will help to explain my capacity for endurance, it may as well go in as a prelude to my journal on the ice-floe.

"I was born in the State of New Jersey; but in early infancy my parents removed to New York city, where I received my early education, and when of suitable age I commenced work in an iron-foundry—my parents, like nearly all others, desiring to keep their sons upon the land. But my heart was always on the seas, and particularly I longed to see something of the Arctic world; the names of Ross and Parry and Franklin had seized upon my imagination, and I longed to follow in their track. To witness the novel scenes, and to share in the dangers of Arctic travel, was at that period the height of my ambition; and while watching the fiery liquid ore that was presently to appear in the shape of grates and fenders, my fancy was off among the icebergs; and, despite the dicta of Shakspeare, I sometimes almost managed to cool my heated brow with

'Thinking of the frosty Caucasus,'

I was disgusted with shop labor; and as no opportunity offered for joining any Arctic exploring party, I concluded to do the next best thing, and ship in a whaler, which at least would bear me a few degrees toward the coveted regions of perpetual ice.

"In execution of this intention, I shipped on board a New London whaler, the bark *M'Clellan*, Captain William Quayle, in 1850, when I was about twenty-one years of age. The *M'Clellan* was bound to Greenland and adjacent seas. It was Captain Quayle's intention, however, to take the sealing on the coast of Labrador first, and for that purpose sailed very early in the season, leaving New London on the 7th of February, 1850.

"After being out a few weeks, one of our shipmates died, and was buried at sea. This seems a slight thing to record, because the poor man has left no historic name behind him; but no one who has not experienced it can realize the great solemnity of a burial at sea, especially when witnessed for the first time by a young person unhardened to the vicissitudes of life. When we deposit our friends in the ground, there seems something left of them; we can at least visit their graves, and adorn them with flowers, and fancy that they know we still care for them; but when the poor discarded body slides over the ship's side, and strikes the water with that heart-sickening *thud*, it appears as though we were giving up our late friend to a more certain and eternal separation. The imagination follows it, indeed, for a while, along the known currents which set to or from the ship, but beyond that we know not its journey, or whither it is carried—whether it ever comes to rest, or is ceaselessly borne about by the ever-shifting waters, until the continual friction first denudes the body of its covering, and then the bones of its flesh, or, perhaps, that it is destined to furnish a ghastly meal to some monster of the deep.

> "'A plunge and a splash, and our task was o'er;
> The billows rolled as they rolled before,
> And many a rude prayer hallowed the wave
> As it rolled above his ocean grave.'

Nevertheless we know that at the last they shall not be forgotten. We have the promise that 'the sea shall give up its dead.'

"We had a rough passage to Newfoundland; but on getting in the ice off the coast of Labrador, we found much smoother sailing, but very cold. Here we met a vessel in the ice (I forget her name); she was out of provisions, and a part of her crew came over the broken ice to solicit aid, and not in vain. Captain Quayle gave them all the food they could carry, with which they joyfully returned to their own vessel.

"Finding no seals, we soon left for the Greenland seas, arriving off Resolution Island in the early part of March; and I shall never forget the terrible weather we experienced in this vicinity—lat. 62° N., long. about 60° to 62° W., just north and east of Resolution Island. I have been much farther north since then, but I never remember feeling the cold more, except while on the ice-floe; and I don't think I should then, only for lack of suitable clothing and sufficient food.

"We were now compelled, through stress of weather, to run into the pack of ice for protection. There we found the wind less violent, and no sea—the unusually heavy floe entirely destroying the force of the huge waves as they beat against it. In the ice here we found many seals of the 'bladder nose' species—so called from the bladder, or hood, which they have on their head; this hood, when they are excited or angry, they can expand to a great size, and then you may know they are ready to defend themselves. It was from this early experience off the west coast of Greenland that I derived my knowledge of the haunts of this kind of seals; and it was this which gave me hope through that dismal Arctic winter, and enabled me to hold out encouragement to the rest of the party while we were on our long drift, knowing that once we could reach this latitude, all fear of starvation would be at an end.

"On this my first voyage we killed numbers of these seals, and many a good battle we had with them. The male, in particular, will defend his family to the death. These seals are quite large, and are taken, like the whale, mainly for the oil.

"Finding no whale off Resolution Island, we next bore away

CAPTURING THE SEAL.

for the coast of Greenland, it being the captain's intention to work his vessel up the coast, through Baffin and Melville bays; and so, getting north of the ice-pack, go to the westward and cross Lancaster Sound, to Pond Bay, where we were sure to find whales. The whalemen, I may as well here explain, always work their ships up the east coast of Greenland in the early summer, finding less ice than to the westward.

"What is called the 'middle-ice,' or 'pack-ice,' is usually encountered between the northerly part of Baffin Bay and Cape

York. Sometimes the belt is wider than at other times; there is a difference in seasons. Sometimes it extends as far south as Cape Walsingham. North of this 'middle-ice' is what whalemen call the 'north water,' which is always free in summer.

"The whalemen always work their ships up the coast of Greenland between the fast bay-ice and this middle or pack-ice, when making for the north water, because there is less ice there than on the westerly side of Baffin Bay; then, having reached the north water, and crossed to the west, make their catch of whales, and start to return home in the latter part of August or early in September, so as to avoid being caught in the fall 'pack,' which at that time of the year is coming down. There are always two packs: one in the spring, caused by the early breaking ice, and the other in the fall, caused by the breaking up of the ice, which has required the accumulated heat of the whole Arctic summer to start it from the sounds, fiords, and inlets, which are less influenced by the warmer under-currents of the open bay.

"But to return to our voyage. We first sighted the land at Holsteinborg, a settlement in the south of Greenland. The land in that vicinity, and, as I afterward discovered, nearly the whole coast, is high and mountainous, presenting a most desolate appearance. If I had not known the fact, I could scarcely have believed that it contained any inhabitants. But we soon had ocular demonstration of this; for, getting the ship in the ice, we were almost immediately 'beset,' and very shortly after a number of these hardy sons of the North were seen coming over the broken ice to pay us a visit; and their first appearance convinced me how much more one is impressed by *seeing* than by reading or hearing. Of course I had always heard of the small stature of the Esquimaux, and thought I knew just how they looked; but when I saw these little creatures approaching—the men less than five feet, and the women not more than four—I realized the difference of race in a way one can not do without seeing. I have often thought since that nature made them so small that they might travel the more easily over the thin ice and the snow, as they often have to do, in pursuit of seals and other game. If they had been made as large and heavy as many of the white race, they would be far worse adapted for the mode of life which that barren country forces them to adopt. As many of them

UNIV. OF
CALIFORNIA

CONGRESS AND POLARIS AT GOODHAVN.

were dressed in seal-skins, with round seal-skin caps on their heads — and, when laid horizontal on the ice, about the length of the smaller kind of seals — I could not help thinking but that God had made them thus, with their brown faces, so that they could imitate the creatures, and so decoy and catch them, which they often do. Holsteinborg, I afterward found, consisted of about a dozen huts, or houses, and less than fifty inhabitants.

"The ice did not detain us long here. It soon opened, and we proceeded northward to Disco, where we were again stopped by the ice. Disco is a regular rendezvous of the English whaling-fleet, as well as being frequently visited by the American whalers. It is on an island, and is a larger settlement than Holsteinborg; the place of the anchorage is called Goodhavn. There are over twenty houses here, and, I was told, seventy or eighty people. I have since ascertained that the Danes who come out here in the governor's suite, and others who visit the country for commercial purposes, and stop any length of time—especially those who intend to make it their home—not infrequently marry native women; so that at some of the settlements you may see a family where the children have the light, flaxen hair of the Dane, and the dark, bronzed cheek of the native. This mixture makes a curious physiognomy.

"The highest point to which we sailed on my first voyage was called the 'Devil's Thumb,' in Melville Bay. This 'Thumb' is a large pointed rock, like an immense Bunker Hill Monument, that rises perpendicularly to the height of five or six hundred feet; and as it stands on a very high, rocky island, its topmost point is probably fifteen or sixteen hundred feet above the level of the sea. There is a great deal of superstition about this 'Devil's Thumb'—partly, I suppose, from its name; but to those who know the difficulty of steering among the icebergs which abound here, and the cross-currents which swirl the bergs about, and of course treat ships the same, there is no need to go beyond nature for objects of terror.

"The 'Devil's Thumb' is so named from its supposed resemblance to a gigantic thumb; it is in latitude about 74° 30' N. When off this point we were again beset in the ice, and, in consequence, the early whaling season was nearly lost to us—as it is necessary to get to Pond Bay by the 1st of July to be sure of

THE "DEVIL'S THUMB."

whales; and as July was now on us, Captain Quayle concluded to turn south before crossing to the west, so as to take the whales as they came from the north, which they commence to do in August.

"When near the Duck Islands, we saw two strange vessels, with colors set; found them to be Americans; and, on speaking, learned that they were the two vessels sent out by the Government to aid in the search for the brave and lamented Franklin, under the command of Lieutenant De Haven, a brave and energetic naval officer. They had been drifting in the pack all winter, and were now slowly working their way north again.

Captain Quayle supplied them with potatoes, and whatever else he could spare. De Haven returned home that fall.

"After parting with the explorers, we took the ice off Disco about lat. 69° 14' N., long. 53° 30' W., and endeavored to get west by taking advantage of every opening in the ice, and soon after sighted the west coast of Davis Strait. Then a thick fog set in on us. At this time we were surrounded by whales; but it is almost impossible to take them when the ice is loose and broken, on account of their running under the large heavy floes to escape, taking the line with them. But we tried our luck, and fortunately captured two.

"There are several species of whales; but those most sought within the Polar Circle are usually either the 'right whale' or the 'white whale;' the former is much larger than the latter. There is also the 'bow-head' whale, and 'sulphur-bottoms.' All the large whales of this region are 'balleeners;' that is, the mouth and upper jaw are furnished with the balleen, or whalebone, of commerce. When a whale is fastened to the ship, and the cutting and stripping of the blubber is going on, the head is usually first severed from the body for convenience in getting at the balleen; but a boat *can* enter the mouth of the whale, and, if necessary, several men could at the same time stand upright and be at work, removing the whalebone from the upper jaw, the head of the whale being about one-third of the bulk of the creature. The whales change their haunts frequently. When they are too closely followed in one sea, they go to other grounds. In 1810, and a little later, whales were plenty in Baffin Bay and along the west coast of Greenland; then, being too sharply followed, they migrated to Hudson Bay; and when they were followed there they became, after a while, very scarce, and the next we heard that they were very plenty off Behring Strait, on the other side of the continent. Is it possible that the whales have the only practicable North-west Passage to themselves? I remember that brave old whaler Scoresby tells in his book about several whales being found in Behring Strait with harpoons in them bearing the mark and date of ships which sailed only in Baffin Bay; and later sailors' yarns have revived such stories, which I always doubted. But Professor Maury, in his 'Geography of the Sea,' puts it down for fact without any question. If the whales do have an 'underground railroad' to the Pacific, they undoubtedly come and go in both directions; for not many years after they were reported plenty in Behring Strait they were back again to the north-west part of Hudson Bay and Davis Strait. But certain it is, that no whale has ever been found on this side of the continent bearing any evidence that it had traveled from the Pacific. I think the scarcity and plenty, and, within certain limits, the changing of their haunts, is explained by the fact that when they become unprofitably scarce in one location the whalers go to another; and thus give them, for two or three seasons, a chance to breed again.

"The right whale is often fifty or sixty feet long, but the white whale does not average more than fifteen—from twelve to twenty. The blubber produces a very superior kind of oil, and its texture is more gelatinous and less gross than that of the larger whales. In the water this fish is a brilliant, shiny white. A common harpoon is scarcely fit for this fish, for it is necessary to penetrate through the blubber to the flesh to have it hold. The Esquimaux consider the flesh of the white whale excellent eating.

"I once had, when I was boat-steerer, quite an adventure with a whale which was determined not to die. It was a large and valuable balleener. Soon after the boat was lowered we got alongside. As I rose to heave the harpoon, it seemed, almost in an instant, that the whale had plunged down to the bottom of the bay; as the rope uncoiled and went over the gunwale it fairly smoked with the intense rapidity of the friction, and I had to order it 'doused' to prevent its taking fire. It came, too, within a hair-breadth of capsizing us. Fortunately the line was over seventy fathoms long, and of the strongest kind. After she plunged we followed on, it taking all our strength to bring the boat near enough to her to keep the line slack. She staid under water the first time so long that we thought she was dead and sunk. It was nearly an hour before she rose; and when she did, the jerk almost snapped our strong line, already weakened by the friction and unusual tension.

"As soon as she appeared she began to beat the water with her flukes, and swirled around so that it appeared impossible to get a lance in her; and, while I was endeavoring to do this, our line parted, and away she went, carrying the harpoon with her. We followed with all the speed we could force, and at last, after several hours' hard pull, came up with her. She seemed to know we were following, and several times disappeared, and then coming up to blow, perhaps half a mile off; but we were bound to have her. On and on she went, on and on we followed. The moon was shining, and the Arctic summer night was almost as light as day, and deep into the night we followed her. Down she went, for the sixth or seventh time, but fatigue was getting the better of her. She was weakening, while, with all the fatigue, our spirits and strength too were kept up by the excitement. At last, when we had been nearly twenty-four hours on the chase, I got another harpoon in her. This seemed to madden her afresh. Another plunge, which had nearly carried us with her; but this time she did not stay down more than ten or twelve minutes. Up she came once more, the water all around covered with blood, and we knew she was done for. Three or four lances were hurled into her ponderous bulk, and at last our exertions were rewarded by seeing her roll over on her side. She was dead. We bent on another strong line, and soon towed her to a floe. But we found ourselves, with our prize, a good nine miles from the ship. We could not, therefore, save the blubber, but we made a good haul of balleen, with which we loaded our boat to its utmost capacity, and then dragged her, with her heavy cargo, the whole distance over the ice to the ship, which is what I call a fair day's work.

"Sailors have a rough life of it, but they often contrive to amuse themselves in circumstances which most landsmen would consider very miserable. Often when the ice was too thick to bore through with the vessel, and we had to lie to, awaiting a break up of the ice, there would be discovered openings here and there, around which would be gathered flocks of aquatic or semi-aquatic birds. On the small islands, and inshore about the base of the rocks, they chiefly congregated. There were the eider-ducks, the little dovekies, the beautiful ivory gulls, and the voracious, thieving, burgomaster gulls—so named by the Danish settlers, which appears to be a reflection upon the unamiable traits of their oppressive burgomasters at home.

"The last-named birds the sailors are always fond of tricking, because they get so much of their living by snatching food from other birds, sometimes even out of their very mouths, and they also steal the eider-ducks-eggs. But there is another bird equally or more voracious.

"One day I saw a messmate fixing a lot of strings about six

feet long, to the ends of which he affixed a bait of seal blubber; then, tying all the strings together at the other end, and also across the middle, he flung the baited ends overboard. Presently a lot of mollimokes espied the food, and one and another seized a morsel, when, suddenly, jerk went Jack's arm, and out flew the blubber from the beaks of the 'mollis.' Over and over they tried it, until at last, baffled and disgusted, away they flew. But to return to our voyage.

"The fog clearing, we pressed our way along through the

EIDER-DUCKS.

broken ice till we got near shore, where we found clear water, and went into harbor, some sixty miles north of Cape Walsingham; but after a few days, finding no whales, we steered for the cape, where we found most of the English and Scotch whaling-ships. Here we were also unfortunate, and soon left for Cumberland Gulf, in the latitude of Cape Mercy, being at the north side of the gulf, 64° 45' N., long. 64° 30' W.

"It was in the early part of September, 1850, when the *M'Clel-*

lan arrived in Cumberland Gulf, and there had never been but a few ships in those waters, except Captain Penny a short time before, and some few Scotch and English whalemen. But little was known of the gulf at that period.

"Captain Quayle, hearing from the Esquimaux that early in the spring, before the ships were able to get in the bay, there were plenty of whales there, called for volunteers to go ashore and stop through the winter, concluding that the vessel should now go home and return for the shore party when they came up again next year. Twelve men volunteered, of whom I was one. We took our traps out of the *M'Clellan*, went ashore, and pre-empted a section of land whereon to build a hut or house. The captain gave us what provisions he could spare; but it was not much, for the vessel had only been provisioned for the usual trip, and the owners had not anticipated that twelve men would require food for eight or ten months longer than was customary. There was very little lumber either that we could get from the ship, so we built the house of stones, filling the crevices with earth and moss, and making the roof by laying poles across and covering these with canvas; inside we built berths, or bunks. Before winter was over we got very short of food, and could not have survived if it had not been for the game we shot and the seals we caught. We had to learn the Esquimaux ways of eating and cooking, and before spring I was pretty well acclimated; and though the life was so rough and so different to what I had been accustomed to, having lived all my previous life in New York city, yet my health was good; in fact, the whole party kept well.

"We had not many opportunities of making pets of any thing out there; the dogs were too fierce, and small animals of any kind were scarce; but one day I saw a young seal; it looked so pretty, with its pure white coat (the young of the Greenland seal is entirely white) and bright hazel eyes, that I took it up in my arms like a baby, and carried it along, talking and whistling to it by the way. The little creature looked at me, turning its head round to look up in my face without any apparent alarm, and seemingly soliciting me to give it something to eat. I thought I should take a great deal of comfort with my little pet, for I had not then got accustomed to seeing the young ones killed, much less eating them myself.

"Arrived at our house, I carefully deposited it outside in a suitable place and went inside to get my supper, hurrying through my meal to get out and look after my treasure. I looked around, but it was not where I had left it. I began to suspect mischief, and, sure enough, there it was, a little way off, *dead*, with its back broken by the heavy heel of a whaler's boot; one of the men, with a malignancy impossible for me to understand, had pressed the life out of my only pet simply to gratify a brutal nature. Had I been quite sure who was the perpetrator, my indignation would have found other vent, I suspect, than words.

"In the spring we had the satisfaction of knowing that we had not wintered there in vain, as we killed seventeen whales; and, had we been more experienced, we could have captured many more; but this was the first season that any whalemen had passed the winter in that region, and we had every thing to learn.

"As summer approached, we began to look anxiously for our ship. All our original stock of provisions had been long consumed, and we had to hunt hard to get enough to eat; and I scarcely believe we should have succeeded in securing enough to sustain so large a party if it had not been for the help of the friendly Esquimaux.

"While we were busy whaling in the spring, and before we had learned to eat whale-meat—for whalemen only strip off the blubber, and abandon the carcass (having also taken the valuable portions of the bone)—the natives would seize upon the latter and strip off all the meat. What they could not eat they put in seal-skin 'drugs,' or bags, and these they stowed away for future use, hiding the bags by covering them up on the various islands in the gulf or inlet. Subsequently, in our hunting excursions, we often came across these 'drugs;' and if our chase had been unsuccessful, and ourselves very hungry, as was frequently the case, we helped ourselves to these reservoirs of old whale-meat; and as much of it had been lying under the stones for several months, it was not particularly savory; but we were often very glad, indeed, to get it.

"It was not until the month of September—a whole year having passed—that we were rejoiced by the sight of a vessel. On boarding her, we found that she belonged to Hull, in Yorkshire, England, and was named the *True Love;* her captain's name was

Parker. She had formerly been a privateer in the American war of the Revolution, and was at the time I speak of about *ninety years old;* and the good old bark was still afloat but a few years ago, and Captain Parker was still in her as late as 1860, and is nearly as old as the vessel. She has since been lost. The fact is, no vessel will last so long as a whaler, unless accident destroys her; for once get a ship soaked with whale-oil, and it is impossible for her to rot.

"On board of her we were surprised to find our old captain, Quayle. He had lost the *M'Clellan* in Melville Bay; and having put his crew on board of different whaling-ships, and sent them home, *via* England, he, with his boat's crew, was taken up by the *True Love*, Captain Parker kindly consenting to come round and pick us up too; and right glad we were to get a good keel under us again, and some civilized food to eat. But still we could not get home. The *True Love* was bound for Hull, and on the 4th of October sailed for England. Nothing worthy of special notice occurred on the voyage until we reached the Scottish coast, where we encountered a terrific gale, which the good old *True Love* weathered; but another whaler which was in our company went ashore and was lost. Being anxious to get home, I went to Liverpool, and sailed from thence in December, in the *Charles Holmes*, Captain Crocker, an American vessel bound for New York.

"When forty days out, we experienced a heavy gale from the north-west; we were still to the eastward of the Banks of Newfoundland. Having had nothing but westerly gales the whole passage, and our vessel having lost nearly all her sails, and, though a new ship, having made considerable water, and there being nearly three hundred passengers on board, some sick, some dying, and all in a most wretched condition, the captain finally concluded to turn back; and though we had been forty days beating to the east of the Banks, the strong westerly gales carried us back in eight days to Queenstown, where the vessel put in for repairs. I staid by her until she was ready for sea again, which was not until March, and once more set sail for home, where I arrived in the ensuing April.

"Having had such a hard experience, and my friends strongly urging the point, I concluded to give up going to sea, and returned to my old business in the manufacture of iron-ware, but very soon grew tired of it, and again longed for the sea. It has its hardships, but it has its compensations too: at least I was sure that I could never spend my life in the stifling atmosphere of an iron-factory; and so, in the spring of 1855, I went again to New London, and shipped as 'boat-steerer' in the bark *George Henry*, Captain James Buddington (uncle of Captain S. O. Buddington, sailing-master of the *Polaris*).

On arriving once more on the scene of my old adventures off the entrance to Cumberland Gulf, where we were bound, we encountered an extraordinary heavy pack of ice. It extended over a hundred miles to the eastward, and it was impossible to get into the gulf while this pack remained along the coast. So, to pass away the time until the ice cleared away, we sailed for Disco Bay, where we were pretty sure to find the 'humpback' whale, which we did, making a good catch. In August we sailed again for Cumberland Gulf, expecting, of course, by that time, that the pack would be gone; but, to our surprise, it was still there. Never in all my experience have I seen any thing equal to it; but, forbidding as it was, we must 'take it' to get into the gulf, though it was so compact and heavy that the July and August suns seemed to have made no impression upon it. But 'nothing venture, nothing have.' We took the ice off Cape Walsingham; and on penetrating the pack about forty miles, it closed on us, and we were regularly 'beset,' our drift being to the southward.

"In the latter part of August I sighted a vessel, which at first we all supposed to be a whaler, as we knew there were several trying to get in the gulf. This vessel remained in sight several days. At times we imagined she had all sail on, and was working through the ice. No one for a moment thought that she was an abandoned vessel, but there was something about her which aroused my curiosity; I seemed to *feel* that there would be a story to tell if I could only get at her; and when she had been in sight about two weeks I asked the captain for leave to go over, with two or three companions, to see what she was made of. He objected at first; thought 'we should never get there' (she was about ten or twelve miles off); and if we succeeded in reaching her he was sure 'we would never get back;' but I was determined, and so at last, in company with the mate, John Quayle, the second mate, Norris Havens, and Mr. Tallinghast, a boat-steerer like myself, we started off for the phantom ship.

"It was early morning when we left the *George Henry*, for we knew we had at least ten, and perhaps more, miles to walk. The task we had set ourselves was no light one; the pack was very rough, and every little while we came to patches of open water; and as we had no boat with us, we were obliged to extemporize a substitute by getting on small pieces of ice and making paddles of smaller pieces; and thus we ferried ourselves across these troublesome lakes and rivers. We were all day on our journey, it being nearly night when we reached the stranger. As we approached within sight we looked in vain for any signs of life. Could it be that all on board were sick or dead? What could it mean? Surely, if there were any living soul on board, a party of four men traveling toward her across that hummocky ice would naturally excite their curiosity. But no one appeared. As we got nearer we saw, by indubitable signs, that she was abandoned.

" ' Toward the shape our steps are bending,
 Northward turns our eager gaze,
Where a stately ship appearing,
 Slowly cleaves the misty haze.
Southward floats the apparition ;
 "Is it, can it be the same?"
Frantic cries of recognition
 Shout a long, lost vessel's name!'

By this time Mr. Quayle was so tired that I had to assist him in boarding the ship, myself and the other two following. We found the cabin locked and sealed; but locks and seals did not stand long. A whaler's boot vigorously applied to a door is a very effective key. We were soon in the cabin. This was no whaler, that was plain; neither was she an American vessel, it was soon discovered. English, no doubt of that. Every thing presented a mouldy appearance. The decanters of wine, with which the late officers had last regaled themselves, were still sitting on the table, *some of the wine still remaining in the glasses*, and in the rack around the mizen-mast were a number of other glasses and decanters. It was a strange scene to come upon in that desolate place. Some of my companions appeared to feel somewhat superstitious, and hesitated to drink the wine, but my long and fatiguing walk made it very acceptable to me, and having helped myself to a glass, and they seeing it did not kill me, an expression of intense relief came over their countenances, and they all, with one accord, went for that wine with a will; and there and then we all drank a bumper to the late officers and crew of the *Resolute*.

"It was now too dark to attempt to travel back that night over the broken ice, and we prepared to stay where we were. Possibly the wine we had taken, being at that time unused to it, partly influenced us to this conclusion; but sleep in the vessel we did.

"In the morning we found it snowing, and blowing very heavy

"EVERY THING PRESENTED A MOULDY APPEARANCE."

from the south-east. We could not hope ever to find our way back to the *George Henry* in such a storm, and so, having made a fire, we were prepared to pass the time as comfortably as possible. Among other things, we found some of the uniforms of the officers, in which we arrayed ourselves, buckling on the swords, and putting on their cocked hats, treating ourselves, as *British officers*, to a little more wine. Well, we had what sailors call a 'good time,' getting up an impromptu sham duel; and before those swords were laid aside one was cut in twain, and the others were hacked and beaten to pieces, taking care, however, not to harm our precious bodies, though we did some hard fighting—*we, or the wine!*

"The storm continued for three days, during which we had ample time to investigate the condition and inspect the contents of the good ship *Resolute*. We found food on board, and were enjoying ourselves so well that we should not have cared if it had lasted six. But the weather cleared up, and we saw that the *George Henry* was still at about the same distance from us; so we took all we could carry on our backs, and started to return, arriving at our ship all safe, though some of us got a good ducking by jumping into the water while attempting to spring from one piece of ice to another. Being so heavily laden, we often fell short of the mark, and went plump into the water; but we were in such good spirits that these little mishaps, instead of inciting condolence, were a continual cause of merriment.

"On arriving at the *George Henry*, we made our report to Captain Buddington, describing our treasure-trove in glowing terms. After a good rest, we again started for the *Resolute*, and staid several days on board. At this time the two vessels were nearing each other—the one voluntarily, the other drifting, as she had already done, for a thousand miles. We did not know this at the time, but learned afterward that the *Resolute* had been abandoned, by Sir Edward Belcher's orders, on May 15, 1854, near Dealy Island, and had drifted all the way to Cape Mercy.

"At last the two vessels were only about four miles apart. We were still having a nice time, when, one morning, we saw several persons coming over the ice, and, to our discomfiture, they proved to be the captain, with several of the crew. We very soon got orders to return on board the *George Henry*, while, to our chagrin, the captain took possession of the *Resolute*.

"We had now drifted as far south as Cumberland Gulf, Cape Mercy bearing west about twenty-five miles distant; but the ice was still close and compact. Had we now caught a good gale from the south-east, we were just in the right position to have been drifted where we wished to get, up into the gulf. But no; instead, we got a gale from the north-west, blowing us directly out of the gulf, and away we drifted past it, and once more to the south of it.

"There was no hope of getting back while the pack-ice lay

along the coast; and our only hope now was to get out of the pack as quickly as possible, and return home. The *Resolute*, getting a lead through the ice, got out on the 14th of October, but the *George Henry* was still fast, and drifting slowly southward; though we too were soon to be released from the pack, but such a release as one would wish to see but once in a lifetime.

"On the 25th of October a strong gale commenced to blow from the north-east, and continued with great violence. On the 26th there was a very heavy sea running under the ice; all through the night, and to the morning of the 27th, it was dark and stormy, with danger all the time of drifting upon great icebergs. Many heavy spurs, rough and jagged, projected from these bergs, cutting fearfully into the vessel, and finally she pounded her keel off, tore her rudder, and injured her stern-post.

"On getting clear of the bay, we went to the pumps, and found the vessel making a great deal of water. On the abatement of the gale, we repaired our rudder as best we could, and then started for home, short of men, of course, as more than half of the crew was with Captain Buddington, on board of the *Resolute*. But with pumping day and night in heavy weather we could not keep the vessel free of water—it would gain on the pumps hourly; but, when the weather was moderate we could keep her nearly free. After a most laborious passage, we made out to keep her afloat until we reached New London, in forty days from the start. The *Resolute* did not arrive until some time after—her passage being sixty days.

"I next went as second mate, in 1856, with Captain James Buddington, in the *George Henry*, having the *Ameret*, a top-sail schooner, as tender, and wintered in Cumberland Gulf, lat. 65° 25' N., long. 67° W.; returned in August of 1857, and sailed again as first mate, and arrived in the gulf October 14; and this season passed another winter there, returning in the fall of 1858.

"I had now become so accustomed to the northern climate that it seemed more natural to me than a more southern one. Sailed again in the spring of 1859, as first officer, but started to return home in November of the same year, as our vessel was dismasted off Cape Charles, and we got into St. Johns, Newfoundland, where we repaired, and arrived home in February.

"In the spring of 1860 sailed as master of the brig *Georgiana*. Previous to my departure, I made the acquaintance of Captain

C. F. Hall. He was then writing to the papers and lecturing through the country, endeavoring to interest and stimulate the public on the subject of his projected expedition. I lent him the model of an Esquimaux kayack, which he used in several of his lectures. . I afterward met him just north of Frobisher Bay. I was in that vicinity at the time he lost the schooner *Rescue*, and at that time I came near losing my vessel, as reported in his work in 'Arctic Researches;' but though she was beating on the rocks, during a violent storm, for twenty-four hours, myself and crew having to get ashore on spars to save our lives, I finally saved both vessel and cargo. Returned home in the fall of 1861.

"Sailed again, in the spring of 1862, in the bark *Orray Taft*, of New Bedford; wintered once more in the North, and returned home in the fall of 1863. The *Orray Taft* was wrecked and lost near Marble Island, in September, 1872, while I was in the *Polaris* at Thank God Harbor; and a short time after the *Ansel Gibbs*, a whaler known to many Arctic explorers, was lost at the same place. The men had to winter on Marble Island, and were not rescued till August, 1873, at the time I was on board the *Tigress* searching for Captain Buddington. Many of the men were lost; some at the time of the wreck, and fourteen from scurvy, brought on by exposure and insufficient and improper food.

"Sailed again, in the spring of 1864, in the bark *Antelope*, of New Bedford, and on this voyage staid out two winters—one in Hudson Bay, and one in Cumberland Gulf. On this trip I took my vessel farther north than any of the whalers had been before. I sailed right ahead into Repulse Bay, and *took the first whale there that was ever caught in those waters*—the whalers having previous to that limited themselves to the latitude of Wager River. Since, however, they have freely visited Repulse Bay. This bay probably offers, on its north shore, more and better harbors than any place within the whaling regions; but the south shore is clean and level, without harbors; and there is a peculiarity about this locality which I have never found elsewhere so near the Arctic Circle, and that is the frequency of thunder-storms, accompanied by vivid lightning.

While I was in winter-quarters in Hudson Bay, Captain Hall visited the bark *Monticello*, which had brought him out, and also other vessels wintering there, including the *Antelope*. I then had long talks with him about getting up another expedition after he' had found out all he could about Sir John Franklin's expedition, and he always wound up by saying he wanted me to go with him. He was badly off for boats at that time, and I let him have one of mine. The *Antelope* was lost in a severe storm in the year 1865, and I returned to St. Johns, Newfoundland, in the steamer *Wolf*, Captain Skinner, and from there got home.

"Sailed again, in the spring of 1867, in the top-sail schooner *Era*, on which voyage the schooner broke out of winter-quarters in December, and drifted out to sea. We had two vessels in company caught in the same drift; one was abandoned, the other run ashore. The *Era*, finally drifting in among some bergs, was frozen in for the winter. During this voyage I met Captain Hall again. He was living with the Esquimaux; in 'training,' as the sportsmen would say, for the great work which he even then had in mind. I supplied him with provisions of various kinds, and he, when he had opportunity, sent the natives with fresh meat to the ships.

"Sailed again, in the *Era*, in the spring of 1869, returning in the fall of 1870.

"In referring to my old log-books, as well as in recalling the events themselves, I find that the experiences of whaling are not essentially different from those of the Polar exploring parties—so far, I mean, as the exposures and dangers are concerned. We were in continual risk of getting 'beset,' and often were closed in, and unable to move for days or weeks, and sometimes compelled to remain and winter, being unable to break out or bore our way through. The old log-books are full of such entries as these:

"'*Schr. Era, July* 17, 1867. Beset in the ice, North Bluff bearing E.N.E.

"'*July* 18. Laying by; ice-anchor out; all sails furled; no water in sight.

"'*July* 23. Working slowly through the ice to the westward.

"'*Sept*. At anchor at Black-lead Island.

"'*Oct*. 29. Bay full of drifting ice.

"'*July* 6, 1869. All hands employed breaking out vessel.

"'*July* 27. Working in the pack, Cape Misery bearing N.N.E.

"'*July* 29. Beset; no water visible.

"'*Nov*. 18. At winter-quarters at Niountelik Harbor. All hands employed sawing ice; eight ships in company. 8 P.M.; blowing hard; ice commenced to break up. Worked all night to try and save the vessel. At 12, midnight, let go the starboard anchor, and got the larboard chain ashore; thick snow.

"'*Nov*. 19. Thick snow; one anchor down; ice all broken up; expecting to go ashore. If wind hauls N.W., we are saved; otherwise the chance is small.

"'*Dec.* 8. Ice on the move, and forcing the schooner inshore, broadside to, through ice nine inches thick.

"'*Dec.* 9. Drove in between grounded icebergs; took out provisions; took ashore square-sail and mainsail to make a house. Two teams of dogs from Niountelik helped haul our things; 20° below zero.'

"On this last occasion the ship remained frozen in until February, and myself and the crew lived ashore in the house or hut we had built with stones and covered with the sails taken from the ship, watching anxiously all the time for a break up, which might either relieve the ship or crush her to pieces. I could not tell what would happen; but, fortunately, in February the ice began to break, and I got over to my ship, found she was still sea-worthy, repaired damages, got our provisions and other articles aboard again, and, getting a lead out, finished my intended trip, making, after all, a very fair voyage.

"On arriving at home, New London, in October, 1870, Captain Hall called to see me. He informed me that he had succeeded in getting an expedition started for the North Pole, and wished me to go with him in the capacity of sailing-master and ice-pilot; but at that time I had a project of my own on hand, and had opened negotiations with a party, expecting to get a vessel for the white whale-fishery, and I so stated to Captain Hall. He called on me several times to persuade me to go, but I felt obliged to decline, having commenced negotiations with other parties. I then heard that he had engaged S. O. Buddington.

"As I did not succeed in effecting an agreement about the whale-fishing which I had had in view, I concluded to remove, with my family, from New London to Brooklyn; and shortly after, the *Polaris* coming to the Navy Yard there, I called to see Captain Hall. He again requested me to join the expedition, making me many promises: at that time all the positions were filled; but he was not to be denied, and he declared he would *make* a position for me, for go I must.

"At last I consented to go; and in forty-eight hours from the time I agreed to accompany him, I had made all my arrangements, procured my outfit, bade farewell to friends, and was on my way to the North Pole. The rest of my Arctic experiences will be found narrated in the history of the *Polaris* expedition, the ice-floe voyage, and in the journal of my trip in the *Tigress*, in search of Captain Buddington and party."

CHAPTER III.

THE POLARIS EXPEDITION.

The North Polar Expedition authorized by Congress.—Captain Hall's Commission.—The *Periwinkle*, afterward *Polaris*, selected.—Letter of Captain Hall's.—Description of the Steamer *Polaris*.—Liberal Supplies.—A patent Canvas Boat.—Books presented by J. Carson Brevoort.—A characteristic Letter of Captain Hall's.—An Invitation to visit him at the North Pole.

THE *Polaris* expedition, or, in official language, the "United States North Polar Expedition," which sailed from the Brooklyn Navy Yard on Thursday evening, June 29, 1871, was under the general command of Commander Charles Francis Hall, whose previous explorations in the Arctic and high northern latitudes will be found summarized in the sketch of his life, to be found in its appropriate place, as descriptive of the chief officer of the *Polaris*.

On the 8th of March, 1870, Hon. Mr. Stevenson, of Ohio, introduced into the House of Representatives a bill authorizing the President to appoint Captain Hall to the command of an exploring expedition to the Arctic regions, reciting the facts of his previous successful journeys, experience, and acclimatization. An identical bill was introduced into the Senate by the Hon. Mr. Sherman, of Ohio, on the 25th of March, and which, after being twice read by its title, was referred to the Committee on Foreign Relations, of which the late Hon. Charles Sumner, of Massachusetts, was chairman.

On the 19th of April Mr. Sumner reported back the bill with an amendment striking out all personal reference to Captain Hall, and substituting the phrase "one or more persons." The bill, as amended, finally passed both Houses of Congress on the 11th of July, 1870, and was signed by the President on the 12th. See p. 101.

On the 20th of the same month Captain Hall received his commission from President Grant, of which the following is a copy:

CAPTAIN HALL'S COMMISSION.

EXECUTIVE MANSION, WASHINGTON, D. C., July 20, 1870.

Captain C. F. HALL:

DEAR SIR,—You are hereby appointed to command the expedition toward the North Pole, to be organized and sent out pursuant to an Act of Congress approved July 12, 1870, and will report to the Secretary of the Navy and the Secretary of the Interior for detailed instructions.   U. S. GRANT.

By a section of this Act the President was authorized to fit out one or more expeditions, and dispatch them toward the North Pole, appointing one or more persons to the command, and also to detail any officer in the public service to take part in it; likewise to give the use of a public vessel suitable for the purpose (*vide* Sec. 9 of the Act, in Appendix, page 428).

THE POLARIS.

The scientific operations connected with the expedition were to be directed by the National Academy of Sciences, of which the well-known and widely esteemed Professor J. Henry is president.*

Captain Hall was allowed to inspect a number of United States vessels, and it was at his desire that the *Periwinkle*, to which the name of *Polaris* was subsequently given, was selected.

The *Polaris* was partially fitted out at the Washington Navy Yard; but it being, on some accounts, more convenient for her

---

* See Appendix, p. 431.

to be completed at Brooklyn, she left Washington on the 10th of June, 1870, and arrived at the Navy Yard, Brooklyn, N. Y., on the 14th. Here she received the last alterations deemed necessary, completed her outfit, received her stores, and shipped her crew.

While lying at Brooklyn, Captain Hall addressed the following letter to the Secretary of the Navy:

Steamer Polaris, Navy Yard, Brooklyn, N. Y., June 23, 1871.

I have the honor to apply for the appointment of Captain George E. Tyson, navigator and master of sledges for the North Polar Expedition, after full consultation with Captain Buddington and First Mate Chester, who agree with myself that the services of Captain Tyson, who has been engaged for over twenty years in voyaging to and from the Arctic seas, would be of great value to the expedition. * * * Captain Tyson is well known to me, and to the whaling-houses of New Bedford and New London, as an experienced, trustworthy navigator and dog-sledge traveler in the Arctic regions......* I have the honor to be,

Yours, respectfully,
C. F. HALL,
Commanding U. S. North Polar Expedition.

Hon. GEO. M. ROBESON, Secretary of the Navy.

The *Polaris* was a screw-propeller of only three hundred and eighty-seven tons; but, in addition to her steam-power, she was fitted with the rig of a foretop-sail schooner, so that, as circumstances dictated, she could be propelled by steam or wind. Did any irreparable accident happen to her machinery, she could still make fair headway under canvas. To guard against accident to the propeller by contact with the heavy ice it was known she must encounter, the screw was so arranged that it could be unshipped and raised to the deck through a shaft in the stern of the vessel, which, as will be narrated hereafter, was done on the 1st of September, 1871. Extra blades were also provided, with which to replace the originals, should they be broken. Her engine was considered exceptionally good, and was the product of Neafies & Levy's establishment in Philadelphia. For its size, it was a powerful worker; and space, in this case, was a prime consideration, as so much room was needed for coal and other stores.

In regard to her boilers, there was an arrangement unique as to United States vessels, one of them being fitted for the use of whale or seal oil as a steam generator; and this was expected not only to be used for the general purpose of propulsion, but

---

* See extract of letter from Captain Edwin W. White, in Appendix, p. 423.

also as a means of heating the vessel when in winter-quarters. This boiler, as will be subsequently described, was willfully destroyed just after leaving Disco.

The hull of the *Polaris* was specially prepared for her Arctic voyage by being planked all over with solid six-inch white-oak timber, the bows being made almost solid, and then sheathed with iron which terminated in a sharp prow, with which to bore her way through the ice. Another peculiarity was a new style of life-preserver, in the shape of a buoy, to be kept slung over the stern, but which could, when occasion required, be instantly detached and lowered to the water by means of a connecting spring which could be reached from the pilot-house; and, by another spring conveniently placed, an electric light, kept secured to the buoy, and rising above it between two and three feet, could be instantly lighted by means of a galvanic battery in the cabin.

In a region where parties are subject to the constant liability of being separated from their ship by the breaking up of the ice, and especially during the dark months from October to February, a contrivance of this kind for forming a beacon-light to those separated by any cause from the ship, might well be termed a life-preserver; and that it was not called into requisition when Captain Tyson and party were separated from the *Polaris* on the ice-floe, shows either that due care had not been taken of the apparatus, and that it was unusable from neglect, or that those in command did not take the trouble to give this aid to their imperiled companions.

Extras of all kinds likely to be needed were amply supplied; every sort of running gear, cordage of different sizes, spare sails, spars, and even an extra rudder. She also carried one small howitzer.

She had also four boats—similar to whale-boats—one flat-bottomed scow, and a patent portable folding canvas boat, intended for the use of transglacial parties. This boat was about twenty feet long, four feet wide, and two deep; and though it weighed only two hundred and fifty pounds, had an estimated carrying capacity of four tons, and was expected, in case of necessity, to carry twenty men, though five or six would be her complement on a surveying trip.

The skeleton, or frame-work, was constructed of ash and hickory, and over this was affixed a water-proof canvas cover, something on the principle of the *oomiaks*, or seal-skin "women's boats," of the Esquimaux. The boat could be readily disjointed and folded, so as to occupy but a small space, and could thus be laid upon a sledge for portage when no longer needed as a boat; and, on reaching water again, could be as suddenly retransformed into its original shape.

Theoretically, it was perfect; but, practically, it was found of little use, being excessively slow. It was used on an exploring trip, and finally abandoned by Mr. Chester, being left at Newman Bay, Mr. Chester and party walking back to the *Polaris*, then distant from the ship about twenty miles.

In the cabin, in addition to the small but select library which Captain Hall always had with him, was a cabinet organ, which had been generously presented to the late commander by the "Smith Organ Company," with the hope that its sweet strains would not only assist the regular Sunday service on board the *Polaris*, but that on other occasions it would help to while away the tedious hours, when prevented from the exercise of more active duties, during the long Arctic night.

Some very valuable books were lost when the *Polaris* foundered. That generous and long-tried friend of Arctic exploration, J. Carson Brevoort, of Brooklyn, New York, had, among other volumes of interest and value, placed on board of that vessel for Captain Hall's use, an entire set of the British Parliamentary Blue-books relating to the English Arctic exploring expeditions. There was also a copy of Luke Fox's "Arctic Voyage of 1635," much valued by its owner,* partly from its bearing the following indorsement in Captain Hall's own handwriting, it having been loaned to him also in 1864:

This book belongs to my friend, J. Carson Brevoort.

To-morrow, March 31, myself and native party, consisting of 13 souls, start on my sledge-journey to King William Land.

C. F. HALL,
29th (Snow House) Enc't., near Fort Hope, Repulse Bay,
Lat. 66° 32′ N., long. 86° 56′ W.

Friday, March 30, 1866.

Part of his library Captain Hall saved—a few books—by leaving them in Greenland with Inspector Karrup Smith, but

---

* See interesting letter of Mr. Brevoort's in Appendix, p. 467.

many others went down with the good ship *Polaris* in sight of Life-boat Cove, while others were mutilated, destroyed, or abandoned.

To show the watchful interest which Captain Hall took in the proper outfit of his vessel, we introduce the following letter, which is characteristic of the man, and explains itself. It was addressed to a friend who had previously discussed the subject of provisioning the *Polaris* with him:

WASHINGTON, D. C., May 28, 1871.

DEAR SIR,—Your letter of the 26th came to hand, and then I telegraphed you as follows: "Do not purchase any Texas corned beef. Don't like it. Letter by mail."

The letter promised I purposed to write and send by last night's mail, but somehow I was so completely hemmed in by callers that I couldn't do it.

I simply say here, relative to the Texas beef, * * * that it does not bear so favorable a reputation among some that have used it as to justify my having it on so important an expedition as the one we are preparing. Indeed, from an examination and use of the article, I am not favorably impressed with it.

I think it quite advisable to have half the bread made of the Graham flour. Am pleased that the same can be done. * * * Am quite sure that the flour will not "heat" when over in the North Polar country. Should there be any danger that it would, before getting out of this melting weather, then it would be wise to have the wheat *kiln-dried* before it is ground. This process could not cost more than one-half cent or so per pound. But some miller ought to be able to tell us whether Graham flour will keep or not. I think it will.

You state as follows: "I do not notice in the list any salt beef." Of course, you can not find such a rank scurvy-breeder in any list I have prepared. It may be that just as we get about ready to start, I shall get you to order the putting up of a few barrels of *slightly* corned beef. This article, put up in this way, will keep in the climate the expedition is going to, and not give us any scurvy, while the ordinary salt beef of the market will. * * *

I thank you for your kindly offer to make your house my home when I come there. In response, let me say that *when I get well settled down at the North Pole,* which I hope and believe will be about the middle of next May (1872), then I may send down word to you to come and see me, and make that objective point *your home.* Methinks I see you shiver at this suggestion. * * *

The Secretary of the Navy is going to let the Polar Expedition have the storeship *Supply*, now on its way from France, as a transport. [United States steamer *Congress* substituted.—*Ed.*] All the senators I have spoken with are quite in favor of his doing so. Senators Morton, Patterson, Nye, Thurman, Sumner, Fenton have each indorsed my written request for a transport to Greenland by adding favorable written sentiments they entertain in behalf of this movement.

If it would help the case at all, *I could get nearly every Senator and Representative* to indorse so reasonable a petition. But, really, yours and your brother's action, before you left Washington, [sufficiently] contributed to show the Secretary the expediency of sending along a transport.

I am blessed with having so noble a soul as is Secretary Robeson to aid me in accomplishing the great work of my life. *I love the man,* and therefore his most intimate friends from early days, as are you and your brother, I profoundly respect. I rejoice that the Secretary has selected you to attend upon me, for now, without difficulty, I can have whatever, in reason, I shall require to help me to make geographical discoveries from lat. 80° N. up to the North Pole—a feat that has baffled the civilized world for more than three centuries. The President has promised to visit the *Polaris* on Wednesday next. Yours, etc.,

C. F. HALL.

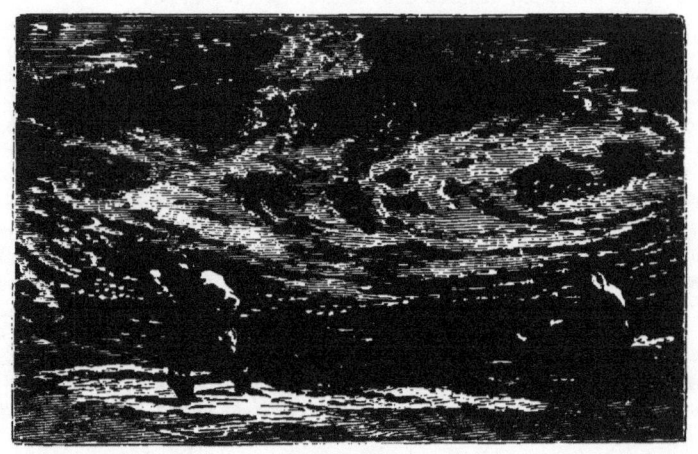

A SNOW-SQUALL.

CHAPTER IV.

The *Polaris* put into Commission.—Official Instructions to the Commander.—Scientific Directions.—Letter of Captain Hall's.—List of the Officers and Crew.

PREVIOUS to sailing, the Secretary of the Navy, Hon. George M. Robeson, put the *Polaris* formally in commission, placing the command of the expedition, the "*vessel, officers, and crew*," under the orders of Captain Charles Francis Hall. This point it is well for the reader to observe, as on its subsequent interpretation the welfare and success of the whole expedition turned.

Captain Hall was ordered to make the first favorable port on the west coast of Greenland, leaving it to his judgment to stop at St. Johns, Newfoundland, if he thought best for any reason to do so. It was further ordered that, if the first Greenland port made was south of Holsteinborg, the *Polaris* should from thence go to the last-named settlement, and from Holsteinborg to the harbor of Goodhavn, on the island of Disco, where the transport-ship, the *Congress*, was expected to bring him additional stores, and establish a dépôt for future use. At the two most northern settlements, Upernavik and Tossac, dogs and any other Arctic necessity were to be called for.

After leaving the last of these Greenland ports, the Secretary directs that the *Polaris* shall go to Cape Dudley Diggs (about 76° N.), "and thence you will make all possible progress, with vessels, boats, and sledges toward the North Pole, using your own judgment as to the route or routes to be pursued, and the locality for each winter's quarters." The *Polaris* expedition, having been provisioned for two years and a half, it was expected that Captain Hall would continue his explorations for the whole of that period, unless "the main object of the expedition, viz., attaining the position of the North Pole, be accomplished at an earlier period," in which event it was of course expected and ordered that the *Polaris* should return to the United States.

The law of Congress having provided for a Scientific Corps to be attached to the expedition, and also provided that the National Academy of Sciences should prescribe their duties and modes of operation, the Secretary included in his instructions to Captain Hall this clause: "The charge and direction of the scientific operations, will be intrusted, *under your command*, to Dr. Emil Bessel; and you will render Dr. Bessel and his assistants all such facilities and aids as may be in your power."

The Secretary further orders that all objects of natural curiosity collected by any of the individuals of the company should be carefully preserved, and considered the property of the Government.

Again: "You will give special written directions to the sailing and ice master of the expedition, Mr. S. O. Buddington, and to the Chief of the Scientific Department, Dr. E. Bessel, that, in case of your death or disability, * * * they shall consult as to the propriety and manner of carrying into further effect the foregoing instructions—which I here urge must, if possible, be done. * * * In any event, however, Mr. Buddington shall, in case of your death or disability, continue as the sailing and ice master, and control and direct the movements of the vessel."

It was directed that Buddington should consult with Dr. Bessel, but the former alone, the commander being dead or disabled, should decide on the propriety of returning to the United States.

The usual directions are given, "to report at all convenient opportunities to the Navy Department," to erect monuments in proper positions, inclosing records of progress and general condition of the party; and to establish *caches* of provisions, according to judgment. After passing Cape Dudley Diggs, no ordinary mode of communication with the civilized world could be expected. It was therefore ordered that bottles closely sealed, or small copper cylinders, containing a statement of the latitude and longitude, with any other facts of special interest, should be thrown overboard daily, if open water or drifting ice promised to convey them to sea. These papers were provided by the Government, printed in different languages, with blanks to be filled in with the occasion of using.

In addition to the instructions of Secretary Robeson, Professor Henry, through the former, requested that "one point should be specially urged upon Captain Hall, namely, the determination, with the utmost scientific precision possible, of all his geographical positions, and especially of the ultimate northern limit which he attains. The evidence of the genuineness of every determination of this kind should be made apparent beyond all question."

To assist in this being done, the *Polaris* was liberally supplied with all needed instruments of the best quality, as also with charts and books, and whatever else was needed to command success.

The instructions to the Scientific Corps were prepared by Professor Joseph Henry, President of the National Academy of Sciences, assisted by Professors Baird, Agassiz, Hilgard and Meek, at Washington, and are of the most elaborate description, as will be seen by referring to the Appendix.

The most prominent point insisted upon was absolute accuracy. It is ordered, say the instructions, "in all cases, that the actual instrumental readings must be recorded, and if any corrections are to be applied, the reason for these corrections must also be recorded." Again: "The evidence of the genuineness of the observations brought back should be of the most irrefragable character. No erasures whatever with rubber or knife should be made. When an entry requires correction, the figures or words should be merely crossed by a line, and the correct figures written above."

As to the subjects committed to the observation and record of the Scientific Corps, we shall merely indicate them in general terms, referring those who desire to make themselves acquainted with the details to the originals in the Appendix.

The first order relates to the keeping of a full and accurate log-book of all transactions concerning the expedition; and a journal of similar import, to be filled up daily, when on sledge-expeditions.

The astronomical observations were to be made four times a day, and each operation repeated three times to guard against mistakes.

The variations of the compass is to be continually watched and recorded, and on sledge-journeys particular attention to be paid to the dip and relative intensity of the magnetic force; and at winter-quarters, "the absolute horizontal intensity to be determined with the theodolite magnetometer, including the determination of the moment of inertia."

Notes to be taken of the various features of the aurora borealis. Also pendulum experiments, to determine the force of gravity in different latitudes.

Tides, currents, sea-soundings, bottom-dredging, and the density of sea water in different localities, to be tested.

Registers of temperature, the pressure of the air, and the proportion of moisture contained in the latter, are all to be made the subject of careful experiment.

The velocity of the winds, observations on the clouds, the precipitation of water, the form and weight of hailstones, the character of snow, and any peculiarities of crystallization, all to be noted.

Electricity in all its multiform developments, the polarization of light, as also optical phenomena, mirage, halos, parhelia, luminous arches, and meteors of all kinds, to be looked after.

Experiments for the detection of ozone in the atmosphere were provided for, by supplying Dr. Bessel with sensitized paper, and with directions how to extemporize the necessary apparatus.

In natural history and geology, it is only necessary here to observe that the scientists were expected to improve all opportunities to make collections of specimens, and to take the utmost care so to label and arrange them that no false deductions might be drawn through errors of fact.

The course and growth of glaciers being of exceptional interest receives large notice, and every suggestion is made to induce a thorough scientific examination of such portions of any as may be traversed by the sledge-parties. The late lamented Professor Louis Agassiz furnished the remarks on this point, indicating to Dr. Bessel the great importance of comparative examinations of the Greenland or other Arctic glaciers with the known history of the Alpine rivers of ice. He wisely forbears giving utterance to any *dictum* on disputed points. He says: "I have purposely avoided all theoretical considerations, and only call attention to the *facts* which it is most important to ascertain, in order to have a statement as unbiased as possible."

While in Washington previous to sailing, a want of mutual respect was known to exist between Captain Hall and Dr. Bessel; and so far was Dr. Bessel's discourtesy carried, on several occasions, that Captain Hall would have been quite justified in refusing to take him in his company, and calling for a volunteer in his place.

It had originally been the intention of Captain Hall, after reaching the head of Baffin Bay, to strike across to the west and sail through Jones Sound, and thence to the North Polar Sea, which had inspired Sir Edward Belcher with such abject fears; but, after careful consideration and consultation with Arctic experts, he finally concluded to sail as nearly due north as possible through Smith Sound and connecting waters.

The day before the *Polaris* sailed, Captain Hall expressed his gratitude, in the following language, for the thorough manner in which the expedition had been fitted out:

Steamer *Polaris*, Brooklyn Navy Yard, June 28, 1871.

SIR,—I have the honor to report that the steamer *Polaris*, selected by you for the expedition toward the North Pole, under my command, is now ready for sea, and will sail to-morrow.

Before leaving port, I can not forbear expressing my great obligation to you for the intelligent and generous manner in which you have provided for the expedition in all respects. The ship has been, under your directions, strengthened and prepared for the special service upon which she enters in the most approved manner; and is supplied with every appliance to make the expedition a success.

The officers and crew of the ship are all I could desire, and the provision made for the subsistence and protection of all on board is the best that could be devised.

Your generous response to every legitimate request I have made in regard to the ship's outfit demands the expression of my warmest gratitude. The only return I can make now is the assurance of my determination, with God's blessing, that the expedition shall prove a success, and redound to the honor of our country, and to the credit of your administration. * * * With an abiding faith that the results of the expedition will prove the wisdom of Congress in providing for it, and justify the generous manner in which you have performed the duty assigned you, I am,

Very respectfully, your obedient servant,
C. F. HALL,
Commanding U. S. North Polar Expedition.

Hon. GEO. M. ROBESON, Secretary of the Navy.

The *Polaris* sailed from the Brooklyn Navy Yard at seven P.M. of the 29th of June, and made a pleasant run of seventeen hours, and then dropped anchor in New London Harbor at eleven A.M. of the 30th of June.

The especial object of putting in to New London was to get an assistant engineer,* the one engaged in New York, Wilson, having deserted. His place was supplied by a better man, Mr. Odell, who had served in the United States Navy during the late war.

The carpenter of the ship had also been taken sick in New York, and had been sent to the hospital just before the *Polaris*

---

* See letter of Captain Hall's in Appendix, p. 453.

sailed, and it was thought that one might be obtained at New London.

From this port Captain Hall also reported the desertion of three other men—a fireman, seaman, and the cook—and the discharge of the steward for incapacity. These positions were all refilled at New London, except the carpenter, who recovered, and was subsequently forwarded by the tender *Congress*.

The following is the corrected muster-roll, as made out by Captain Hall, on July 2, and forwarded by him to the Secretary of the Navy:

C. F. Hall..............................Commander.
Sidney O. Buddington...............Sailing and Ice Master.
George E. Tyson......................Assistant Navigator.
H. C. Chester.........................First Mate.
William Morton........................Second Mate.
Emil Schuman.........................Chief Engineer.
Alvin A. Odell.........................Assistant Engineer.
Walter F. Campbell...................Fireman.
John W. Booth........................   "
John Herron...........................Steward.
William Jackson......................Cook.
Nathan J. Coffin.....................Carpenter.

Seamen.

Herman Sieman.   Joseph B. Mauch.
Frederick Anthing.   G. W. Lindquist.
J. W. C. Kruger.   Peter Johnson.
Henry Hobby.   Frederick Jamka.
William Lindermann.   Noah Hays.

Emil Bessel........................Surgeon and Chief of Scientific Corps.
R. W. D. Bryan..................Astronomer and Chaplain.
Frederick Meyers.................Meteorologist.

Esquimaux.

Joe....................................Interpreter and General Assistant.
Hannah..............................   "
Puney................................Child.

Added at Upernavik. { Hans Christian........................Dog-driver, Hunter, and Servant.
Wife of Hans.............................
Augustina................................Child.
Tobias....................................   "
Succi......................................   "

CHAPTER V.

BIOGRAPHICAL SKETCH OF CAPTAIN HALL.

Nativity and early Life of Charles Francis Hall.—Leaves his native State of New Hampshire and settles in Ohio.—Takes to Journalism.—Attracted by Arctic Literature.—Unsuccessful Effort to join M'Clintock.—Sails for the Arctic Regions in the *George Henry*, of New London.—The Tender *Rescue* and the Expedition Boat lost in a Storm.—He explores Frobisher Bay and Countess of Warwick Sound.—Collects Relics of Franklin's Expedition.—Returns to the United States.—His Theories regarding the Franklin Expedition.—Sails for the North, 1864, in the Bark *Monticello*.—His Discoveries.—Skeletons of Franklin's Men scattered over King William Land.—Annual Reports.—His Life with the Esquimaux.—Return to the United States.—Physical Appearance.—Mental Traits.—In the Innuit Land he did as the Innuits do.—Persevering Efforts to organize the North Polar Expedition.—President Grant personally interested.—"That Historical Flag."—How he would know when he got to the Pole.—His Premonitions.—His last Dispatch.

CHARLES FRANCIS HALL, though long a resident of Ohio, was born in the township of Rochester, in the State of New Hampshire, in the year 1821. His early life was far from luxurious, though not lacking in the ordinary comforts of country homes. He was early inured to work, and received only the usual common-school education of the period, which was far more limited then than now.

But a lad fond of reading will readily make amends for the limitation of school facilities, and young Hall was omnivorous in this respect, so that not only the books in his own family but those of his friends and neighbors were sought out, borrowed, and read. He thus became possessed of a curious conglomeration of information, over which he brooded, without, it would seem, any proper direction as to systematic study. And thus he plodded his way along, like many another dreamy lad, whose heart and aim is all beyond, and outside of, his every-day occupations.

As his school-days ended the unattractive labor of a blacksmith's shop opened before him; and though not much to his taste, this heavy work assisted materially in developing his muscles and hardening his constitution, thus indirectly helping to fit him for the arduous adventures of his later years.

CHARLES FRANCIS HALL.

While yet a young man, he left his native place, and with it the blacksmith's trade. Setting his face westward, after some experiments elsewhere, he settled in Cincinnati. Here he made arrangements to learn the seal engraving, and in this business he continued for some years; but he had not yet found his forte. This sort of work, though more artistic than shoeing horses or welding iron, did not satisfy him.

He took to journalism, and published at Cincinnati, first the

Occasional, and subsequently the *Daily Penny Press;* and both of these periodicals amply prove that, though Captain Hall was not a college graduate nor a professed scientist, he was very far from being an ignorant man. He was well-read, intelligent, thoughtful, and a persevering student of whatever he undertook to make himself acquainted with.

For nearly ten years before he sailed on his first Arctic voyage he had been an enthusiastic reader of Arctic literature. Naturally attracted by the subject, which has fascinated so many brilliant minds, he searched out, read up, and carefully studied every thing relating to Polar affairs which he could get hold of; and by the time that England and the United States were fully awakened to the necessity of sending relieving parties to search for Sir John Franklin, young Hall was fully aroused, eager and anxious to join in the search. The first Grinnell Expedition especially excited his enthusiasm, but no way then appeared open to him by which he could join it. Disappointed in that, he made another unsuccessful effort to go out with M'Clintock, in 1857.

At this time his mind was so unsettled between his desire to go on a Polar expedition and the necessary claims of his family —for he had married in Cincinnati—that his business, never very profitable, became more and more embarrassed. To his eye, the Polar regions had all the attraction of a terrestrial paradise; its glistening icebergs and snow-clad plains were as enchanting to his imagination as the fairy-tales of younger days; and, above all, he had that impression of fatalism, that inspiration of a personal mission, which looked to some of his friends like a mania, but which was a convincing voice to him that success was possible, and that he was the person to succeed.

But he had no money, no means whatever of fitting up a private expedition, no influence at Washington, at that time, by which he might hope to attain his purpose, nor, after De Haven's return, did the Government appear inclined to invest further in that direction. But here and there our enthusiast gained friends; touched the heart of one man by his pictures of some stray wanderer of that fated expedition dragging out an isolated and half-savage existence among the Esquimaux; interested the imagination of another by narrations of the wonderful scenes which had met the wondering gaze of preceding explorers; reached even to the pockets of others, who believed that such devotion would accomplish something, if the right start was given; and offers of aid at last cheered and encouraged him to hope that his heart's desire might yet be fulfilled.

But it was not until the year 1860 that he was at last enabled to put his long-cherished plans in operation. In pursuit of information among practical men, who knew the modes of life among the Esquimaux, and the resources for living on the shores north of Hudson Bay, and north and west of Cumberland Sound, Captain Hall visited New London, Connecticut. Here he was fortunately introduced to the firm of Williams & Haven, who generously tendered him a free passage in their bark, the *George Henry*, to which was attached as tender the famous *Rescue*, a schooner once known as the *Anaret*, and which had been consort to the *Advance* in 1850–51, in the De Haven Arctic expedition.

A fund was raised by his friends in New York, Cincinnati, New London, and elsewhere, to provide the necessary outfit; and on the 29th of May, 1860, he had the inexpressible pleasure of at last finding himself sailing toward the goal of all his hopes.

An "expedition boat," a fine large sail-boat, had been expressly built for him, and in this he expected, by portage and otherwise, to reach King William Land, to prosecute his researches in shallow waters which the bark could not enter, and farther to the north and west than the whaler was destined to go.

He reached Cyrus Field Bay without special incident, and made some interesting trips in his expedition boat; but his hopes were fearfully dashed by its wreck and entire loss, during a violent storm, which occurred on the 27th of September, at which time the *George Henry* was endangered, and the *Rescue* went ashore and became a total loss. Captain Tyson, then master of the brig *Georgiana*, was involved in the same storm, and, though for some time expecting the certain destruction of his own vessel, he sent a portion of his crew to try and save Hall's precious expedition boat; but their efforts were ineffectual. Though sadly disappointed by the loss, it did not wholly dishearten him. He could not do as he had predetermined, but he decided to stay and do what he could. His aim had been to proceed north and west in search of possible survivors of the Franklin expedition, through the connecting waters north and west of Fox Channel; but the loss of the *Rescue* and his expedition boat completely frustrated this intention. He was for a moment cheered by the promise of a stout whale-boat by old Captain Parker, of the British whaler *True Love*, but accident prevented his receiving it. He then resolved to make what explorations he could with dog-sledges, and subsequently, with the aid of these and an "old, rotten, leaky, and ice-beaten boat" obtained from the *George Henry*, made that thorough examination of both shores and the terminus of Frobisher Bay and Countess of Warwick Sound which has since become a part of Arctic history. The account of the reliquiæ belonging to the visits of that ancient voyager, collected and brought by Captain Hall from that region, is fully detailed in his graphic and entertaining work entitled "Arctic Researches."

On this expedition he was absent two years and a half, returning with accurate charts, and much other valuable information regarding the inhabitants and resources of the country. His crowning geographical discovery on this trip was that of proving the water named by Frobisher as a strait, and which had been so designated on the maps for two hundred and eighty-four years, to be a bay. But still mindful of his original object, Captain Hall had no sooner returned to the United States, than he set about planning another journey to the north-west. He had brought home with him two of the natives, Ebierbing and Tookoolito—the "Joe" and "Hannah" of the Polaris Expedition. These Esquimaux, or Innuits, as they prefer to call themselves, had been taken to England in 1853; and the woman especially had acquired many of the habits of civilization, spoke sufficiently good English to act as an interpreter, and could read a little. "Joe" was an excellent pilot, and could also speak some English.

From what Captain Hall had learned from the natives during his sojourn and explorations around Frobisher Bay, he had become fully convinced that the Esquimaux held the secret of the fate of Sir John Franklin's company, and that by living with them long enough to gain their confidence he should be able to extract all the truth from them.

These ideas were so inwrought in his mind that he determined to return to the Arctic regions as soon as a new outfit could be secured, and to remain there expatriated for half a decade; to live with and among the natives, making himself completely one of them—all with the benevolent hope that he might be able to

JOE, HANNAH, AND CHILD.

find, relieve, and bring back to civilization some possible survivor of the lost expedition.

All of his old and many new friends came forward to his support. The winter of 1863–64 was principally spent by him in lecturing, writing, and visiting where he could create or renew an interest in his new expedition.

New London again furnished him a passage. This time it was made in the bark *Monticello*, of which Richard W. Chappel was agent. In the log-book we find the following entries:

"*June* 30, 1864. The Arctic expedition; Mr. Hall and his two natives take passage for Repulse Bay." [The two natives were Joe and Hannah.—*Ed.*]

"*Aug.* 21. Anchored close to Dépôt Island, and landed Hall's expedition.

"*Dec.* 14. Two sledges, with eight natives on one, and Mr. Hall's Joe and a white man on the other, came to the ship and brought 225 lbs. of deer and musk-ox meat and some suits of native clothing; they were five days coming. They also brought five ox, six bear, and ten deer skins. Mr. Hall sends a letter saying he has heard of six men living with Esquimaux north of Repulse Bay. He will visit the ship in January. 56° below zero.

"*Dec.* 19. Natives returned to Mr. Hall. Will come again at full moon.

"*Jan.* 13, 1865. Mr. Hall and Esquimaux, seventeen in number, came to the ship. Five days on their journey down. Mr. Hall is not looking very well, but I believe he is enjoying good health.

"*Feb.* 10. Mr. Hall, two natives, and one woman left for their homes at the north. Three hearty cheers from ship's company, and all the good wishes that could be expressed for his welfare. I think it will be doubtful if he will be able to reach Repulse Bay before midsummer, as the natives don't want to get there before July or August. Think Mr. Hall will not go without them.

"*March* 22. Mr. Hall sent a dog-team to the ship, with 600 lbs. of ox-meat, to divide among the ships.

"*March* 25. Albert and Jack left for Mr. Hall. Took a boat from the *Antelope* [Captain Tyson] for Hall. Albert loath to leave, as he does not know when he would see ships again, as he was going to travel with Hall."

HALL'S DISCOVERIES.

Hall pursued his investigations in pursuit of information regarding Franklin's party with unsparing devotion and great success. He gathered, from evidence collected among the natives, that one of Franklin's vessels had actually made the North-west Passage while yet five living men remained on board; also that, when abandoned by the crew, the vessel had been left in perfect order, and was found by the Esquimaux, in the spring of 1849, near O'Reilly Island, lat. 68° 30' N., long. 99° 8' W., where it had been frozen in.

Captain Hall said that the skeletons of Franklin's men were scattered over King William Land; and he explains that the Esquimaux of that region are very different in character and disposition from those of Repulse Bay; and that, with one or two exceptions, they had refused to render the lost explorers any assistance, though they were well able (having enough food themselves) to have saved their lives; but instead of aiding them, not only allowed them to perish, but also plundered them of every thing they could make use of, and even suffered their dogs to eat them. He also heard and believed the native statement that

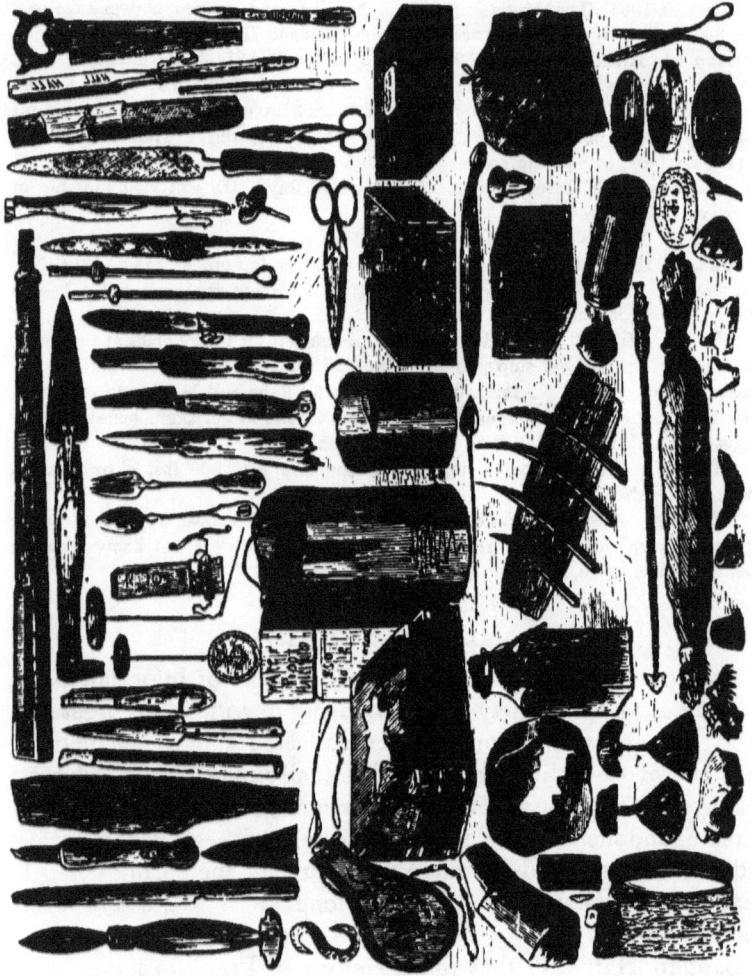

RELICS OF FRANKLIN'S EXPEDITION.

some of Franklin's men had eaten their companions. Captain Hall believed also that the original records of the *Erebus* and *Terror* are still in existence, and can probably still be found in a vault eastward and inland from Cape Victory. He was unable to reach that point himself, on account of a war which was then in progress between the native tribes of that region.

During the five years which Captain Hall spent in these explorations, he made no definite attempt at purely Polar discoveries, though incidentally making many.* He did not go above 70° N.; but he searched very thoroughly Melville Peninsula, Fury and Hecla straits, Pelly Bay, Boothia Peninsula and Gulf, and part of King William Land, connecting these with his Frobisher Bay explorations. He procured from the natives as many as one hundred and fifty relics of the Franklin Expedition, consisting of articles which had belonged either to the officers or ships of the lost expedition, and, with full records of his sojourn and journeys in the Innuit land, returned to the United States September 1, 1869.

CAPTAIN HALL'S ANNUAL REPORTS.

At various times, through the courtesy of whaling captains, Captain Hall sent home news of his progress. In 1865 he had learned that Captain Crozier of the *Terror*, with Parry, Lyon, and one other whose name he could not obtain, had survived the others, who earlier succumbed to cold and starvation; and that the three latter lived for some time on the flesh of their dead comrades. One Innuit (Hall uses this word to express Esquimaux, as generally understood) had sheltered and fed one of the Franklin party until he died. The evidence gathered went to prove that Crozier and one other was alive in the fall of 1864. It was also affirmed by the natives that Franklin's men had engaged in a battle with a tribe of Indians—not Esquimaux—near the estuary of Great Fish River; that none of the whites were killed, but many of the Indians; and that after this Captain Crozier and two others had gone to the south-west by land, to try and get to Fort Churchill or York Factory, and that then they had food, and either skin or rubber boats.

In the fall of 1866, Captain Hall wintered on Repulse Bay; and during the winter of 1866–67 he made a six weeks' journey with dog-sledges to the north-west, mainly to secure an ample supply of dogs for the next season. He took with him on this journey five white men—selected from the crews of whalers then lying at Repulse Bay—the two natives, Joe and Hannah, and thirty dogs.

On this occasion he met a small hostile tribe, but finally conciliated them by presents of old knives and tin pans, and re-

---

* See letter to Judge Daly, in Appendix, p. 424.

ceived in return forty dogs. He experienced very cold weather, and many hardships and inconveniences, but all kept their health through the journey. These natives said that some of the white men had been with them, and one had died, and was carefully buried.

On his return to Repulse Bay, Captain Hall expressed himself so certain of finding some of Franklin's party still alive, that he offered five hundred dollars in gold to each white man who would accompany him in the search during the season of 1867–68. Five seamen from the whaling fleet then in harbor offered to go, and, commencing their engagement in August, employed about two months in hunting, so as to secure a stock of provisions for their journey, while awaiting the hardening of the snow until it was fit for sledging. Captain Hall seemed to feel the pressure of a special call in this direction. To the captains in Repulse Bay he frequently remarked, "If I die, I shall die doing my duty."

In 1868, Captain Hall had procured additional evidence that Captain Crozier and one man had survived until 1864, and heard of others in King William Land; but he was not made happy by finding any white man alive who had belonged to that unfortunate company. Convinced at last of the fact that there were no survivors to rescue, Captain Hall returned to the United States in the bark *Ansel Gibbs*,* of New Bedford, with the full determination to secure a vessel and outfit to go in search of that geographical point which has so long eluded the efforts of Arctic adventurers, the North Pole.

PHYSICAL APPEARANCE.

Physically Captain Hall was well adapted for his chosen work. He was a well-proportioned, powerfully built man, about five feet eight inches in height, muscular rather than stout, though weighing not far from two hundred pounds. Life and vigor seemed inseparable from the thought of him. His head was well shaped, large, and covered with a profusion of wavy, brown hair; the beard also was thick and heavy, evincing still more of an inclination to curl. A phrenologist would have admired the ample de-

---

* The *Ansel Gibbs* was lost on the 19th of October, 1872, near Marble Island, her crew being forced to winter on shore, and were not rescued until August of 1873.

velopment of the coronal and temporal regions, and the broad, ample, reflective forehead, while a poet would have found in his expressive blue eye the very counterpart of the good knight Arthur,

"Of the
Frank and azure eyes."

His general expression was pleasant, but somewhat dreamy withal, when in repose, but kindling into a brilliant enthusiasm when his favorite topics were discussed; but possibly a skilled physiognomist would have descried too much of a poetical temperament in our Polar knight-errant, to have had much faith in him as a discriminator of, or successful commander of men.

MENTAL TRAITS.

Much—far too much—has been said in disparagement of Captain Hall, on account of his lack of what is technically called a "liberal education." He had all the education which was needed to have carried him to the end of his enterprise, had he not been thwarted by the cowardice of one, and the jealousy of others. One of the most intelligent and generous friends of Arctic research, who for years befriended Captain Hall, has assured the writer "that Hall knew 'Bowditch's Navigator' by heart—every line of it, and was perfectly competent to navigate a vessel;" while another friend of many years' intimacy, and perfectly competent of judging, writing to the *Cincinnati Commercial* (October 16, 1871), says, "Self-taught, to a considerable extent, as was Franklin, nevertheless Hall excels in the exactness and precision of his field-work, in the determining of latitude and longitude, and in his careful, conscientious record of magnetic and astronomical observations; for these, as well as for his accurate and reliable charting out of coast-lines, Hall has been complimented by the British Admiralty, and his work has stood the severest tests of our own Coast Survey Office, and of the Smithsonian Institute."

If Hall had not the advantage of a severe and systematic training in mathematics or science, he had at least thoroughly fitted himself for the work he had set himself to do. He did not pretend to be a scientific naturalist, but he was thoroughly competent to make and record geographical discoveries, and that was *the* object of the *Polaris* expedition; all else should have been subordinate to that. He was, despite his temperament, energetic, persevering, and courageous, and, above all, unselfish. The extent to which he was able to overlook the insolence and impertinence of those who owed him duty and allegiance is something marvelous to consider. Indeed, he carried this too far. Had he dealt more sternly with the beginnings of insubordination, we might have had a far different story to tell; but every other feeling and sentiment seemed swallowed up in the absorbing desire to get north, and it is somewhat difficult to realize how, for instance, he could, under the circumstances, have used such language as the following, which he did on the occasion of a public reception given him by the Geographical Society of New York: "I have chosen my own men—men that will stand by me through thick and thin. Though we may be surrounded by innumerable icebergs, and though our vessel may be crushed like an eggshell, I believe they will stand by me to the last." Here, certainly, was either a sad lack of discrimination, or a wonderful power of ignoring disagreeable facts when their recognition threatened to interfere with the progress of the expedition.

One pleasant feature of Captain Hall's character was his ever-abounding gratitude. His heart overflowed toward those who assisted him in any way, even the most trifling. Words seemed all too tame to express his sense of obligation to those who had forwarded his Arctic exploration schemes; and mingled with this recognition of friendly human help, was the ever-present sentiment of gratitude to God that he had been permitted to do something toward elucidating the fate of Franklin's expedition, and to add to the geographical knowledge of the Polar regions.

During the period of his stay in the land of the Innuits, Captain Hall had become so habituated to the usages of that people that he had actually learned to enjoy his residence among them. He liked them, believed in them, and thought them more intelligent and trustworthy than most Arctic travelers are willing to concede. The truth is, that they trusted and believed in him more than other travelers had permitted them to do, and hence the feeling of trust and confidence became reciprocal. He learned to like the repulsive food they live upon; fasting, when it was scarce, with the *sang froid* of "one to the manner born," and relishing the blubber, when it came, with the best of them.

It will thus be seen that the experience of Captain Hall in the

Arctic regions justified the hopes of his friends and the confidence of Government in intrusting him with the command of the *Polaris* expedition. He was thoroughly acclimated, was well known, and had friends among the natives at many stations, particularly on the western bays and shores of Davis Strait and Baffin Bay; having lived with them in their "igloos," or snow huts, in winter, traveled with them on their dog-sledges, shared their summer tents, and used, when necessary, their "oomiaks," or family boats. In the "Innuit land he did as the Innuits did."

After his return, in September, 1869, until the spring of 1871, Captain Hall was indefatigable in his endeavors to enlist public sympathy, and Congressional and other aid, for his new enterprise. In his previous voyages private parties had furnished, by subscription and donations, both vessels and outfit; but now the national aid was to be secured, the sympathy of the Government enlisted. Hitherto the United States had not officially done much in the way of Arctic explorations. When the liberality of Mr. Henry Grinnell had presented two vessels ready furnished for an Arctic voyage, the Navy Department ordered to the command Lieutenants De Haven and Griffith, to officer the *Advance* and *Rescue*. The second, under the auspices of the United States, and this wholly at the national expense, was the relief searching expedition under Lieutenant Hartstene, who went to the aid of Dr. Kane, and brought him back. The third is that which we have under consideration. Others wholly American, though unaided by the National Government, will be found recorded elsewhere. The *Polaris* expedition may truly be described as Captain Hall's, in its inception as well as execution; for, though the expense was borne wholly by the United States, and its late commander was most faithfully assisted by devoted and energetic friends, yet had it not been for the untiring perseverance with which the late commander of the *Polaris* canvassed the country, lecturing, writing, and by personal interviews with intelligent men of all classes who could by any possibility bring influence to bear in forwarding his plans, the expedition would never have sailed.

To those who saw him in Washington during the winter of 1869-70, he appeared like the embodiment of a single idea, and that idea was how to get aid for his projected Polar expedition. Here, fortunately, as in all large cities, there are always found men of brains and comprehensive intellects, to whom appeals of this nature come with the certainty of a sympathetic response. It is not always so in country places. Great men may be born in the rural districts, but they are almost certain to gravitate to cities.

Among those most ready to listen to the Arctic enthusiast—for such Captain Hall was, nor did he take any pains to conceal the fact—were such men as Senator Sumner, of Massachusetts; Sherman, of Ohio; Fenton, of New York; and others of that stamp, some of whom we may have occasion to refer to hereafter. The President, also, was personally interested in the project, and favorable to it; but though Senators and Representatives may feel great sympathy with many appeals and plans which come before them, they do not always consider themselves at perfect liberty to vote appropriations of the public money for carrying them out. A certain respect is usually paid to the supposed wishes and opinions of constituents. Hence it was no light task for Captain Hall and his immediate coadjutors to secure the vote of $50,000, which was finally granted by Congress for the outfit of the expedition.

A few days previous to the sailing of the *Polaris* expedition, the "American Geographical Society" of New York tendered a public reception to Captain Hall and his officers. On this occasion Mr. Henry Grinnell presented to Captain Hall the historical flag which, in 1838, Lieutenant Wilkes, of the United States Navy, had borne nearer to the *South* Pole than any American flag had ever been before. This flag had also gathered interest in every fold by subsequently being carried by Lieutenant De Haven to a higher northern latitude than any flag had ever been; next it passed into the hands of Dr. Kane, who bore it, still onward and upward, beyond De Haven's highest; and then Dr. I. I. Hayes, "'mid snow and ice," mentally, if not vocally, shouting "Excelsior," pitched it thirty-seven miles nearer the pole than his lamented predecessor had attained. It was again exhibited at the "reception given by the Geographical Society of New York to the officers and crew of the *Polaris*" on the eve of Feb. 16, 1874.

Captain Hall, in accepting this already glorified bunting, said he "believed that this flag, in the spring of 1872, would float over a new world, in which the North Pole star is its crowning jewel." In continuing his remarks, he declared that he did not expect during the first winter to reach above 80° N., and from thence in the spring to make sledge-journeys toward the pole.

Anticipating the criticism which some astronomers have made, "that he would not know when he got to the pole, even if he should really reach it," Captain Hall remarked: "On reaching that point called the North Pole, the north star will be directly overhead. Without an instrument, with merely the eye, a man can define his position when there. Some astronomers tell me I will find a difficulty in determining my position. It will be the easiest thing in the world. Suppose I arrive at the North Pole, and the sun has descended. Suppose there is an island at the North Pole; around it is the sea. I see a star upon the horizon. If I were to remain a thousand years at the pole, that star will remain on the horizon without varying one iota in height. Then, again, when I am at the pole, on the 23d of June I take the latitude of the sun; just $23\frac{1}{2}°$ high at one, and all hours. Five days before the 24th of June, and five days after, with the finest instruments we have, you can not determine one iota of change. Therefore you will see that it is the easiest thing in the world to determine when you arrive at the North Pole. The phenomena displayed there will be deeply interesting, provided there is land there; and I am satisfied, from the traditions I have learned from the Esquimaux, that I will find land there."

Had Hall reached the pole, it would be safe to say he would have been the happiest man alive. His enthusiasm was boundless; this was his spirit, as expressed in his address: "Many who have written to me, or who have appeared to me personally, think that I am of an adventurous spirit and of bold heart to attempt to go to the North Pole. Not so. It does not require that heart which they suppose I have got. The Arctic region is my home. I love it dearly—its storms, its winds, its glaciers, its icebergs; and when I am among them, it seems as if I were in an earthly heaven, or a heavenly earth!"

It was indeed a cruel fate which baffled and prevented the execution of the cherished hopes of such an exuberant spirit as this.

CAPTAIN HALL'S PREMONITIONS.

Captain Hall had made voluminous notes of his second expedition, 1864–69; but these records have not yet been published; he was too busy, during his short stay at home, to write them up, being occupied the whole time in securing the necessary influence and means of organizing the *Polaris* expedition. But the manuscript notes of his "Franklin Search" journey he took with him on board of the *Polaris*, intending to prepare them for publication; but some sad premonition of evil—

> "A strange and weird and phantom-seeming thing,
> Which stood dim outlined in a sable shroud,
> Though shapeless, as in noonday hangs a cloud"—

hovered before him, hinting at the insecurity of the fate which might be his; and he left these valuable papers with the inspector-general of North Greenland, Mr. Karrup Smith, at Disco.

A very fortunate providential monition, it would seem, as, had he taken them with him, they would probably have gone down in the foundered *Polaris*. It was that he might have time to prepare these records for publication that he sought the assistance of Mr. Meyers in writing up the journal of the day; but as Mr. Meyers refused his aid, Captain Hall foresaw that he must postpone his literary work; and this was one of the things which seriously annoyed and troubled him. The account of his last illness and death will be found in its appropriate place in Captain Tyson's journal, as part of the history of the *Polaris* expedition.

The last dispatch which Captain Hall indited for transmission to the Secretary of the Navy (a fac-simile of a portion is given below) was written at lat. 82° 3′ N., at Cape Brevoort, and a copy of it was placed in a copper cylinder, and left there, protected by a covering of stones. It will be found entire in the Appendix.

FAC-SIMILE OF CAPTAIN HALL'S WRITING.

CHAPTER VI.

Dr. Emil Bessel.—Sergeant Frederick Meyers.—Mr. R. W. D. Bryan.—Sidney O. Buddington.—Hubbard C. Chester.—Emil Schuman.—William Morton.—Letter of Captain Hall's.—The *Polaris* sails.—Disaffection on Board.—Meets the Swedish Exploring Expedition.—Favorable condition of the Ice.—United States Ship *Congress* arrives at Disco with Supplies for the *Polaris*.—Insubordination on Board.—Captain Hall's Idiosyncrasy.—He "bids Adieu to the Civilized World."

DR. EMIL BESSEL, the chief of the Scientific Corps, was a native of Germany. Though a comparatively young man, he had already established a reputation in his native country. He was a

DR. EMIL BESSEL.

graduate of the famous university of Heidelberg. His researches in zoology and entomology had brought him into close friendly

SIDNEY O. BUDDINGTON.

relations with Dr. August Peterman, of Gotha, who appointed him as scientist on board the *Albert*, belonging to M. Rosenthal, a walrus-hunter, which sailed in the year 1869 to the north of Spitzbergen. He was considered an accomplished surgeon and an enthusiastic naturalist, and was also esteemed in every way intellectually fit for the position.

In person he was slight, somewhat delicately built, and of quick, nervous temperament. The shape of the head indicated predominant mentality, while the features were regular and pleasing in their contour; hair and beard dark and heavy, with a bright, dark eye, which was susceptible of very varied expression, completed the *tout ensemble* of what would pass for a handsome man, built on rather too small a scale.

Mr. FREDERICK MEYERS, the meteorologist, was a native of Prussia, and a sergeant of the United States Signal Corps, established in Washington. He had received a thorough education, and was considered especially fitted for the position to which he was appointed.

HUBBARD C. CHESTER.

Mr. R. W. D. BRYAN, astronomer and chaplain to the expedition, was a young man of very superior talents, a graduate of Lafayette College, in Easton, Pennsylvania.

SIDNEY O. BUDDINGTON, sailing-master, was a native of Groton, Connecticut, and had been captain of various whalers for over twenty years. He was familiar with the perils of ice navigation, and was generally esteemed a capable navigator.

HUBBARD C. CHESTER, first mate, is a native of Noank, Connecticut, and made a good officer.

EMIL SCHUMAN, the chief engineer, was a German, regularly educated for his profession, and a draughtsman of considerable skill.

WILLIAM MORTON, second mate, has spent most of his life at sea, having served thirty years in the United States Navy as yeoman, and having made at least one memorable Arctic voyage with Kane; on which occasion he crossed the great Humboldt Glacier in company with the native Hans, and, looking out upon the waters since visited by the *Polaris*, thought he had discover-

EMIL SCHUMAN.

ed the Polar Sea. He was an efficient helper of Kane's; and though he made his Arctic reputation upon a visual mistake, in that he has good company; nor does it detract from his faithful service to Kane and Hall. He is a native of Ireland, but has resided for many years in Jersey City, in the State of New Jersey.

Just before sailing, Captain Hall addressed the following letter to a friend on whom he greatly relied:

Left New London Harbor, 4 A.M. this day,
July 3, 1871.
On board the *Polaris*, now stopped to let off the pilot.
This my last letter.

MY DEAR FRIEND,—Inclosed please find a document that will explain itself. I wish you to see the Secretary of the Navy in person, and have all done for Captain Tyson that can be done. Try and have a position given him such as will command a commission. In proof that Captain Buddington is anxious to have all done that can be for our old friend, he will sign this letter with me, which will have an earnestness that can not be better expressed.

To Captain Tyson I have paid $———— in advance out of my own pocket. * * * Government now allots me $———— per month for my family; I ask if I am to pay Captain Tyson for his valuable services. * * *

I have neglected to report to Government that James M. Buddington, of Pequan-

WILLIAM MORTON.

nock, New London, Conn., will be ice-pilot to the transport to Disco, Greenland, for $——— per month. Much more I will soon write you, my noble and attentive friend; but now I cease my incoherent, hasty work.

Do all you can for Tyson. Let as good a position as possible be given. * * * The ship's company is now of the best material. Glorious is the prospect of the future. C. F. HALL.*

The *Polaris* left New London at 4 P.M., on the 3d of July, and proceeded without especial incident, except a severe storm, which occurred on the evening of the 3d, to St. Johns.

Mr. Noble, one of the companions of the artist Bradford, when he went "after icebergs," made the remark that "there was stupidity somewhere," because the beauties of St. Johns harbor and vicinity had not been written up by summer tourists; and the same feeling occurs to all who have an eye for natural beauties who visit Newfoundland for the first time.

The harbor is completely land-locked, and is about a mile long, and half that distance in width, and is entered through a narrow

---

* Captain Buddington's signature is not attached to the letter.

channel flanked with high rocks, picturesque in the extreme. Here, as elsewhere, the "west end" is the fashionable part of the town, which is terraced along the hill-side, showing its whole extent to the approaching visitor; while the east side is devoted to business, and that business principally the storage of oil, in bright red warehouses.

While lying in this pleasant port, a point of discussion arose as to the authority of the commander over the Scientific Corps. Strong feeling was mutually exhibited, which extended to the officers, and even the crew, among whom was developed an unmistakable feeling of special affinity on the score of national affiliation. At this point it really appeared as if the foreign element were far more in sympathy with each other, as fellow-countrymen, than they were with furthering the hopes of Captain Hall, and the main object of the expedition. However, matters were smoothed over; the Scientific Corps were left free to follow their own course, and the threatened disruption of the party avoided.

The vessel laid at St. Johns for a week, some repairs being found necessary for her machinery. Efforts were also made to secure a carpenter, but without success; and Captain Hall notified the Secretary of the Navy that he should leave without one, deeming longer delay inexpedient.

Touching first at Fiscanaes, and then at Holsteinborg, Greenland, Captain Hall was agreeably surprised to meet at the latter port the Swedish Arctic exploring expedition, consisting of a brig and steamer, under commander Frederick W. Von Otter. This expedition had been up as far as Upernavik, but was now homeward bound, and the commander kindly offered to take and forward dispatches for Captain Hall.

Commodore Von Otter reported the navigation in Baffin Bay to be exceedingly favorable, he having seen no ice, except occasional bergs, between Disco and Upernavik. He also furnished Captain Hall with a tabular statement of the results of some thirty deep-sea soundings and sea temperatures, obtained on board H. S. M. steamship *Ingegera* between Holsteinborg and Upernavik (for which see Appendix, p. 457), with other pleasant courtesies, duly acknowledged and thoroughly appreciated by the commander of the *Polaris*.

The favorable reports of the Swedish expedition inspired Captain Hall with new hopes and increased enthusiasm, as the prospect of making good progress northward during the present season was thus confirmed. Governor Elberg, of the Holsteinborg district, also very kindly offered to aid the purposes of the expedition in any way that he could. Hall and he had met before, and they now greeted each other as old friends.

There were two objects in putting into Holsteinborg; one was to get a supply of coal, and the other, a stock of reindeer furs, to make up into winter garments; but neither of these could be obtained. The supply of coal on hand was only fifteen tons; and though the governor generously offered to let Captain Hall have two-thirds of it, yet the latter very properly declined to strip the settlement, and concluded to await the tender which he knew was to bring a supply, and for which he was now every day hopefully looking. No reindeer skins could be had either, for the reason that none of these animals had recently been obtained. The deer formerly visited that part of Greenland in large herds, but of late years they had totally disappeared; and hence no skins could be found for sale.

The *Polaris* lay at Holsteinborg until the 3d of August, awaiting the tender *Congress*, and then sailed for Disco, reaching the harbor of Goodhavn on the afternoon of the 4th, twenty-four hours' sailing-time.

On inquiry it was found that the inspector of the district, Mr. Karrup Smith, was away on his annual tour through the district, and might probably not return for two or three weeks; and though his lieutenant was very cordial and kindly disposed, he hesitated to assume the necessary responsibility. In this dilemma, the wife of the inspector, Mrs. Smith, came to the rescue; and expressing herself warmly in behalf of the expedition, and the propriety of the Danish officials doing all in their power to assist the party from the "great and glorious country of the United States," she suggested that a boat-party be sent off to seek the inspector and request his return to Goodhavn. Captain Hall promptly adopted the advice, and detailed his chief mate, H. C. Chester, for the duty.

After a search up and down the coast, involving a distance of one hundred and seventy-five miles, almost entirely by oars, Mr. Chester was fortunate enough to find the object of his inquiry at Rittenbek, a station to the north of Disco, and the inspector, instantly acceding to the request, returned with his boat, in company with Mr. Chester, on the 11th instant. During the absence of the boat, however, the United States steamer *Congress*, Captain Davenport, had arrived at Disco (August 10), and thus relieved Captain Hall of a "mountain-load of anxiety," and making the presence of Mr. Smith less necessary, though no less agreeable. It was, however, desirable to have his sanction for the use of the Government store-house, as a dépôt for the extra supplies brought out by the *Congress*, for the future use of the expedition; and this was readily granted by Mr. Smith, who also engaged to have the stores carefully preserved till called for, and this gratuitously. The *Polaris* was here replenished with all the extra stores and coal which she could carry, and the balance was placed in the offered store-house belonging to the Danish Government.

Besides the stores and dispatches, the *Congress* had also brought out Captain Tyson's commission as assistant navigator to the expedition, and he was henceforth a regularly enrolled officer. Captain Hall still wanted two men to thoroughly complete his Arctic party. Of these, one was a resident Dane, named Jansen, and enjoying the grand title of "governor" at Tossac, and the other a native Esquimau, Hans Christian. Both of these had been attached to Arctic exploring expeditions before, with Drs. Kane and Hayes, and it was naturally thought that their knowledge and experience of Arctic resources would add to the efficiency of the expedition, especially as Hans was reputed a good hunter and dog-driver. The last accomplishment is not easily learned, while at the same time it is essential to the success of sledge-traveling. In addition to these, there were still dogs and furs to be obtained. Jansen was to be found at Tossac, and Hans at Proven, a little to the south of the former.

Leaving Goodhavn on the 17th of August, where also the *Polaris* parted with the *Congress*, the former made all sail for Upernavik, which was reached on the 18th; thus making two hundred and twenty-five miles in little more than thirty-three hours.

While lying at Goodhavn, it leaked out that furtive raids had been made on the liquors and other stores by unauthorized parties, and for a while it seemed that an open quarrel would be the outcome of the discovery. But Captain Hall's patience and forbearance were equal to the emergency, and the misdemeanor was condoned.

Here he had hoped to meet Dr. Rudolph, the late Governor of Upernavik, who had been thirty years in Greenland, but who was now expecting to return on a visit to his native place—Denmark—and by whom he expected to be able to send to the United States his latest dispatches.

Doctor, or now ex-Governor, Rudolph—for his successor, Mr. Elberg (son of the Governor of Holsteinborg), had arrived—immediately offered every facility within his power to aid Captain Hall. He sent two men in kyacks—small native boats—one to find Jansen, and the other to find Hans. The former was unsuccessful, as Jansen declined to come; but Hans, who was at Proven, fifty miles to the south of Upernavik, was secured. A boat from the ship had to be sent for Hans, and, under the pilotage of Mr. Chester, he made the *Polaris* on the 20th of August. Mr. Chester made this trip of a hundred miles, rowing, between noon of August 19th and eight P.M. of the 20th.

Captain Hall appears to have had very decided premonitions of disaster, from the fact that he left here in charge of Inspector Smith a quantity of valuable papers relating to his second expedition, and particularly to his search for Sir John Franklin—an extraordinary step to take under the circumstances, as his object in taking them with him was to write them up for publication on his return.

Why Captain Hall studiously avoided all allusion to the dissensions on board—why he even compliments the "material" of the expedition, when he was all the while suffering from insolence and disaffection—can only be explained by his idiosyncrasy, which *enabled him to sink every thing else in the one idea of pushing on to the far north.* He dreaded nothing so much as being delayed, or compelled to return. He was willing to die, but not to abandon the expedition. Not being able to get deer-skins as he expected, Captain Hall procured supplies of seal and dog skins for extra clothing, which answers nearly or quite as well, except for boots.

The last dispatch from Greenland received by the Secretary of the Navy from Captain Hall was dated at Tossac, and forwarded, through the courtesy of the Danish governor, by a vessel bound to Copenhagen. It was, however, too late for that year. The communication with Denmark is by an annual visit of a ship in the early summer, which remains on the Greenland coast for a few weeks only, touching at different settlements, and then returning. When the following dispatch was written the vessel had sailed; and there being no other means of sending it, a whole year elapsed before it could be forwarded, which was done on the next annual visit of the Danish ship. It was not received in

UPERNAVIK.

Washington until August of 1872, at which time Captain Hall had been nine months in his solitary Polar grave. The following is the dispatch:

Position by my observation, lat. 73° 21' 00" N.; long. 56° 5' 45" W.—HALL.

United States Steamship *Polaris*, Tossac (or Tes-su-i-sak), Greenland, August 22, 1871.

SIR,—I have the honor to report my proceedings since the date August 20 and 21, of my last communication written at Upernavik.

It was 8.30 P.M. of August 21, when we left the harbor of Upernavik, having on board Governor Elberg, of whom I made previous mention, and several of his people, bound for this place on a visit. After steaming twelve miles to the northward and westward, we hauled up in front of a small island settlement, called King-i-toke, where Governor Elberg and myself, with a boat's crew, went ashore to purchase dogs, furs, and other requisites for the expedition.

Not a little was I chagrined, at this place, to find the resident governor perfectly immovable in his purpose not to comply with my desire to purchase some of his dogs, although he had many of the best sledge and seal dogs I had ever seen. All my efforts in liberal offers and otherwise, combined with the persuasive language of Governor Elberg, failed to induce the honorable official of King-i-toke to sell even a single one of his fine dogs. However, I was able, after considerable difficulty, to get from his people eleven dogs, to add to the number already possessed by the *Polaris*.

Having spent two hours at King-i-toke, we returned aboard, and at once (1 A.M. of August 22), resumed our voyage for Tossac, threading our way, by the aid of good native pilots, among the numerous reefs, rocks, and islands with which Upernavik and vicinity abound. At 5.30 A.M. of the 22d we arrived at Tossac, lat. 73° 21' 18" N., long. 56° W. At once I called on Jansen, and, to my astonishment and disappointment, found that a mistake had been made in any one of us expecting that his consent had been, or could be, obtained to leave his home at the present time; * * * but at the same time he has the desire to do all he can in supplying the expedition with dogs, furs, etc. This desire, however, I find to be combined with *a face of brass*, for he charges unheard-of prices for his dogs, and will not deviate a hair, knowing as he does, and as I do, that this is the last place, and the only place, I can now depend upon with any hope of getting the supply to make up the number needed for our expedition.

By the consent and co-operation of the Government authorities of Denmark resident in Greenland, I have concluded to contract with Hans Christian, by which he enters into the service of the United States North Pole Expedition, as dog-driver, hunter, and servant, at a salary of $300 per annum. * * * His wife and three children are to accompany Hans.

The prospects of the expedition are fine; the weather beautiful, clear, and exceptionally warm. Every preparation has been made to *bid farewell to civilization for several years, if need be*, to accomplish our purpose. Our coal-bunkers are not only full, but we have full ten tons yet on deck, besides wood, planks, and rosin in considerable quantities, that can be used for steaming purposes in any emergency. Never was an Arctic expedition more completely fitted out than this.

The progress of the *Polaris* so far has been quite favorable, making exceedingly good passages from port to port. * * * The actual steaming or sailing time of the *Polaris* from Washington to New York was sixty hours; and from the latter place to this, the most northern civilization settlement of the world, unless there be one for us to discover at or near the North Pole, has been *twenty days, seven hours, and thirty minutes*. Had I known, on leaving New York, that the United States transport would be at Goodhavn, Greenland, as soon as the *Polaris* could reach that port, the *Polaris* would have been here in advance of the present time just eighteen days; but there is no cause of any regret—indeed, there is every reason to rejoice that every thing pertaining to the expedition, under the rulings of high Heaven, is in a far more prosperous and substantially successful condition than ever I have hoped or prayed for. Every effort we are making to get ready to leave here to-morrow. I will at the latest moment resume my place in continuing this communication.

Evening, August 23, 1872.

We did not get under way to-day as expected, because a heavy, dark fog has prevailed all day, and the same now continues. The venture of steaming out into a sea of undefined reefs and sunken rocks, under the present circumstances, could not be undertaken. The full number of dogs (sixty) required for the expedition is now made up. At the several ports of Greenland where we stopped we have been successful in obtaining proper food for the dogs.

August 24, 1 P.M.

The fog still continues, and I decide we can not wait longer for its dispersion, for a longer delay will make it doubtful of the expedition securing the very high latitude I desire to obtain before entering into winter-quarters. A good pilot has offered to do his very best in conducting the *Polaris* outside of the most imminent danger of the reefs and rocks. * * *

November, 1.30 P.M.

The anchor of the *Polaris* has just been weighed, and not again will it go down till, as I trust and pray, a higher, a far higher, latitude has been attained than ever before by civilized man. Governor Elberg is about accompanying us out of the harbor and seaward. He leaves us when the pilot does. Governor Lowertz Elberg has rendered to this expedition much service, and long will I remember him for his great kindness. I am sure you and my country will fully appreciate the hospitality and co-operation of the Danish officials in Greenland as relating to our North Polar expedition.

November, 2.15 P.M.

The Polaris bids adieu to the civilized world.

Governor Elberg leaves us, promising to take these dispatches back to Upernavik, to send them to our minister at Copenhagen by the next ship, which opportunity may not be till next year. God be with us. Yours ever,

C. F. HALL.

Hon. GEO. M. ROBESON, Secretary of the Navy, Washington, D.C.

As we thus bid farewell to Captain Hall, as he sails, with his bright hopes clouded by insubordination, but with a brave, undaunted heart, northward to unknown seas, we drop the general narrative, and now take up Captain Tyson's notes of events as they occurred, going back a few days to recover some details of interest occurring between New York and the Greenland port.

CHAPTER VII.

NOTES BY CAPTAIN TYSON ON BOARD THE POLARIS.

Captain Tyson's Soliloquy on leaving Harbor.—A Thunder-storm.—Arrive at St. Johns.—Icebergs in Sight.—Religious Services on board the *Polaris* by Dr. Newman, of Washington.—Prayer at Sea.—Esquimau Hans, with Wife, Children, and "Vermin," taken on board.—Firing at Walrus.—The Sailing-master wants to stop at Port Foulk.—The *Polaris* passes Kane's Winter-quarters.—An impassable Barrier of Ice.—Misleading Charts.—The open Polar Sea recedes from Sight.—Afraid of "Symme's Hole."—*Polaris* enters Robeson Channel.—Surrounded by Icefields.—Council of Officers.—Puerile Fears.—Sir Edward Belcher.—The American Flag raised on "Hall Land."—Seeking a Harbor.—Repulse Harbor.—Thank God Harbor.—Providence Berg.—Housing the Ship for Winter-quarters.

"*June* 29, 1871. As we left the Navy Yard and steamed toward the Sound, the vision of friendly faces from which I had just parted seemed to follow me with an intensity I have not always had time to realize. When acting as master the thoughts must be quickly withdrawn from all on shore, and concentrated on the business of the ship. But for once I find myself sailing without a designated position and toward unknown seas, and with leisure to think of the past and to anticipate the future. To observe others, instead of commanding them, is a new sensation on leaving port, and I gave myself up to the novel employment.

"As we passed through the East River, with the great city of a million souls on my left, and half a million on the right, I could not help thinking how few of all these took any interest in our peculiar mission; and of those who knew of the outfit of the *Polaris*, the majority, no doubt, thought we were wild and reckless men, willfully going to our own destruction. But some there were who bade us God-speed—some large souls who could look through the danger to the honor, and who sympathized with that mysterious attraction which ever draws us on to seek the unknown. I hope this expedition will repay the cost and trouble of its getting up.

"On we sail—the sunset behind us, the bright summer night

beckoning us to the familiar waters of Long Island Sound. Hell Gate is passed, and all looks well for a good night.

"*June* 30. At 11 A.M. dropped anchor in New London harbor—so recently my home. Some of my old friends here will think I have started on a wild-goose chase. But as to that all depends on good management. If we have that, I think we can get farther north than any one has been yet. In the evening some members of the Baptist congregation came on board, and a religious service was held. I have heard that Captain Hall always favored the cultivation of a religious sentiment among his ship's company.

"We remained at New London until Monday, July 3. When we left it was fine weather, but the next day we were surrounded with fog, which continued for three days. On the evening of the 3d, during the first watch, a great change came over the sky; dark clouds, changing every moment to a deeper blackness, were massed above the horizon to the south-west, and in almost less time than it takes to describe it the entire sky was covered as with a pall; a sudden rain-squall, with violent thunder and brilliant lightning, quickly succeeded each other. The lightning seemed almost continuous, so incessant were the flashes. The very firmament was in a blaze from horizon to zenith, while peal after peal of deep reverberating thunder echoed and re-echoed across the sky like the cannonading of contending armies. But the *Polaris*, undisturbed, moved serenely on her way. The storm continued until midnight, and then gradually subsided.

"The 9th being Sunday, service was held in the cabin, Captain Hall taking occasion to remark that it was his intention, whenever weather permitted, to hold Sunday services.

"The next day we sighted the coast of Newfoundland, encountering some loose floating ice as we approached St. Johns. In the harbor, which we made on the 11th, were two good-sized icebergs.

"I see there is not perfect harmony between Captain Hall and the Scientific Corps, nor with some others either. I am afraid things will not work well. It is not my business, but I am sorry for Hall: he is fearfully embarrassed.

"*July* 19. Bade farewell to St. Johns. In the evening a very fine show of northern lights commenced about 10 P.M., and continued till past eleven. The sailing-master talked of resigning and going home, but matters have been smoothed over.

"On the 26th we passed a heavy piece of ship's timber; it looked as if it had been a long time in the water—a piece of some wrecked whaler, I suppose. If it could speak, who knows what a romance it might have to tell? We are now well up toward Fiscanaes; ought to see the coast to-morrow.

"*Thursday*, July 27. Many icebergs in sight; a great many to the southward, and some to the east. Can see the coast plainly, lat. 63°. In the afternoon, about three o'clock, a native pilot came out in his kyack from Fiscanaes and boarded us; these fellows are very daring, and risk themselves in their little, dancing, feather-like boats far out of sight of land.

"Only remained at Fiscanaes until the 29th, leaving at 3 A.M.,

THE FISCANAES PILOT.

not finding Hans here, as expected. Weather delightful. A few hours later we passed Lichtenfels, a missionary station. In the latter part of the afternoon the weather changed; a fresh breeze sprang up, and it commenced to rain. A heavy gale of wind from the south-west created much sea, and, darkness coming on, it was deemed prudent to stop the engine, as there was danger of running on bergs in the uncertain light; also shortened sail. The storm lasted about four hours.

"*July* 31. Reached Holsteinborg. Like most of the Greenland settlements, this is a small place. You can stand on the deck of the *Polaris* and count not only all the houses, but almost all the people, for every one that can walk gets out to look at a vessel in the harbor. Captain Hall thought he might perhaps find Hans here, but he was up to the northward. Left Holsteinborg on the 3d of August, about 2 P.M., and next morning sighted Disco. Icebergs in plenty here. Many of them are from one hundred and fifty to two hundred feet high. To those who see icebergs for the first time the size seems to make the greatest impression, but afterward the beauty of many of them keeps the eyes fascinated; but to the Arctic sailor the permanent feeling is— look out! there is danger! Early in the afternoon we received a pilot, and at 3 P.M. of the 4th of August we cast anchor at Goodhavn.

"Some one has been at the stores. Captain Hall told me he would not have any liquor on board; but Dr. Bessel procured an order for some for medical purposes, and the "thirsty" have found out where it is stowed.

"On Sunday, August 6, went to church, which, considering the size of the place, was well attended both by resident Danes and Esquimaux. Here, I suppose, we shall wait for the *Congress*, United States store-ship. Mr. —— told me that my commission would be sent out by her. After seeing what I have, it would suit me just as well if it did not come, for then I should have a decent excuse to return home. There is nothing I should like better than to continue the voyage if all was harmonious, and if each person understood his place and his proper duties.

"*Aug.* 10. United States store-ship *Congress* arrived from New York with provisions and coal. After storing the *Polaris* to her utmost capacity, the rest was landed at Disco, as a dépôt, in case the expedition should need it hereafter. Captain Davenport and Rev. Dr. Newman, who came up in the *Congress*, have had their hands full trying to straighten things out between Captain Hall and the disaffected. Some of the party seem bound to go contrary anyway, and if Hall wants a thing done, that is just what they won't do. There are two parties already, if not three, aboard. All the foreigners hang together, and expressions are freely made that Hall shall not get any credit out of this expedition. Already some have made up their minds how far they will go, and when they will get home again—queer sort of explorers these!

"*Aug.* 17. Captain Hall has purchased a number of dogs for our sledge-excursions. The Rev. Dr. Newman, of Washington, came aboard the *Polaris* and held a service, using the following prayer, one of three which he has written expressly for the expedition:

PRAYER AT SEA.

"'O God of the land and of the sea, to Thee we offer our humble prayers. The whole creation proclaims Thy wisdom, power, and goodness. The heavens declare Thy glory, and the firmament showeth Thy handiwork. Day unto day uttereth speech, and night unto night showeth knowledge. There is no speech nor language where their voice is not heard. And we thank Thee for the clearer and fuller revelation of Thyself to man in Thy precious Word. Therein Thou hast revealed Thyself as our Sovereign and Judge. Thy law is perfect, converting the soul. Thy testimony is sure, making wise the simple. Thy statutes are right, rejoicing the heart. Thy commandment is pure, enlightening the eyes. Thy fear is clear, enduring forever. Thy judgments are true and righteous altogether. Although far from home and those who love us, yet we are not far from Thee. We are ever in Thy adorable presence; we can never withdraw from Thy sight. If we ascend up into heaven, Thou art there; if we make our bed in hell, behold, Thou art there; if we take the wings of the morning and dwell in the uttermost parts of the sea, even there shall Thy right hand lead us, and Thy right hand shall hold us. Oh, help us to be ever conscious that Thou seest us, and knowest us altogether. Though the darkness may cover us, yet the night shall be light about us; for the darkness and the light are both alike unto Thee. While on the mighty deep, be Thou our Father and our Friend; for they who go down to the sea in ships, that do business in the great waters, see the works of the Lord and his wonders in the deep. It is Thee who raiseth the stormy wind which lifteth up the waves; it is Thee who maketh the storm a calm, so that the waves thereof are still.

"'Oh, hear us from Thy throne in glory, and in mercy pardon our sins, through Jesus Christ our Lord and Saviour. Give us noble thoughts, pure emotions, and generous sympathies for each other, while so far away from all human habitations. May we have for each other that charity that suffereth long and is kind, that envieth not, that vaunteth not itself, that is not puffed up, that seeketh not her own, that is not easily provoked, that thinketh no evil, but that beareth all things, hopeth all things, endureth all things; that charity that never faileth.

"'May it please Thee to prosper us in our great undertaking, and may our efforts at this time be crowned with abundant success. Hear us for our country, for the President of the United States, and for all who are in authority over us. And hear us for our families, and for all our friends we have left at home; and at last receive us on high, for the sake of the great Redeemer. Amen.'

"After the service we weighed anchor and steamed out of the harbor. The men on board of the *Congress* cheered us as we went off, and the most of us returned it. The weather is fine, but many icebergs are all around; some nice steering is required to avoid running afoul of them.

"UPERNAVIK, *Aug.* 18. Captain Hall, being disappointed about getting the deer-skins at Holsteinborg, has now to try and buy dog-skins. They make up into very warm clothing, and are a great deal better protection against cold winds than almost any amount of our woolen clothing. Civilized clothing will keep off *still* cold, but it takes skins to protect against the searching Arctic winds.

"Hall now tells me that Davenport was prepared to take one party home in irons for his insolence and insubordination, but that another said he would leave if he did, and that then all the Germans of the crew would leave too; and that would break up the expedition. Was ever a commander so beset with embarrassments, from which there seems no way to free himself except by giving up all for which he has worked so long and so hopefully?

"P.M. Mr. Chester has got back with a nice load—not only Hans, but his wife and three children, with all his household goods, and skins alive with vermin. Hans, it seems, would not come without his family.

"Yesterday several sections of our 'oil boiler' were ordered to be thrown overboard; so, when our coal gives out, we shall be unable to raise steam.

"KINGITUK, *Aug.* 21. We left Upernavik at 8 A. M., and arrived at this little settlement at 11. The captain could not get dogs enough at our last port, and hopes to find some here. The Governor of Upernavik, who came up on the *Polaris*, intending to go as far as Tossac, accompanied the captain ashore—I suppose to influence the natives to sell their dogs. Succeeded in getting about a dozen.

"*Aug.* 22. Arrived at Tessuisak, or Tossac; here Captain Hall completed his supply of skins and dogs—have got about sixty altogether; but the man Jansen, whom he hoped to get, would not come. He had grown too big; he is governor (!), or some such thing, of this collection of huts. He was a good man with Dr. Hayes.

"*Aug.* 23. Can't get away for the fog.

"*Aug.* 24. Sailed to-day. This is the last settlement we expect to stop at. Now we may say we are at the entrance of our work. Only a few days more, and, if the ice does not beset us, we shall be through Smith Sound.

"*Aug.* 27, *Evening.* We have reached lat. 78° 51' N., past Kane's winter-quarters. It was on Thursday, three days ago, that we left Tessuisak. We steamed out after dark, and almost ran afoul of an iceberg, and afterward encountered a great many of them, and also considerable ice; but the *Polaris* worked through very well. Yesterday saw a party of walruses on the ice; fired at them, but they got out of the way. It is almost impossible to kill a walrus with a ball fired toward the front, unless the eye is hit. The skull is very thick, except on the crown of the head, which is a difficult point to strike.

"The sailing-master wants the *Polaris* to go into Port Foulke and lie up; then he can stay there and take care of the ship, and the others can go up north in sledges if they want to.. But I am glad to see that Hall perseveres, and will have his way about that; and indeed there is nothing to hinder.

"One revelation after another. Seeing Captain Hall very constantly writing, I asked him if he was writing up his Franklin search-book, about which he had often talked to me. He said, 'No; I left all those papers at Disco!' I did not like to ask him, but I looked 'Why?' A sort of gloom seemed to spread over his face, as if the recollection of something with which they were associated made him uncomfortable, and presently, without raising his head, he added, 'I left them there for safety.' I saw the subject was not pleasant, and I made no further remark; but I could not help thinking it over.

"It was about half-past three o'clock to-day when we came up to Kane's winter-quarters of 1853-55, and this evening to the point where he abandoned the *Advance*. But he went much higher by land.

"*Aug.* 28. Last night, just before midnight, at which time it was my watch, Chester came down and reported that an 'impassable barrier of ice' lay ahead of us. I went up; found the vessel had been slowed down; met our sailing-master, who was in a fearful state of excitement at the thought of going forward. I went up aloft, and looked carefully around. There was a great deal of ice in sight, which was coming down with a light northerly wind. It looked bad; but off to the westward I saw a dark streak which looked like water, and I believed it was; went down and reported to Captain Hall that the ship could skirt round the ice by sailing a little to the south, and then steering west-north-west. At this time the sun set about 11. P.M., and rose again by 1 A.M., so that it was nearly light all the time.

"During my watch I got the vessel over to the west side, and found a passage of open water varying from one to four miles in width. During my watch below we reached Cape Frazier, in latitude about 80° N. The obstructing ice which we sailed round to avoid was very thick—from ten to forty feet—showing that it had been formed on the shore plateaus, or shoals, as no such ice ever forms in the open water.

"We are now on the west side of the sound. It was from about here, I think, that Dr. Hayes traveled with dog-sledges up to 81° 35' N. I hope we shall get much farther. But there is one, at least, on board who thinks we have come too far already. Out on such cowards, I say! I keep aloft much of the time. All the points about here are *wrong on the charts.* Here Captain Hall went ashore and coasted around a little in the boat.

"Moved on again; passed Carl Ritter Bay; and when we got to Cape George Back the ice led us off to the north-east, and we crossed Kennedy Channel, and then over again to Cape Lieber, where we brought up in a fog about fifteen miles from land. Here a copper cylinder was thrown overboard, containing a record of our progress. We have now gained lat. 81° 35' N. *Can't make any thing out of the charts.* As old Scoresby says, 'They are more of a snare than a guide.' But we are now at the head of Kennedy Channel, and ought soon to see *Kane's open sea!*

"Still sailing north; some trouble working through the ice. *Here should be the open sea,* but there is land on both sides of us! To-day we have sailed into a bay which Morton and Hans must have mistaken for a sea; this bay lies to the eastward, inclining to south. Captain Hall calls it Polaris Bay. We are not deceived; *we have sailed right across it.* It is not surprising that Morton made this mistake if it was foggy, for on many days one could not see across it; it is about forty-five miles wide. The land, though, is plain enough to be seen, for it is high, quite high land.

"Still sailing on. We have now got into a channel similar to Kennedy's, only wider, and must be part of the water mistaken for the open sea. This channel is seventeen or eighteen miles wide, and obstructed by heavy ice. I hope we shall be able to get through, but it don't look like it now. I see some rueful countenances. I believe some of them think we are going to sail off the edge of the world, or into 'Symme's Hole.' But so far we see no worse than I have seen scores of times in Melville Bay

—ice. Captain Hall has called this new channel, after the Honorable Secretary of the Navy, *Robeson Channel*—a good name: without the good-will of Secretary Robeson we should not have been here; and if the *Polaris* should get no farther, her keel has plowed through waters never parted by any ship before.

"*Aug.* 29. Surrounded by ice-fields, and a thick fog has settled down on us; all last night working through the ice, but fear we shall get no farther. Have had to fasten to a floe until about 7.30 P.M., when we made a lead near the eastern coast, in hopes to find a harbor here. A second cylinder has been thrown overboard. Went ashore in the boat with Captain Hall, and examined a bight inshore to see if it would do for a harbor. No protection; would not do. The ice is pressing heavily upon the sides of the *Polaris*. The captain has ordered a quantity of provisions to be taken out and put on the ice.

"*Aug.* 30. Put the provisions aboard again; drifting out of Robeson Channel to the south-west, wind from the north-east. Steamed in under the land, and came to anchor behind some bergs. It is blowing a gale.

"*Sept.* 1. Unshipped the propeller to save it from injury.

"*Sept.* 2. Captain Hall requested Captain Buddington, Mr. Chester, and myself to come into the cabin; wanted to consult about attempting to proceed farther north. Mr. Chester and I wanted to go on as far as it was possible to get, but the senior officer was opposed. I could have told that before. He was very set, and walked off as if to end the discussion. Captain Hall followed him, and stood some time talking to him. After a while Captain Hall come toward us, and ordered us to see to the landing of some provisions. I said nothing more, neither did Chester.

"These puerile fears remind me of Sir Edward Belcher's expressions when discussing the possibility of M'Clure and Collinson having ventured into the Polar Sea north of the Victoria Archipelago. He says: 'If they entered the Polar Sea on the range of these islands, with comparatively open water for one hundred miles, they might drift to and fro for years, or until they experienced one of those northern nips which would form a mound above them in a few seconds! The more I have seen of the action of the ice—the partially open water and the deceitful leads into the pools—the more satisfied I am that *the man who once ventures off the land is in all probability sacrificed!*' That man, and all like him, ought to have staid at home. The channel was at that very time open to the north-east, and we could have gone on.

"*Evening.* This afternoon Captain Hall spoke to me again about our going north. He seemed to feel worried. I told him that I should gain nothing by it, but that it would be a great credit to *him* to go two or three degrees farther. He appears to be afraid of offending some one. I don't speak all my mind; it might be misapprehended, and mistaken for self-interest. God knows I care more for the success of the expedition than I do for myself. But I see it's all up, and here we stop. Have ascertained that the highest latitude made by the ship was, by dead reckoning, 82° 29';* but we have drifted nearly a degree since then.

"*Sunday.* Our usual Sabbath service in the forenoon. Snowing to-day. Can scarcely see the land, though so close to it.

"*Sept.* 4. This morning there was water making all around; every thing to be got aboard again; and the shipping of the propeller again was not so easy, as it was all frozen over; but at last it was got into position, and in the evening got up steam, and tried to work nearer inshore to the water. About 11 P.M. we had got through, and free of the ice. Lowered a boat, and I went with Captain Hall ashore to examine the place for a harbor.

"*Sept.* 5. At midnight, last night, Captain Hall raised an American flag on this land—the most northern site on which any civilized flag has been planted. When it was run up, Captain Hall pronounced that he took possession 'in the name of the Lord, and for the President of the United States.' He then returned on board, and we let go the anchor at half-past twelve, low meridian, on the 5th of September. This place, which we had examined, was only a bend in the coast, and afforded no protection as a harbor; we therefore steamed through the open water, and resumed our search to the southward; but not finding any better place, we returned to our former anchorage, and began immediately to land provisions again; snow still falling. Captain Hall named the bight we examined 'Repulse Harbor.'

"On the 7th of September we weighed anchor and steamed in nearer to the shore. There was some discussion as to going over to the west side to look for a harbor, but the sailing-master de-

---

* Subsequently corrected. Real latitude, 82° 16'.

clared she should not move from there, and so Captain Hall gave up. We have now brought the ship round behind an iceberg, which is aground in thirteen fathoms of water. This iceberg is about four hundred and fifty feet long, three hundred feet broad, and sixty feet high. Our latitude, by observation, 81° 38′ N., long. 61° 45′ W. We had been, I should think, nearly fifty miles farther north, but the current had set us down. We are now preparing to put permanent stores on shore, so that if the vessel gets nipped we shall have something to depend on.

"*Sunday.* After service this morning, Captain Hall announced that he would name our winter-quarters *Thank God Harbor*, in recognition of His kind providence over us so far. He also named the iceberg to which we fastened 'Providence Berg.' Came near having an explosion the other day, when coming into harbor, by the fireman neglecting to feed the boilers; just discovered in time.

"*Sept.* 11. Commenced housing the ship with canvas, and, after the ice becomes strong enough, we shall bank her up.

ICE BREAKING UP.

CHAPTER VIII.

A Hunting-party.—A cold Survey.—Description of Coast-hills.—A Musk-ox shot.—Landing Provisions.—Arctic Foxes.—Captain Hall prepares for a Sledge-journey.—Conversation with Captain Tyson.—Off at last.—Captain Hall "forgets something."—Twenty "somethings."—The Sun disappears.—Banking the Ship.

Dr. Bessel and Mr. Chester, with Joe and Hans, went off to-day on a hunting-excursion. Shot a musk-ox and some hares. These hares in winter are almost entirely white, having but a small black spot near the ears, and they can cover even that spot

POLARIS AT CAPE LUPTON—WINTER-QUARTERS, 1872–'73.

with their ears. In consequence of their being colorless like the ice and snow, they can not be readily distinguished if they keep still. The ice, which had been broken up by a southerly gale, is beginning to pack.

"*Sept.* 16. To-day has been the most wintry-feeling day we have had. This morning Mr. Bryan, Mr. Meyers, and Mauch went out to make a survey. They got to a mountain about fifteen miles off in a south-east direction, intending to start a base-line from it; but they returned, between one and two o'clock, nearly frozen, as they had broken through some thin ice and got wet; there was a sharp wind, which helped to discomfit them.

Have been out to see what this country is like where we must spend our long winter nights. The coast-hills are very high—from nine to thirteen hundred feet or more in height—and the great scars and cracks in the rocks look as if wind and weather, frost and ice, and sudden changes of temperament, had done their worst with them; at the base of these rocks there is a large amount of débris—stones and sand, great scales from the rocks which have been split off by the frost. Off to the south there is a large glacier, which sweeps round in a wide circuit and falls into the bay north of us. There was no snow on these hills when we came to anchor here; what fell the first few days ran off and dried up fast. The mountain ranges which we can see in the interior, so far as I can tell, were also clear of snow; and the land, this 20th of August, is bare, except what can be distinguished of the distant glacier, which is white. The soil very rapidly absorbs any moisture. The hot summer has heated the ground so that the snow does not lie. The soil is a light clay.

"This bay of Thank God Harbor is about twelve miles long and nine wide. The *Polaris* lies at lat. 81° 38'.

"Esquimaux have evidently lived here; saw their traces to-day; circles of stone, indicating where their tents had been placed; but we have seen none of them. Perhaps they used to come here in the summer, and have now emigrated permanently to the south. Have found some spear-heads made of walrus teeth, some pieces of bone, and other little things which only Esquimaux use. All we pick up we give to Dr. Bessel. The landscape is all of a dull neutral tint—a sort of cold gray. It will soon be all of one color, and that white. The frame-work of the observatory which was set up had to be strengthened and braced; the wind almost blew it down.

"*Sept.* 23. A large halo round the sun. We shall probably have a change of weather soon.

"*Sunday.* Mr. Bryan read a sermon and one of Dr. Newman's prayers. Mr. Bryan is quite a favorite aboard, and deservedly so. He never makes any trouble.

"Dr. Bessel and the natives have returned from their hunting excursion. They took with them a sleigh and team of eight dogs. Hans is driver. The first part of the week they had fine weather, but encountered a stiff gale on Friday—the same which almost shook down the observatory. They have brought home the greater part of a musk-ox which they killed. The dogs which draw the sledges are taught to bay these oxen, and keep them at a stand as they do the bears. One of the dogs had been thrown, but not much hurt. The musk-ox has been reported by some writers to be extinct, but it seems they are not here; and Captain Hall certainly found them plenty west and north of Hudson Bay in 1865. The flesh of this one was good, and did not taste of musk in the least—very much like other beef. The meat, head, and skin weighed over three hundred pounds. As they stand, some of them will weigh twice that. The Labrador musk-ox is so strongly permeated with the flavor of musk as to be scarcely eatable.

"*Sept.* 24, *Evening.* Captain Hall has a very pleasant way of getting along with the men; they were highly pleased with some remarks he made to them on Sunday, and have got up a letter of thanks, which they sent into the cabin. It was very well worded. I have no copy of it, but this was Captain Hall's reply:

"'United States Steamship *Polaris*, C. F. Hall commanding.

"'SIRS,—The reception of your letter of thanks to me of this date I acknowledge with a heart that deeply feels and fully appreciates the kindly feeling that has prompted you to this act. I need not assure you that your commander has, and ever will have, a lively interest in your welfare. You have left your homes, friends, and country; indeed you have bid a long farewell for a time to the whole civilized world, for the purpose of aiding me in discovering the mysterious, hidden parts of the earth. I therefore must and shall care for you as a prudent father cares for his faithful children. Your commander,

"'C. F. HALL,

"'United States North Polar Expedition,
In winter-quarters, Thank God Harbor,
Lat. 81° 38′ N., long. 61° 44′ W. Sept. 24, 1871.'

"*Sept.* 27. At 11 A.M. commenced a violent snow-storm, which has continued all day. This afternoon at 4 P.M. the ice broke up and packed. We must expect winter weather soon.

"*Oct.* 1. The snow-storm which commenced on the 27th of September continued for thirty-six hours, and the following day the pack-ice crowded against the ship badly. She ought to be sheared up and banked. All day yesterday a strong gale from the north-west. Weather fine to-day (Sunday). Hour of service changed from 11 A.M. to 8.30.

"*Oct.* 2. In consequence of the pressure of the ice, a considerable quantity of provisions have been taken ashore. To-day they were all covered up with snow, and some of the men are ordered to haul them off the flat ground and place them under the lee of a hill. There ought to be a house built to shelter them. Some seals have been seen, and some of the men have been out hunting for them, but they got none; and a white fox which they saw also escaped them. These Arctic foxes seem the most cunning animals I ever saw. It is very difficult either to shoot or trap them.

"Try to do a little reading and writing, but the light is very weak now. The sun makes us but short visits. The ice is now so well hardened that Captain Hall is talking of preparing a sledge-party to go north.

"*Oct.* 3. Captain Hall is feeding the dogs up, and looking over his things to decide what he will take. Had a conversation with Captain Hall. He told me that he would like to have me go with him, and then he stopped, and, pointing to the sailing-master, said, 'but I can not trust that man. I want you to go with me, but I don't know how to leave him on the ship. I want to go on this journey, and to reach, if possible, a higher latitude than Parry before I get back.' I told him 'I would like to go; but, of course, I was willing to remain and take what care of the ship I could.' I did not tell him how much I wanted to go.

"*Oct.* 6. Preparations still going on for the sledge-expedition. Captain Hall told me to-day that he would take Chester with him instead of me, giving as his reason 'that, if the vessel should break out, it would be better for me to be aboard to assist the sailing-master.' He has been having every thing weighed, so as to know exactly what weight the dogs will have to carry, and what rations to allow themselves. Been examining the dogs' harness, and preparing extra lines. It seems slow work.

"*Oct.* 10. Every thing ready at last; they will get started to-day at 12 M. There are two sledges; each sledge has seven dogs: Captain Hall and Joe in one sledge, and Mr. Chester and Hans in the other. This journey, I understand, is merely preliminary to a more extended journey in the spring. He wants to get a general idea of which will be the best route; he hopes to find some better way than over the old floes and hummocks of the straits.

"*Evening.* Saw Captain Hall well on his journey; went with some of the men, and helped haul the heavy laden sledge up the steep hill. They drove off the plains to the eastward, a little north by east. I watched them as long as I could see them, and

CAPTAIN HALL'S SLEDGE-JOURNEY.

hope he will have a safe and successful trip; but after all the time spent in preparing and packing, I have no doubt he has forgotten something; he is rather peculiar that way.

"*Oct.* 11. As soon as this snow-storm is over, I shall try and get material to build a house ashore to put our stores in; otherwise we shall have to dig them out of the snow whenever they are wanted. Hans has returned with a letter from Captain Hall; it seems he has forgotten, not one thing, but several, and is now waiting five miles off for Hans to return with them."

[A copy of the above letter or dispatch was found in the writing-desk of Captain Hall, and was preserved on the ice-floe. We give it entire.—*Ed.*]

"'Sir,—Just as soon as possible attend to the following, and send Hans back immediately:

"'Feed up the dogs (14) on the seal-meat there, giving each 2 pounds.

"'In the mean time order the following articles to be in readiness:

"'My bear-skin mittens;

"'3 or 4 pairs of seal-skin mittens (Greenland make);

"'8 fathoms lance warp;
"'20 fathoms white line, for dog lines;
"'1 pair seal-skin pants, for myself;
"'12 candles, for drying our clothing;
"'Chester's seal-skin coat;
"'1 candlestick, 1 three-cornered file, 4 onions;
"'1 snow-shoe;
"'1 cup, holding just one gill;
"'1 fire-ball, and the cylinder in which it is (this hangs up in my office);
"'Have the carpenter make, quick as possible, an oak whip-handle, and send the material for 2 or 3 more;
"'A small box that will hold the 1 pound of coffee which I have;
"'A small additional quantity of sinew;
"'Try and raise, if possible, 2 pairs of seal-skin boots that will answer for both Chester and myself.

"'The traveling we found very heavy yesterday, the snow being very deep, and just hard enough to allow ourselves, the dogs, and the sledge to break through at every step. We were *three full hours* in making the first two and a half miles from the ship. The dogs, being poor and weak, were more disposed to lie down and take a nap than to work, but the whip, swung by the energetic arm of our excellent dog-driver, "My Joe," at length warmed them up, so that after a fashion we accomplished a hard day's work, but only the distance of five (5) miles. These drawbacks and obstacles, however, are nothing new to an Arctic traveler. We laugh at them, and plod on, determined to execute the service faithfully to the end.

"'Have Mr. Bryan compare my watch with chronometer D; then, by a good watch-guard attached to it, suspend it to the neck of "Hans," having the watch next his warm bosom.

"'Do not omit sending my bear-skin mittens, which I left behind by mistake.

"'Have Hannah make a small watch-bag to suspend to my neck, then place the same on Hans's neck, with the watch in it.

"'Tell Dr. Bessel to be very mindful that the chronometers are all wound up at just the appointed time every day.

"'While Hans is absent, we are to go on a hunt for musk-cattle. Hasten Hans back without the loss of a moment.

"'I should have sent Hans back last night, but I desired to first know that our company apparatus was all complete. The "Conjuror"* works well. May God be with you all. Respectfully, C. F. HALL,
"'Commanding North Polar Expedition, in snow hut 5 miles east of Thank God Harbor, on sledge-journey toward the North Pole, and on a musk-ox hunt. October 11th, 1871, 6h. 25m. A.M.

"'You will preserve this carefully, as I have not the time to copy it now. Tell Hannah and little Puney to be good always. H.

"'S. O. BUDDINGTON, Sailing and Ice Master, North Polar Expedition.'"

"*Oct.* 17. Our sun has set behind the mountains, and we shall not see him again this winter from the ship; we might, no doubt, by going to the top of the hills, see the upper limb for two or

---

* A small stove, for use in traveling.

three days longer; but their great height cuts him off from the ship. Have commenced banking up the ship with snow, to keep out the cold. Want to get all snug and taut before Captain Hall returns. Set some of the men to fixing up the canvas housing, covering in a portion of the deck, and making it almost as tight as a room. The twilight deepens, and we have but a few hours which can be called daylight. Yesterday (19th) it was exceedingly cold, and blowing such a gale that the men had to stop work on the banking, and get inside.

"*Oct.* 21. The banking is not finished yet, but we have got the deck housed in, and we now creep in and out through a small opening. Been to work making a house on shore; I made it out of hard-wood poles; tried to get some lumber out of the ship, but could not, though there was some aboard.

GOTHIC ICEBERG.

CHAPTER IX.

Putting Provisions ashore.—Return of Captain Hall.—"Prayer on leaving the Ships."—Captain Hall taken Sick.—What was seen on his Sledge-journey.—Apoplexy?—M'Clintock's Engineer—Death of Captain Hall.—A strange Remark.—Preparing the Grave.—The Funeral.—"I walk on with my Lantern."—Thus end his ambitious Projects.

"ALL the stores on shore, coal, clothing, guns, ammunition, apparel, and a portion of every thing which we should most need, are being packed in my little house. It could have been made stronger if I had the lumber, but the poles do tolerably well.

"*Oct.* 24. Still engaged in banking; it is heavy work, as we are making the bank about ten feet thick.

"*Afternoon.* Captain Hall and the rest returned to-day about one o'clock; all well, and have lost no dogs. Have been gone just two weeks. Captain Hall looks very well. They expected to go a hundred miles, but they only went fifty. I saw them coming, and went to meet them. Captain Hall seems to have enjoyed his journey amazingly. He said he was going again, and that he wanted me to go with him. He went aboard, and I resumed my 'banking.'"

[Among the articles found on the ice-floe was a small private desk of Captain Hall's, which Esquimau Joe took charge of. In this was found a small book of nine pages, containing the three prayers composed for the use of the expedition by Dr. Newman. On the outside it was indorsed:

"C. F. HALL.
"*Thank God Harbor;* lat. 81° 38' N., long. 61° 44' W."

On the upper margin of the first page was penciled, "By Dr. Newman, for the North Polar Expedition."

The second prayer, "on leaving the ships," was indorsed as follows, in Captain Hall's handwriting:

"Read 1st time 6h. 45m. to 6h. 50m. A.M., Tuesday, Oct. 17, 1871, in our snow-house, 5th enct. (encampment) on the New Bay. Lat. —— N., long. —— W.

"*Oct.* 20, 1871. Read A.M., 7h. 0m. at our 6th enct., N. side entrance of what I now denominate Newman Bay, after Rev. Dr. Newman, the author of the three prayers of this book."

The third prayer, which was to have been first used *at the North Pole*, will be found in the Appendix. The second we give entire.—*Ed.*]

PRAYER ON LEAVING THE SHIPS.

"Almighty Father in Heaven, Thou art the God of all ages, climes, and seasons. Spring and autumn, summer and winter obey Thy command. In the tropics Thou dost cause the sun to send forth floods of light and heat upon plain and mountain, until the earth burns like a furnace; and here in this far-off northern clime Thou givest snow like wool and scattereth the hoar-frost like ashes. Who can stand before Thy cold? But Thou art our shelter from the stormy blast, and our cover from the storm.

"We return Thee hearty thanks for our safe and prosperous voyage over the great deep; and now as we leave our ships, be Thou our guide and protection while we traverse these mountains of ice. As unto Thine ancient people through the wilderness so may it please Thee to be unto us, as a cloud by day and a pillar of fire by night. We are here to explore the unknown regions of our earth, to enlarge the scope of human knowledge, and advance the best interests of mankind. Others have perished in the noble but perilous attempt, but may it please Thee to preserve us amidst dangers seen and unseen, and bring our labors to a successful termination. Grant us health of body, vigor of mind, and cheerfulness of soul. Save us from doubts and fears and all misgivings. May our courage never forsake us, nor our resolution falter for a moment. Send us the inspiration of Thy Spirit that will give us warmth of soul and gladness of heart amidst these ice-bound regions.

"Be pleased to suggest to our minds the direction we should take, and point out to us the path which will lead us to the desired destination, that our hearts may be glad and rejoice in the consummation of our plans.

"Help us to be kind and true to the people of this distant land, that they may learn of Thee and of a better civilization by our deportment and example, and especially of that Divine Christianity which is the hope of the world.

"Unto Thee, Almighty Father, we offer our prayers for the health and happiness of the dear ones at home, who are now thinking of us, and may we meet them again in peace and safety.

"Pardon all our sins, we humbly beseech Thee; keep our minds in perfect peace, and at last, when life is over, may we behold Thee in Thy glory in Heaven, for the sake of Christ our Redeemer. Amen."

"*Oct.* 24, *Evening.* I kept at work till it was too dark to see, and then came aboard. Captain Hall is sick; it seems strange, he looked so well. I have been into the cabin to see him. He is lying in his berth, and says he feels sick at his stomach. This sickness came on immediately after drinking a cup of coffee. I think it must be a bilious attack, but it is very sudden. I asked him if he thought he was bilious, and told him I thought an emetic would do him good. He said if it was biliousness it would. Hope he will be better to-morrow.

"*Oct.* 25. Captain Hall is no better. Mr. Morton and Mr.

Chester watched with him last night; they thought part of the time he was delirious.

"*Evening.* Captain Hall is certainly delirious; I don't know what to make of what he says. He sent for me as if he had something particular to say, but— I will not repeat what he said; I don't think it meant any thing. No talk of any thing in the ship but Captain Hall's illness; if it had only been 'the heat of the cabin,' which some of them say overcame him, he could have got out into the air, and he would have felt better. I can not hear that he ate any thing to make him sick; all he had was that cup of coffee.

"Chester has been telling me about this sledge-journey. He says they went up to the large bay we saw to the eastward of Robeson Strait, and they have discovered a lake and a river. They went up on to the top of an iceberg near the mouth of this river, where they could overlook the large bay. From its southern cape to its head it is thirty miles long. He says Captain Hall has named this southern cape after Senator Sumner; the bay itself after Dr. Newman; and the north cape after Mr. J. Carson Brevoort, of Brooklyn." [Here he wrote his last official dispatch.—*Ed.*]

"There was plenty of open water all around, and they could see the seals at play. The ice in the strait itself was moving, so they did not venture on it. They made six camps on the way, and halted at Cape Brevoort, where they could see about seventy miles farther to the north. Captain Hall thought their eyes took in land as far north as 83° 5', but they could not be sure of any thing at that distance, unless it was some very high mountain or a familiar landmark. Where they stopped last, at their sixth encampment, it was too hilly for the dogs to go any further.

"Joe built snow-huts for them, and they did not find it very cold; they saw musk-ox and the tracks of bears and wolves; and have seen foxes and rabbits, geese, partridges, and other birds; so that Captain Hall is quite encouraged, thinking that, when he goes again, their party can depend on getting game to help subsist on. Joe shot some seals at their very last encampment. Mr. Chester says that Captain Hall wrote a record, and put it in one of the copper cylinders, which was left at Cape Brevoort, digging down and covering it up with stones. The weather was warmer than it is here.

"*Nov.* 1. Captain Hall is a little better, and has been up, attempting to write; but he don't act like himself—he begins a thing, and don't finish it. He begins to talk about one thing, and then goes off on to something else: his disease has been pronounced paralysis, and also apoplexy. I can't remember of any one dying of apoplexy in the north except Captain M'Clintock's engineer, and he died very suddenly; went to bed well at 9 P.M., and was found dead in his state-room in the morning. I always thought that might have been heart disease. Hope the captain will rally.

"*Nov.* 3. Captain Hall very bad again. He talks wildly—seems to think some one means to poison him; calls for first one and then another, as if he did not know who to trust. When I was in, he accused —— —— and —— —— of wanting to poison him. When he is more rational he will say, 'If I die, you must still go on to the Pole;' and such like remarks. It's a sad affair; what will become of this expedition if Captain Hall dies, I dread to think.

"*Nov.* 5. No change for the better—worse, I think. He appears to be partially paralyzed. This is dreadful. Even should he recover his senses, what can he do with a paralyzed body?

"*Nov.* 8. Poor Captain Hall is dead; he died early this morning. Last evening Chester said the captain thought himself that he was better, and would soon be around again. But it seems he took worse in the night. Captain Buddington came and told me he 'thought Captain Hall was dying.' I got up immediately, and went to the cabin and looked at him. He was quite unconscious—knew nothing. He lay on his face, and was breathing very heavily; his face was hid in the pillow. It was about half-past three o'clock in the morning that he died. Assisted in preparing the grave, which is nearly half a mile from the ship, inland; but the ground was so frozen that it was necessarily very shallow; even with picks it was scarcely possible to break it up.

"*Nov.* 11. At half-past eleven this morning we placed all that was mortal of our late commander in the frozen ground. Even at that hour of the day it was almost dark, so that I had to hold a lantern for Mr. Bryan to read the prayers. I believe all the ship's company was present, unless, perhaps, the steward and cook. It was a gloomy day, and well befitting the event. The place also is rugged and desolate in the extreme. Away off, as

BURIAL OF CAPTAIN HALL.

far as the dim light enables us to see, we are bound in by huge masses of slate rock, which stand like a barricade, guarding the barren land of the interior; between these rugged hills lies the snow-covered plain; behind us the frozen waters of Polaris Bay, the shore strewn with great ice-blocks. The little hut which they call an observatory bears aloft, upon a tall flag-staff, the only cheering object in sight; and that is sad enough to-day, for the Stars and Stripes droop at half-mast.

"As we went to the grave this morning, the coffin hauled on a sledge, over which was spread, instead of a pall, the American flag, we walked in procession. I walked on with my lantern a little in advance; then came the captain and officers, the engineer, Dr. Bessel, and Meyers; and then the crew, hauling the body by a rope attached to the sledge, one of the men on the right holding another lantern. Nearly all are dressed in skins, and, were there other eyes to see us, we should look like any thing but a funeral cortége. The Esquimaux followed the crew. There is a weird sort of light in the air, partly boreal or electric, through which the stars shone brightly at 11 A.M., while on our way to the grave.

"'Thus end poor Hall's ambitious projects; thus is stilled the effervescing enthusiasm of as ardent a nature as I ever knew. Wise he might not always have been, but his soul was in this work, and had he lived till spring, I think he would have gone as far as mortal man could go to accomplish his mission. But with his death I fear that all hopes of further progress will have to be abandoned.

CHAPTER X.

Captain Buddington passes to the Command.—Scientific Observations.—The first Aurora of the Season.—Sunday Prayers discontinued.—Dr. Bessel Storm-bound in the Observatory.—Meyers to the Rescue.—An Arctic Hurricane.—Fast to the Iceberg.—Sawing through the Ice.—Electric Clouds.—Pressure of Floe-ice.—The Iceberg splits in two.—The *Polaris* on her Beam-ends.—Hannah, Hans's Wife, and the Children put Ashore.

"CAPTAIN BUDDINGTON has passed to the command without question, it being understood by all that such was to be the case if Captain Hall died or was disabled.

"Bad weather. Observations continue to be made by Dr. Bessel and Mr. Meyers on the temperature, the force and velocity of the wind, the deviations of the compass, and whatever else the weather permits. They work in the observatory. On board we make frequent observations on the tides; usually every hour, sometimes more frequently. Dr. Bessel and Mr. Meyers say now that the highest latitude the ship reached was 82° 16'.

"*Nov.* 15. Appeared the first aurora of the season, not very brilliant. The land to the east of Robeson Strait north of Washington Land we now call 'Hall's Land.' When the weather is calm and very cold, the ice in Robeson Channel is partially closed, but with every strong breeze that blows it opens again for miles at a stretch. If we could get through one of these leads, we might find the open water beyond. The weather is very changeable. A change of 20° in a few hours is not uncommon. The last gale was ascertained to have a velocity of forty-seven miles per hour.

"*Sunday, Nov.* 19. After prayers this morning it was announced that the service would be discontinued in the future. It was suggested that 'each one could pray for himself just as well.' I think the Sunday service has a good influence; it seems a pity to discontinue it. Perhaps the cabin is needed for something else.

"*Nov.* 20. Last night Dr. Bessel went over to the observatory, and at 9 A.M. we were all a little surprised to find that he had not come back; but, as he had a fire there, did not think so much of it, only that a fearful snow-storm had come on, which might keep him prisoner there too long. At half-past nine Mr. Meyers concluded to go and see if any thing had happened to him. It was with the utmost difficulty that he made his way up the hill toward the observatory. Several times he was driven back by the force of the storm. Joe and Hans offered to go with him; and finally, after an hour consumed in trying to get a few rods, they succeeded in reaching the house. It was well they went. Dr. Bessel had been without fire for eight hours, his coal having given out; and he had not dared, in the storm and darkness, to try and get to the ship for fear of losing his way. One of his ears was frozen.

"Mr. Meyers got one of his eyelids frozen in the brief time he had been battling the wind and snow, and even Joe's right cheek was touched. However, they all got back to the ship; and after Dr. Bessel had had some warm coffee and food, and his ear attended to, he was all right again. This gale kept veering and backing between east and north, and part of the time attained the immense velocity of sixty miles per hour, the thermometer standing at 24° below; which, if it had been calm, we should think moderate weather.

"One of the men, Herman Sieman, going out to examine the tide-gauge, was lifted up by the storm and carried quite a distance, and then thrown violently upon the ice, which was overflowed with water; he had to give up. The snow-drift was as blinding as the gale was furious; it has shaken up the ice so during the night that between one and two o'clock it began cracking around the ship, and the snow-wall which we had been at such pains to build gave way, and sank fully two feet.

"*Nov.* 22. Yesterday the ice broke all around us, the snow drifting so that we could not see our condition, or how to remedy any thing. However, we put out another anchor; but the ship drifted closer toward the berg. Toward noon some of the men succeeded in getting over the floe to the iceberg, and with the aid of hatchets they fastened three ice-anchors, to which the ship was secured by hawsers. These held her more steady, and in the afternoon the gale abated. When the weather cleared, we found that the water was open all around us.

"To-day the weather continues fair, but very cold; the wind from the east. When we came to see what damage had been done, we found, among other things, that two of our sledges were lost — probably drifted away when the ice broke. Most of the dogs were fortunately on board, having been brought in when the gale came on.

"*Nov.* 25. Sawed an opening through the ice, so as to warp the ship round under better shelter toward the centre of the berg, as in our present position we are only protected in one direction. Moved her along one hundred and twenty feet, to the middle of the berg on its long side.

"Last evening some electric clouds were observed—a shining white, and circular in form. There have been some seen before.

"*Sunday.* There was a brief service held, but the announcement was made that attendance was not compulsory, though it was desirable that all should attend. Mr. Bryan would like the service to be regular, but I don't know who is head of the chaplain's business now.

"*Nov.* 28. Fair weather and moderate temperature yesterday; but the latter part of the day the barometer fell gradually, and in the evening a snow-storm, with a gale, set in from the south—south by west. The floe-ice was pressed against our berg so violently that it parted in two. We swung to our anchors, but the ship was forced upon the foot of the berg, which lay to the southwest of us, shaking and straining the vessel badly. At ebb-tide she keeled over, and lay nearly on her beam-ends; careening so much that it is difficult to keep one's footing on deck.

"The foot of the iceberg is now pushed beneath her, raising her stem two feet and a half. Sent Hannah and Hans's wife with the children to the observatory for safety. Also sent some more stores ashore, in case we have to abandon the vessel. Think the vessel could be hauled off, but no orders are given. If she is left in this way she will get farther and farther on to the spur of the berg, and get such a straining as will set her leaking.

CHAPTER XI.

Thanksgiving.—A Paraselene.—Dr. Bessel's bad Luck.—"It is very dark now."—Oppressive Silence of the Arctic Night.—The Voracity of Shrimps.—"In Hall's Time it was Heaven to this."—A natural Gentleman.—No Service on Christmas.—The *Polaris* rises and falls with the Tide.—Futile Blasting.—The New Year.—Atmospheric Phenomena.—The Twilight brightens.—Trip to Cape Lupton.—Height of the Tides at Thank God Harbor.

"THANKSGIVING was remembered at the table, but in no other way. Fair weather.

"On both Saturday and Sunday evening there was a *paraselene*—three moons showing besides the true one; the four so arranged as to form a beautiful cross. This curious phenomenon is more common in these latitudes than farther south, caused, I presume, by some peculiar state of the atmosphere—a sort of double refraction.

"*Dec.* 6. Dr. Bessel has bad luck. Yesterday was fair until the middle of the day, but in the afternoon a gale sprung up from the south, increasing through the evening. Toward midnight the wind subsided, but a snow-storm came on. About 2 A.M. Dr. Bessel started for the observatory, and, missing his way, wandered about all night until six o'clock this morning. It has just occurred to him that it would be well to run a rope or a wire from the ship to the observatory, so as to guide him in the darkness; and it will be put up to-day.

"There are occasional displays of northern lights, but not very brilliant; have seen a number of shooting-stars. No doubt we miss more than we see. The last aurora, December 10th, appeared in the form of an arc over the hills, embracing about 20° from south-east to north-west. Whatever there is of a peculiar nature shows very plainly, because there is nothing all around us to intercept the view, except in the direction of the hills, and most of the atmospheric and electrical phenomena are above them.*

---

* The French expedition, which went north of Spitzbergen to make auroral observations, came to the conclusion that their nearest approach to the earth was forty-six miles.

It is very dark now, but not totally dark. We shall soon reach the shortest day, and can then look forward to a blessed change from this gloomy, everlasting twilight; we can hardly tell day from night, and, if it was not for our time-pieces, should get sadly mixed up—the more so that there is so little regularity observed. There is no stated time for putting out lights; the men are allowed to do as they please; and, consequently, they often make night hideous by their carousings, playing cards to all hours.

"Can not get away from the ship much; it is too dark to go out of sight of the vessel, except for an hour or two at high meridian; and, once away from the ship, the gloom and silence of every thing around settles down on one like a pall. There is no whistling of the wind among the trees, for none of these exist here; and out on the open plain the wind strikes you without notice. There is nothing to be ruffled or disturbed by it, so that you *feel* it before you hear it, except you are near a gorge in the hills, between which it sometimes comes roaring loud enough.

"The other evening I had wandered away from the ship, disgusted with the confusion and noise, and longing for a moment's quiet. Once beyond range of the men's voices, there was absolutely no other sound whatever. It was quite calm; no wind, no movement of any living creature; nothing but a leaden sky above, ice beneath my feet, and *silence everywhere*. It hung like a pall over every thing. So painfully oppressive did it become at last that I was frequently tempted to shout aloud myself, to break the spell. At last I did; but no response came, not even an echo.

"'The space was void; there I stood,
And the sole spectre was the solitude.'

The men have had revolvers and other fire-arms presented to them by the commanding officer; what use they are expected to make of them I have not inquired.

"*Dec.* 16. The time drags heavily; shall be glad when we can get out and do something, if it is only to bob for shrimps. What an astonishing number of these creatures there are! Some of the men catch and eat them. I don't like their looks; they appear to be the principal food of the seals. They are very numerous, or very voracious, or both. A musk-ox skull, which Dr. Bessel hung in the water, with some other 'specimens' which he wished cleaned, was not only picked clean of every edible speck, but almost polished by the shrimps.

"*Dec.* 17. Another aurora reported last night; very shifting and changeable.

"*Dec.* 18–24. Nothing occurring that is pleasant or profitable to record. I wish I could blot out of my memory some things which I see and hear. Captain Hall did not always act with the clearest judgment, but *it was heaven to this*. I have not had a sound night's sleep since the 11th of November. Would he had lived till spring!

"Some preparations are being made for Christmas, and all hands, I hear, are to be invited into the cabin this evening. We have passed our shortest day, and that interests me most. If I can get through this winter I think I shall be able to live through any thing. Mr. Bryan does not say much, but I think he feels it as much as I do. He is naturally a gentleman, with the true instincts of right and wrong.

"No service on Christmas. The ship continues to rise and fall with the tide, her stem resting on the foot of the iceberg. At neap-tide the leak might be repaired — gradually, if not all at once — by working a few hours at a time as the tide permits; but no orders are given to attempt it.

"*Dec.* 28. A futile attempt was made to break up the foot of the berg by blasting, but it was too strong. The amount of powder necessary to blast it successfully would endanger the ship, as she lies.

"*Jan.* 1, 1872. The first day of the new year, and eighty days since we have seen the sun. Considering the heterogeneous elements of which this expedition is composed, it is something to be thankful for that we all commence the new year in good health, and without any open and acknowledged disagreement, which, considering, too, some circumstances, I think remarkable. If Dr. Bessel could hear the compliments he gets sometimes, he would be highly gratified! Well, all I can do is to keep silent: my position does not warrant interference.

"*Jan.* 2. Tried the blasting again, with no better result. The ice is very strong and thick; the stars bright at midday.

"*Jan.* 6–7, 12 M. For twenty-four hours there has been a display of the aurora borealis, consisting chiefly of successive bands of light, more or less brilliant. These lights help to make our dark days less gloomy. It is very cold; the thermometer stands at 48° below zero. Some of the atmospheric phenomena are matters of dispute, or rather discussion, between the three members of the Scientific Corps: for instance, on the 10th instant, about 5 A.M., there was a bright arc observed in the sky, extending from the western horizon toward the east, and reaching up to the zenith; it appeared to be about 12° from the Milky Way, and parallel with it. This continued only about an hour; but as it disappeared, there remained three cloud-like shapes of about the same brightness, resting near the zenith. At one time, some narrow, bright stripes were visible. Whether this was to be considered as a true aurora, or as some unique electrical phenomenon, was the query.

"*Jan.* 17. The twilight toward the south-east is visibly increasing. How naturally every eye turns to that quarter, hoping to see the arc of light extended.

"*Jan.* 24. Can see water to the north, and perceive that the ice is not closed, but drifting in the strait. Dr. Bessel has been out this morning with two men and a sledge-team, to ascertain if the open water extended any great distance—for it is too dark to see far from the vessel. He only made nine miles—to the north; he could not get farther on account of the headland, or cape, being covered with smooth ice, over which the dogs could not go nor the men climb. As far as he could see, it was open water. But it is yet too dark to accomplish any thing in the way of explorations. On account of the darkness, they were unable to find a pass. The ice was drifting with the current.

"*Jan.* 26. Yesterday Mr. Chester thought he would try. He took four men with him, and a dozen dogs to draw the sledges. He thought he could get over the mountains, or find a pass through them; started about 10 A.M.; returned about four o'clock, as completely baffled as Dr. Bessel had been. Within the last three days the thermometer has varied from 20° to 35° below zero.

"*Jan.* 31. A violent snow-storm, the wind blowing with hurricane force. The tide-hole is completely chocked with snow, and the storm is so violent that it is impossible for the men to keep it clear; therefore, impossible to make the usual tidal observations.

"*Feb.* 1, 1873. Gale increasing; has attained a velocity of 53.6

miles per hour. To-day I went over to Cape Lupton to see what I could make out. Cape Lupton is a bold headland, eighteen hundred or two thousand feet high. I managed to get to the top, and from that elevation saw that the ice was completely cleared out of the channel; in fact, there was free water everywhere except in the bay, the ship itself being firmly inclosed; but for eighty miles to the north, had we been in the channel, we could certainly have sailed in free water.

"*Feb.* 2. The tide-holes have been cleared out, and the usual observations resumed. The tides vary from two and a half to seven feet eight inches. The highest spring-tide yet observed was seven feet eight inches; the neap are from two and a half to three feet rise.

HEAD AND ANTLERS OF THE ARCTIC REINDEER.

CHAPTER XII.

An impressive Discussion.—Daylight gains on the Night.—Barometer drops like a Cannon-ball.—Four mock Moons.—Day begins to look like Day.—The Fox-traps.—The Sun re-appears after an Absence of one hundred and thirty-five Days.—Mock Suns.—Spring coming.—An Exploring-party in Search of Cape Constitution.—A Bear-fight with Dogs.—New light on Cartography.—Tired of canned Meat.

"LAST month such an astonishing proposition was made to me that I have never ceased thinking of it since. The time may come when it may be proper for me to narrate all the circumstances. It grew out of a discussion as to the feasibility and expediency of attempting to get farther north next summer. My own opinion is that we ought to do all we can to carry out Captain Hall's wishes, and the just expectations of the Government and the country. If the season should prove as favorable as it was last year, there is no reason why we should not reach the pole itself. It would be a lasting disgrace not to utilize to the utmost a ship fitted out with such care and expense. It is enough to make Captain Hall stir in his ice-cold grave to hear some of the talk that goes on.

"The last gale made wild work with the ice and the bergs; at one time clearing the strait of the floe-ice, and driving it in a pack before it. The icebergs have also been driven together, and the hummocks thrown up in heaps as if the very ice demons had been having one grand set-to.

"Another attempt to get through or over the hills to the north; but the last party got no farther than the rest.

"*Feb.* 5. The daylight is beginning to gain on the night a little. In the middle of the day we can see to read without a light, but only for a little while at a time. Joe and Hans have begun hunting for seals, as they can now be heard under the ice making their breathing-holes; but as yet none have been caught.

"*Feb.* 10. Quite mild and pleasant weather—only 13° below zero. More auroras reported, which I did not see.

"*Feb.* 11–14. Several days of pleasant weather. Nothing to record.

"*Feb.* 17. A sudden change; barometer dropped like a cannon-ball; sky suddenly overclouded, with violent squalls of wind from the south-east, backing to the north-west, and then veering again to the south-east; next, snow-drift. It is now blowing at the rate of fifty-eight miles an hour. These gales last two or three days; we have had them alternating with a few days of pleasant weather nearly all the winter.

"*Feb.* 21. The last gale continued with variable winds, and as varying velocities, for three days. Last night there was another paraselene, more complicated than the former. There were the four false moons besides the true—five in all. The true moon was surrounded by a halo, which also embraced two of the false ones; while the other two had a separate halo, making a large circle concentric with the first. The two mock moons nearest to the true showed the prismatic colors. It was a beautiful and curious sight.

"*Feb.* 22. Day is beginning to look like day, or rather dawn. We do not see the stars any more in the middle of the day, but neither do we see the sun yet. For over three months we have seen the stars in the day-time whenever the sky was unclouded, and the moon when it was not stormy. Much of that time the stars were very bright, and the moon also.

"*Sunday, Feb.* 25. No service; walked over to Captain Hall's grave. Always seem to walk in that direction. It is now getting so much lighter that we shall be able to do something, I hope, soon. As yet, the hunting has amounted to nothing; where there is water one day ice is found the next.

"Nothing to record; first a gale, then a snow-drift, then squalls, then fair weather, and repeat. This formula would do for the whole winter, with slight variations.

"*Feb.* 28. A glorious day. The sun has showed himself once more. I happened to be the first to see him. If it had not been for the hills, we should have seen him yesterday, or day before. Never was expected guest more warmly welcomed. It is *one hundred and thirty-five days* since we have seen his disk. Poor Hall! how he would have rejoiced in the return of the sun. His enthusiasm would have broken loose to-day, had he been with us. And to think that there are those who are glad that he can not come back to control their movements!

"March comes in with the temperature 37° below zero.

"*March* 2. 43° below. Variable weather until Wednesday the 13th, which was very pleasant, and in which we had a repetition of the phenomenon of mock suns, instead of moons—a parhelia of three false suns surrounding the true, and in this case, two of the suns exhibited the primary colors.

"The hunters are very persevering, but they bring in nothing. The animals have all migrated apparently either to the north or south; but probably, as the sun gains strength, we shall find something alive to repay our exertions. Since the appearance of the sun our nights are less dark; it can hardly be said to be dark at all, but like a deepened twilight; and soon we shall not have even that.

"*March* 19. Have been removing the canvas housing from the ship. Since the sun has appeared, it has been discovered that lantern light is very injurious to the eyes. It seems a little premature, for I see a storm coming up.

"*March* 20. Last night a violent snow-storm.

"*March* 21. Gale continues.

"*March* 22. Velocity of the gale fifty-three miles an hour.

"*March* 23. Clearing off. Sent to examine the fox-traps; one white fox had been caught by the foot. These traps have to be placed some distance from the ship. The foxes are very wary, and readily scent dogs or men at a long distance. To-day, also, Hans shot a female seal with embryo; both the skins and skeletons of these have been prepared as specimens, and will be presented to the Smithsonian Institution at Washington.

"*March* 27. Weather mild—above zero to-day, 1° to 3°; much warmer than we have had it. A wind from the south-east brings us a light snow, but it does not appear to decrease the temperature. Spring is coming; some birds, thought to be partridges, were seen yesterday. Another sledge-excursion, this time to the south, to examine the southern fiord—a fiord lying about twenty-eight miles south of Polaris Bay—Dr. Bessel, Mr. Bryan, and Joe. They take fourteen dogs, and they intend to get as far as Cape Constitution, and to make surveys and astronomical observations.

"*March* 30. A little snow, but 1° to 5° above zero for the last three days. We shall soon have plenty of game.

"*Sunday, March* 31. Service to-day, but most of the crew off shooting; killed one hare and some partridges. These birds are provided with plumage to match the country; they are unlike the southern birds; feathers all white, so that they are not easily seen, which is a great protection to them, but gives the gunner much trouble.

"P.M. We were surprised to see Mr. Bryan and Joe coming back without Dr. Bessel; thought some accident had happened to him. But it seems they had only broken their sledge, and had come back for another, the doctor remaining on a little island in the mouth of the fiord, where he had been made happy by finding some petrifactions. They had left all the stores with him.

"*April* 2. Yesterday Mr. Bryan and Joe, with Hans also, started off with two sledges to rejoin Dr. Bessel. Weather continues fair, and we are fitting up the boats for an exploring trip to the north next month. I think myself it would be better to go in sledges; for though there is open water there is also much floating ice, and I am afraid the boats will be stove. But I am ready to go any way that promises success, or chance it any way.

"*April* 8. Quite an excitement to-day—the party from the fiord have returned, bringing with them a bear which Joe shot, and also a seal. One of the dogs had been injured in the fight with the bear, and is an object of great attention with the men. This is a very plucky dog, and is called 'Bear;' it seems that it took several severe blows from the polar, which had made a better fight than they sometimes do. Another dog had been thrown with such violence against an ice-hummock that it was left for dead, but the next day showed itself at their camp nearly recovered.

"They have made some curious discoveries in regard to former surveys. Having crossed the southern fiord, which is about twenty-two miles wide, they traveled along the coast for forty miles in search of Cape Constitution, but did not find it. It is farther south than it is placed on the chart; and as Dr. Bessel reports having gone south of the latitude which Dr. Kane gave as Morton's highest without reaching the cape, it shows conclusively that there was a mistake made in the latitude. They could see the island, however, which lies off the cape far to the south. Into the first fiord, which we had called the 'Southern,' they could not penetrate more than twenty miles on account of the numerous icebergs with which it was filled; but having ascended some of these, they could see that the fiord terminated in a glacier extending in a south-easterly direction as far as the eye could reach. The shores were high and rocky. At some points the rocks were almost perpendicular, and from six to seven hundred feet high.

"South of this fiord they found another, smaller—not named—which was also beset with icebergs, and apparently between thirty and forty miles long. Several glaciers were seen, but, I believe, not examined. It was in this most southerly fiord that the bear was killed. They went as far south as 80° 45′ N. Hans, who was with Morton when he supposed he had seen the open Polar Sea, perceived on this occasion that he was farther north than he had been then. The party would have pushed on to Cape Constitution, but could not get their sledges over the steep hills along the coast, and the shore-ice was so invaded by open water that they could not venture to trust to it. As it was, they had to carry their sledges at several points.

"*April* 16. Mercury has fallen several degrees—getting much colder; the strait now choked with ice, which has drifted in, filling it as far as the eye can see. From the summit of Providence Berg there is no open water visible. The weather is fine and calm, but cold. The sun is continually above the true horizon now, though at times hidden from our view by the hills. We may now say it is always day.

"*April* 19. The two Esquimaux have gone off on a hunt. There is a general desire for some fresh meat. We have every thing on board but this. There was nothing spared in the outfit of the *Polaris*. All the canned food has turned out good, and some of it is really excellent; but a change of diet is desirable, and fresh meat is necessary for health in these regions.

SEALS.

CHAPTER XIII.

Sledge *vs.* Boat.—What Chester would do when he got Home.—Photographing a Failure.—Off on a Sledge-journey with Mr. Meyers, Joe, and Hans.—Habits of the Musk-cattle.—Peculiar strategic Position.—Encounter a Herd.—How the Young are concealed.—Dull Sport.—Newman Bay.—Preparing for Boat-journeys.—What does he mean?—Climatic Changes.—Glaciers.—Wonderful Sportsmen.—The Ice thick and hummocky.—A dangerous Leak.

"Dr. Bessel wishes to go in sledges to the north. He has made a formal request to the captain to that effect. I believe he got for answer that 'he' (Bud.) 'intended to take the boats and go north himself.' But no one thinks he will go. I wish to go with a sledge myself, and shall soon make the proposition.

"*April* 23. Joe and Hans have returned from their hunt. Have had good success, having shot seven musk-oxen. They had to leave three in the igloo they had built, as the dogs could only haul four. A stiff breeze from the north-east.

"Had a talk with Chester about the astounding proposition made to me in the winter. We agreed that it was monstrous, and must be prevented. Chester said he was determined, when he got home, to expose the matter.

"*April* 27. Mr. Chester, with the two Esquimaux and one of the men, took a couple of sledges and went off for the musk-oxen left in the hut; also to find open water, if they can.

"*May* 1. Had a talk with Captain Buddington about a sledge-journey to the north; he did not think it practicable. Why he thinks thus, I know not; it could not do any harm for me to try. Perhaps he may change his mind yet.

"Dr. Bessel has been trying to get some photographs; so far, not much success. He tried to photograph Captain Hall's grave; but it was too dark then.

"Three days ago, a gale with snow-drift sent the ship against the berg, causing her to careen considerably. Have set men to cut away the ice so as to free her before high water, or worse damage will come of it.

"P.M. Mr. Chester and party returned; they had started in the direction of Newman Bay, but got only about twenty miles. Could not tell about the water, whether it was open or not, as they had taken an inland course. They had met with some musk-oxen, and killed two. Bear-tracks have been observed within half a mile of the ship, and consequently no one cares to go far unarmed.

"*Sunday.* Yesterday a sharp storm from the north, with furious snow-drift, which continued until this morning. The velocity of the gale was rated at fifty-three miles per hour.

"*May 9.* I have at last got a couple of sledges, to try and get to the northward; Mr. Meyers, Joe, and Hans will accompany me. We start now (4 A.M.); shall get to Newman Bay, and farther, if possible.

MUSK-OX.

"*May, Evening.* Got back to the ship, having been gone six days. We took an east-north-east direction inland, and succeeded in reaching Newman Bay, and from thence went on in the same general direction, inclining more to the north, until we reached lat. 82° 9′. Mr. Meyers surveyed the shores of Newman Bay. At the end of this there is a glacier, such as is found in so many of the bays and fiords of Greenland. I paid my principal attention to getting game, as the ship's company was in want of fresh meat. There was plenty to be found; I soon noticed the tracks of musk-oxen, all showing that they had come from the south-east. Newman Bay runs south-east and north-west, and the oxen came from the interior—from the head of the bay. One day we came upon a large herd of them. They act very curiously when attacked. They all form round in a circle, stern to stern, and so await an attack. The dogs surround them and keep them at bay. Now and then a dog gets tossed. Joe and I fired and reloaded as fast as we could; the animals made no rush at us. We killed eight, and the rest ran off. Having secured our slain, we hauled one of them to the encampment. We had a heavy piece of butchering to do to skin it, and cut up the best pieces to save for the ship.

"The next day we followed up the trail, and came up with them; bagged four more; but we were too far from camp to get them there without more help. These cattle develop their great weight on what looks like very slender diet; their food is the moss and lichens which grow on the rocks, and to obtain it they have first to scrape away the snow with their hoofs. I forgot to mention that there were some calves with the herd, three of which were killed. We did not see them at first, for at the approach of danger the young ones get under the parents' body, and the hair of the musk-ox is so long that, almost touching the ground, it hangs like a curtain before the young, completely concealing them from view. The musk-ox is a very heavy creature; several of those which we shot would weigh from five to six hundred pounds apiece. Their legs are very short in proportion to their size and weight. It is not very exciting sport, for there is no more chance of missing them than the side of a house. When they have been checked by the dogs, and got themselves in a circle, there is nothing to do but to walk up and shoot them.

"Saw a few white foxes; they are very different from the musk-ox, and will lead one a fine chase; and it requires a skillful marksman to hit them, for they are so swift in their movements, and so cunning, you think you have them, and the next moment they are out of sight. We saw no reindeer nor wolves. These Esquimaux dogs are wolves enough for me. We saw very little open water—a few leads in the floes, and that was all.

"Newman Bay would average seven miles in width, and is sixty or seventy in length. There appear to be two small islands near the head of it.

"Sent off the two Esquimaux and two of the men, to bring home the rest of the musk-meat. We brought eight, and four calves.

"A few lemmings (the *M. torquatus*) have been seen. One of

the men caught a live one, and the carpenter found a dead one. These lemmings are small gnawing mammals. Sometimes called the Arctic mouse, they differ considerably from the common, in having sharp sickle-shaped claws, the two middle ones of the fore feet being extraordinarily long for a creature only about five inches in its whole length. These creatures inhabit the southern as well as the north Polar regions, but are not found elsewhere. They do, however, work down a little below the Arctic circle toward Hudson Bay. They burrow in sphagnous swamps in summer, and between stones and rocks in winter, where they feed on roots and moss. When they travel they make a perfectly straight course, and only an absolutely insurmountable obstacle is sufficient to turn them aside.

"*May* 20. One of the boats has been sent over the ice on sledges to Cape Lupton, four miles to the northward, as that was the only place to launch her. Between here and there, the channel being narrower, the ice is still packed; above Cape Lupton open water appears earlier than in our vicinity. The ice up there looks weak. Weather very pleasant now; the snow disappearing from the mountains, and the pack-ice softening. Another sledge-expedition, with Joe and Hans, has gone to bring to the ship the remaining musk-oxen, also the snow-tents and sleeping-bags, which were left on the preceding trip.

"*May* 24, 25. Another boat has been transported to Cape Lupton, and provisions and stores are now being forwarded. Nothing will do but explorations by boat. I am going in one, though I have expressed my opinion plainly that we could do nothing with boats at this time of the year. The ice begins to move and the water to show between the cracks, but the strong current packs it in the narrow channel.

"P.M. Hans found a piece of an Esquimau sledge on some elevated ground about three miles from the ship, to the northward of our position, toward Cape Lupton. It had doubtless been there a long time, but shows that the Esquimaux have formerly inhabited this region. It may be that it was only a summer encampment, but every thing indicates that they have been here, and most likely they came from the west coast. They would hardly come overland from Southern Greenland, because they could not pass the glaciers which pour down the eastern coast, particularly the great Humboldt Glacier, which flows from an immense inland *mer de glace*. There are many things which indicate that the climate of North Greenland has altered since its early history, and that it is much colder now than it was then.

"*April* 30. Two of the men who had been off on a long walk reported that they had seen tracks of bears near some fresh-water ponds, and also musk-oxen. Two of the latter were resting on the snow at the foot of a hill, near one of these ponds. At a distance of several hundred yards the creatures perceived the men, and suddenly rose, when the men fired at them. While Sieman was reloading, Kruger called his attention to the fact that one of the animals was making for him in the rear, and was coming on at all speed. Sieman retreated, to get an opportunity to reload. In the mean time the valorous creature was joined by the others, when they halted, and assumed their particular tactics, heretofore described. The men, both of whom had retreated to a considerable distance, being surprised at the animal showing fight, fired again, killing the female, which instantly fell, while the male, with the calf, took to flight. We afterward ascertained that in accomplishing this feat they had expended *three hundred balls!* One of the men admitted he had fired *seventy shots*. They did not pursue the fugitive. After hearing their story, five of the crew volunteered to go out with a sleigh and bring home the game, and also to try and find the two which had escaped.

"*May* 31. The men have returned with the three animals, having discovered and killed the male and calf. Some hares were seen, but none taken. Some of the men go out every few days hunting. A partridge and snow-bird were shot to-day. The Esquimaux have been off to Newman Bay, and brought back two seals.

"*June* 3. The ship has made so much water that the donkey-engines have been started; after four hours' work she was pretty well freed for the time, but unless the leak is stopped it will get worse.

"*June* 4. At the request of Captain Buddington, I went to-day about twenty-five miles northward to examine the state of the ice, and to report to him upon the prospects of success for a boat expedition. Joseph Mauch accompanied me. I found the ice closely impacted, very rough, thick, and hummocky; not only filling the channel, but crushing against the land. I examined carefully for leads, but could see none; continued on the search for about twenty-eight hours, and then returned to the ship. I sent in my report in writing to Captain Buddington, expressing my opinion that nothing could be done with boats at that time.

"*June* 5. The continued warm weather is telling on the snow and ice which surrounds the vessel; the ice is loosened, and the vessel, feeling the water, is rising steadily.

"Discovered to-day a dangerous leak on the starboard side of the stern, at the six-foot mark; two planks were badly split. No wonder; the strain has been tremendous, with her stem resting on the foot of the iceberg above the level of the stern all winter. In the men's quarters, in the forepart of the ship, they say they can hear the water entering at flood-tide.

"*June* 6. Some attempt made to stop the leak; have not yet succeeded. Chester has been out, and reports that there is now a practicable opening for the boats.

ESQUIMAU DOG.

CHAPTER XIV.

Two Boat-parties arranged.—A Disaster.—Chester's Boat crushed in the Ice.—The "Historical Flag" lost.—Chester takes the patent Canvas Boat.—Captain Tyson's Boat-party. — Reach Newman Bay.—Dr. Bessel's Snow-blindness.—Drift-wood. —Extinct Glaciers.—Unfavorable Condition of the Ice.—A Proposal rejected.— Return to the Ship.

"OUR parties are now arranged. Mr. Chester will take charge of one boat, and with him will go Mr. Meyers and four men: Fr. Jamka, F. Anthing, H. Sieman, and Kruger. I shall take the other boat, and will be accompanied by Dr. Bessel and four men: H. Hobby, F. Jansen, William Lindemann, and G. Lindquist.

"*June* 7, 8 P.M. Mr. Chester's party left the ship with a sledge-load of things for his boat, which is awaiting him at Cape Lupton.

"*June* 9, P.M. Chester's party have all returned, having had the misfortune to lose their boat, and nearly their lives. This boat was named the *Grant*, after the President. It happened in this way: they got launched about noon yesterday, and after rowing about a mile were stopped by a large floe, on which they halted, and drew up the boat. Discovering open water ahead in the afternoon, about a quarter of a mile off, they hauled her over the floe, and again launched her in open water; but the ice closing about them, they only succeeded in getting a little over a mile, when they were compelled to pull up again on to a floe— this time between two icebergs grounded on the shore — a very dangerous position; for at this time of the year—indeed, at any time during the continuance of warm weather—icebergs may explode at any moment.

"Here they set up their tent and prepared to spend the night. F. Anthing had the watch; Mr. Chester and Mr. Meyers had lain down about twenty yards from the boat; three of the men were lying in the tent close to the boat. Suddenly the watch shouted out, '*The ice is coming!*' All immediately sprang to their feet and made for the shore. They had hardly cleared the icebergs, when the heavy floe, full of hummocks, came on with such force as to shatter one of the bergs, which fell with a great crash, crushing the boat to pieces; at the same time the pack-ice crowded in blocks, lapping and overlapping each other; and for a time it seemed that they would lose all; but after a while the pressure ceased, and they managed to get out on the ice and save some of their things. Mr. Meyers was very fortunate in saving his, but the others lost every thing. Mr. Chester lost even his journal, and the historical flag which Mr. Grinnell had presented to Captain Hall just before the *Polaris* sailed." [This appears to have been subsequently recovered.—*Ed.*] "When this accident happened they were only about seven miles from the ship, so there was no difficulty in walking back; but they had to take another boat.

"I called the cape near which they lost the boat Cape Disaster, and the bay they were on, beyond Cape Lupton, Folly Bay, which I believe was rather displeasing to Mr. Chester. He is going to take the patent canvas boat and start again in a day or two.

"*June* 10. I start this P.M., with Dr. Bessel and four men.

"NEWMAN BAY, *June* 12. Arrived here without much trouble, working at intervals as I could get a lead through the ice; we are stopped here by a heavy pack to the northward. The channel is open southward to the ship, but completely closed ahead. Must camp here, and have hauled up the boat until the ice opens and gives us a lead. Have set up our tent, and await the movement of the ice. Dr. Bessel's eyes are bad, and he can do nothing; he is troubled with snow-blindness. I take my gun in hopes of finding game, but dare not go out of sight of the boat, lest I should miss a chance of working up, should a lead open. The boat is heavy, and all hands are needed to handle her. Shot some eider-ducks, gulls, and dovekies; saw some brent geese.

"*June* 17. Mr. Chester has at last arrived in that patent boat; she is dreadfully slow—makes about three miles an hour. They got started two days after I left, and have been all the week getting up here.

"They started at noon on the 12th from north of Cape Lupton, to which they were brought in dog-sledges by Joe and Hans, who returned to the ship; but that day could not find a lead. The next day they got launched, and worked along through a narrow opening for between two and three miles, when they had to haul up, and draw the boat ashore on account of the ice. Twenty-four hours later they got another start, worked her through a mile and a half, and then drew up on a floe, not being able to reach the shore on account of heavy pack-ice and bergs, which kept them off. During that night a strong north wind began to drift the ice they were on; and being unable to escape, they were drifted back on their course all night, and in the morning found themselves south of Cape Lupton. About 7 A.M. the ice separated, and they got a lead to the north, and joined us.

"At present we can get no farther, as there is no open water; but as the ice is setting south, we hope the channel will be cleared before long. Weather is pleasant. Little willows, that are more like a vine than trees, running only a few inches from the ground, are found in the ravines. Mosses and flowers are now to be seen everywhere. The doctor has been suffering from snow-blindness, and will return to the ship at the first opportunity. He is quite discouraged. A little drift-wood has been picked up on the bay shore, apparently small branches of trees, very much worn and wave-tossed. All the pieces are quite small—the largest two or three feet long, and two and a half or three inches thick; most of the pieces much smaller. The men burned some to make a fire, but Dr. Bessel has saved some specimens. It was not easy telling what they were. The experts thought they were black-walnut, ash, and red pine. There is nothing which grows around here any thing like it.

"Mr. Meyers brought with him a copper cylinder, and having made a record of Captain Hall's death, with latitude and longitude of the place of deposit, a hole was dug in the ground, into which the cylinder was put, and then stones piled over it in such a form as would at once be recognized as artificial, if any other human eyes are destined to gaze around on this solitary place.

"There appear signs around Newman Bay of extinct glaciers, as the moraines may be seen on the shores; but at present there is only one at the head.

"*June* 24. The ice-floes still fill the channel. One compact field of ice stretched quite across, grazing the shores both east and west, as it went on its southerly course at the rate of nine or ten miles a day.

"Chester's party are not very comfortable in their canvas boat. She is not fit for such rough sailing as we have to encounter; it is square fore and aft, and the slowest craft I ever saw. She would do for a party of children to paddle about on a calm and placid lake; but you might as well put an egg-shell in the way of an ice-pack as this patent contrivance.

"*June* 27. Have made several attempts to get farther to the north, but have not been able to force a way through the pack. If we had sledges and the dogs, we might have done something. We have not been above 81° 57′ 26″. The last two days, strong northerly breeze, with snow-squalls.

"It becoming daily more apparent that nothing could be done with the boats, I proposed to Chester to unite our two crews and organize a pedestrian exploring-party. My plan was to go on ahead, either alone or with one or two companions, and divide all the rest of the company into squads, or rather couples, who should follow in a given direction—as nearly due north as the lay of the land permitted; and that each party following should make *caches* at certain described intervals, so that we should have had something to eat on the return journey. In this way, taking our guns with us to assist in procuring food, we could have walked to the pole itself if the land extended so far, without any insuperable difficulty during the Arctic summer, when game of various kinds is so abundant; but I could get no one to join. Some were indisposed to the exertion of walking, and some did not know how to use the compass, and were probably afraid of getting lost; and so that project fell through. I then consulted with Mr. Chester about waiting and trying to get farther north. He, as well as I, was anxious to get up to the 83°, if possible. If the ship would have waited for us, there might have been an opportunity; but Captain Buddington had told me plainly, if he 'got a chance to get out he would not wait;' so that there seemed no other course to pursue but to get back to the *Polaris*.

"*July* 4. Received orders from Captain Buddington to return to the ship. Mr. Chester says 'he won't go.' He had sent two of his men for additional provisions, and Buddington wished to detain them, but, at their urgent request, allowed them to return, and sent the order by them. They report the ship leaking badly.

"Overheard two of the men talking; they 'thought, if the captain got a good chance, he would sail south without waiting for any one.' They were very cool, and said they 'didn't care,' appearing to think they could get down the coast in boats before cold weather set in.

"*July* 6. Hans came up on a sledge, bringing a written order to Chester to return, which he passed to me. Dr. Bessel took the opportunity of going back on the sledge with Hans. The state of the ice is such that we can get neither north nor south with the boats. It has turned out just as I said it would; with sledges we might have done something. As there was no using the boat, I concluded to haul mine up in as safe a place as I could find, and then walk back to the ship. Our boat was so heavy, with its contents, that my crew was not sufficient to haul it over the rough hummocky ice, and Chester let me have some of his crew to assist. We were between Cape Brevoort and Cape Sumner when we started, and it took us nearly forty-eight hours of most fatiguing labor to reach a ravine near Cape Sumner, where the boat could be partly protected, and there we hauled her up, placing the tent, and what stores I was obliged to leave, as well secured as possible, and, this accomplished, set out to walk to the ship.

"*July* 8. Back again on board the *Polaris*. Stood the walk of over twenty miles very well. So did the men.

ARCTIC WOLVES.

CHAPTER XV.

Engineer's Report.—A new Inscription.—A gentle Awakening.—Providence Berg disrupted.—Having "enough of it."—Lost Opportunities.—The Advent of little Esquimau "Charlie Polaris."—Beset near Cape Frazier.—Alcohol Master.—Interruption of his morning "Nip."—Drifting with the Floe.—Pack-ice in Smith Sound.—The Oil-boiler.—The bearded Seal.—Preparations for spending another Winter in the North.—A south-westerly Gale.

"Mr. Schuman, the engineer, reports that the pumps had become choked, and that some water had got into the lower hold, and injured a quantity of provisions.

GRAVE OF CAPTAIN HALL.

"As there is no probability that we shall be allowed to do any thing more—the captain being in waiting for an opportunity to get south—I have been over to see if Captain Hall's grave had been put in order; when he was buried it was too dark to work, and the ground frozen too hard to do much except cover it with stones for security. There was a board at the head, with the inscription, written in pencil, by the engineer, Mr. Schuman:

"'TO THE MEMORY OF C. F. HALL,
LATE COMMANDER OF THE NORTH POLAR EXPEDITION.
DIED NOV. 8, 1871.
AGED 50 YEARS.'

Some of the men, particularly Sieman, took great interest in securing the grave. Captain Hall was generally liked by the men.

"Mr. Chester has sent in three of his men. They walked the distance in twelve hours; but Mr. Meyers was twenty-eight hours on his way. He got caught in a snow-drift shortly after starting, and, being in danger of losing his way, was obliged to seek the shelter of a rock until the storm abated. The men report that Mr. Chester, with one man, Sieman, remained to try and save his boat. Perceiving that he could not get north, he started to get his boat ashore, which the roughness of the ice made very difficult; on the way they dropped many things, and had to go back for them. After about ten hours' hard labor and travel, they reached the land where I had left my boat, pretty well broken down. However, they got to the shore, and leaving the boat as it was, made the best of their way to the ship, bringing such clothing and other things as they could carry.

"*July* 24. A gale from the north, which I hope will blow the ice out of the bay.

"*July* 25. Cleared off; the bay partially opened; much water in the hold. I wanted the captain to divide the crew into three watches, and so have all hands take a turn at the pumps, to save fuel. Shortly after there was a sudden accession of water in the hold, and it was suggested to the captain that some one in the engine-room had willfully opened the stop-cocks and flooded her, so that those in favor of hand-pumping 'should have enough of it.' Captain Buddington went down to the engine-room to see about it, but had the door shut in his face for his pains! He has now divided the company into two watches for pumping; but after the first flood was got rid of, the ship was easily kept free with from two to four minutes pumping every hour. Some of the timbers appear to have swollen and closed the seams.

"I talked with Chester about fixing Captain Hall's grave; and he got a board and shaped it out properly, and cut the inscription in very nicely; and then we fixed it up, so that the grave now looks, though dreary enough, not quite so neglected as it did.

"The channel still full of pack-ice and heavy floes.

"*Aug.* 1. Still in Polaris Bay. What opportunities have been lost! and the expedition is to be carried back only to report a few geographical discoveries, and a few additional scientific facts. With patience we might have worked up beyond Newman Bay, and there is no telling how much farther. Some one will some day reach the pole, and I envy not those who have prevented the *Polaris* having that chance.

"Several of the men have gone back to Newman Bay to try and recover some valuable instruments of Meyers's, and other things which were left there. They brought back all they could carry, and reported the channel off the bay to be full of ice. The tent they left. Three boats lost on that ill-considered trip.

"*Aug.* 12. The wife of the Esquimau, Hans, has added a male member to the expedition. These natives have not outgrown some of their savage customs. Like the squaws of our Western Indians and other uncivilized people, the women are left alone in the exigencies of childbirth, and free themselves, like the inferior mammals, by severing the umbilical with their teeth. They very soon recover, but in the settlements various customs supposed to conduce to the welfare of the child are deemed necessary. Among other things, the clothes of the mother are always abandoned or destroyed, and are never worn again. The boy has been named, by acclamation, 'Charlie Polaris;' thus combining a remembrance of our late commander and the ship.

"P.M. This afternoon, the ice opening, and a good lead of water appearing, with a northerly wind, we weighed anchor, and steamed out of Polaris Bay about five o'clock.

"*Aug.* 15. Last night, at midnight, in Kennedy Channel, heading for Cape Frazier, during my watch, Mr. Bryan, with Chester, was out on the ice making some astronomical observations. When I went below, instead of the ship being kept to her course, she was allowed to fall off, and thus got beset in the broad waters of Peabody Bay, or Kane Basin, as it is sometimes called. About 8 A.M. we were enabled to get her headed right again, but could make no progress, and anchored to a floe in lat. 80° 2' N.

"We have only ice-anchors now; our ground-tackle has been broken or lost. One was broken when Providence Berg split up, by the weight of a heavy mass of ice which fell on it, and the other was lost under grounded ice when the berg pressed us inshore.

"Fuel we are rather short of. A great deal was used in working the donkey-engines, when hand-pumping would have answered, and the men were willing to pump; some of them even commenting on the waste of coal. There is a quantity of provisions left ashore at Polaris Bay.

"Our floe is drifting, and taking the vessel with it, slowly to the southward; we are at present in charge of the current, as there is no wind. Open water appearing to the south during the night, got up a good head of steam, and tried to force the vessel through the ice which beset us, but did not succeed, and had to tie up to the floe again. Got a start about noon, the ice loosening. We hauled in our ice-anchors, and made good progress. Weather fair, and wind southerly. Passed Cape Constitution, and proceeded until near midnight, when the ice closed around us again, and we anchored to the floe.

"*Oct.* 16. Still drifting with the floe. Ice opening here and there, but we get no chance; but probably shall soon, as narwhals have appeared, and they always breathe through the large ice-cracks. Saw several to-day.

"P.M. Thick and foggy. By observation at noon lat. 79° 59'.

"*Oct.* 19. Fog hung about us for twelve hours, then cleared by a fresh northerly breeze. No lead visible, and we still drift with the floe.

"12 M. Tried to shift the position of the vessel, as we are in danger of being nipped; the ice is very heavy, and strong pressure. There is now a quantity of stores, clothing, and some bags of coal kept on deck, so that they may be at hand to throw overboard in case of necessity. The ice in Smith Sound varies very much in different seasons. This sort of pack-ice baffled Kane in July and August, between Cape Parry and the Cary Islands, but it has been traversed in August by several explorers without difficulty, from the time of old William Baffin down. There seems no rule about it; probably depends on the force and direction of the winds when the ice begins to break up in the north. The weather is fine.

"*Oct.* 23. Made a few miles, then found something was the matter with the boilers, and had to draw the fires. The ship making considerable water, the hand-pumps were kept going: if this vessel had not been very strong she could not have gone through what she has. Boilers repaired; and on the morning of the 26th got up steam again, and pushed forward about half a mile, working toward the shore, as the heavy pack is in the middle, and we hoped to find a lead between it and the shore ice.

"P.M. Towed the ship through a narrow opening for a short distance, but beset again, and obliged to desist. So we pass the days — boring a few rods or a quarter of a mile, then tying up again, to repeat the process, with slight variations, through the days and nights. All around us lies the pack-ice, with large bergs in sight, some grounded, and others like sentinels watching the progress of the floe to which we are fast. Young ice has formed over the open water, strong enough to bear. Snow and rain and fog have succeeded each other. Rain is rather a curiosity; but we have had a little to vary the meteorological changes.

"*Sept.* 1. Our summer is almost over; the days are visibly shortening. A steady, slow drift to the southward. Some seal have appeared. Nothing to record.

"*Sept.* 13. Have reached lat. 79° 21′ 30″. Some walruses have been seen; Hans has shot a seal, and Joe fired at a walrus, but their hides are so thick, and their heads so impenetrable that it is difficult to either kill or secure them without lance or harpoon.

"*Sept.* 21. Our coal is getting so short that I think the destruction of the oil-boiler will yet be regretted. This boiler was made expressly for the use of the expedition, on the presumption that we should be absent two years or more, and of course we could not store coal enough to be absent all that time; and it was expected that blubber or seal-oil could be used for fuel. But what portion of it that did not go overboard between Disco and Tossac was left at Polaris Bay; and that was the end of our 'oil-boiler.'

"The meat of the seal is now very welcome to the men, as a change from the canned food. And the blubber, too, will soon be needed, for cook and steward have used up the kerosene *ad libitum* to kindle fires.

"The last day of September has come, and though the open

water can be seen to the southward, we can not get to it; water is also visible to the north, but we are kept in the pack. But the ice keeps working, and a strong gale, I think, would open it. During the last six weeks we have done little except drift, with now and then a spurt at the engines. In this time we have made about sixty miles—about ten miles a week, and mostly drifting. At 12 M. our latitude was found to be 70° 2′ N.; and the temperature begins to suggest the coming winter: it is a little below zero.

"*Oct.* 4. October came in fair and clear. Have passed Rensselaer Harbor, where Dr. Kane wintered during 1853-55. I am surprised that in the latitude of Rensselaer Harbor he should have found the darkness so intense as he describes it. It was not totally dark with us at high meridian at any time in clear weather, but it was too dark to travel about much, or do any shooting, unless it was full of the moon. The ice keeps groaning, as if a change of some kind was impending. There is little chance of getting home this fall. We shall have to spend another winter here, I expect. If we were heading the other way, I should not mind that; but to go home without having done all we could is galling.

"Yesterday Joe shot a large bearded seal of uncommon size even for this species—eight feet long and six in circumference.

"Mr. Meyers is not very well—looks as though he had a touch of scurvy. Seal meat and blood is a specific against that disease; the Esquimaux rarely have it, and they live nearly altogether on seal-meat.

"Have commenced work on a house in which to store provisions, as there is no telling when the ship may get nipped. I wanted some lumber from the ship to build it, but could get only poles and canvas. Into this house we shall remove a quantity of provisions; and keep clothing, guns, and ammunition ready on deck to heave over at the shortest notice.

"*Oct.* 4, 5. Four more seals killed; house nearly finished; light snow. There is no more doubt that we must winter here. The men are cutting fresh-water ice for the engineer to use in the small boiler which supplies steam for the pumps. This boiler is badly crystallized inside from the use of salt-water.

"A bear has been tracking round the ship, but was not seen or scented by the dogs.

"*Oct.* 9. The bear was seen to-day on the ice-floe, about a mile from the ship; the ice was not in a condition to go in chase of him.

"*Oct.* 12. There has been a large number of seals killed; they become plenty as the water opens. We are now drifting much more rapidly than we did, and within sight of the eastern shore. Symptoms of a gale from the south-west.

THE LUMME OF THE NORTH.

CHAPTER XVI.

JOURNAL OF GEORGE E. TYSON, ASSISTANT NAVIGATOR ON UNITED STATES STEAMER POLARIS, KEPT ON THE ICE-FLOE.

Adrift.—The fatal Ice Pressure.—"Heave every thing Overboard!"—The Ship breaks away in the Darkness.—Children in the Ox-skins.—First Night adrift.—Snowed under.—Roll-call on the Ice-floe.—Efforts to regain the Ship.—The *Polaris* coming!—A terrible Disappointment.—The overladen Boat.—Three Oars, and no Rudder.—The Ice breaks beneath us.—Drifting to the South-west.—Regain the large Floe.—Hope of regaining the *Polaris* abandoned.—Building Huts.—Native Igloos.—Estimating Provisions.—Locality of the Separation.—Meyers's and Tyson's Opinion.—Two Meals a Day.—Mice in the Chocolate.—Too cold for a Watch.—Too weak to stand firmly.—Hans kills and eats two Dogs.—Natives improvident.—Lose Sight of the Sun.—The Dogs follow the Food.

"*Adrift, Oct.*, 1872. Blowing a strong gale from the north-west. I think it must have been about 6 P.M., on the night of the 15th, when we were nipped with the ice. The pressure was very great. The vessel did not lift to it much; she was not broad enough—was not built flaring, as the whalers call it; had she been built so she would have risen to the ice, and the pressure would not have affected her so much; but, considering all, she bore it nobly. I was surprised at her great strength.

"In the commencement of the nip, I came out of my room, which was on the starboard side of the ship, and looked over the rail, and saw that the ice was pressing heavily. I then walked over to the port side. Most of the crew were at this time gathered in the waist, looking over at the floe to which we were fastened. I saw that the ship rose somewhat to the pressure, and then immediately came down again on the ice, breaking it, and riding it under her. The ice was very heavy, and the vessel groaned and creaked in every timber.

"At this time the engineer, Schuman, came running from below, among the startled crew, saying that 'the vessel had started a leak aft, and that the water was gaining on the pumps.' The vessel had been leaking before this, and they were already pumping—Peter and Hans, I think, with the small pump in the starboard alley-way.

"I then walked over toward my room on the starboard side. Behind the galley I saw Sailing-master Buddington, and told him what the engineer said. He threw up his arms, and yelled out to 'throw every thing on the ice!' Instantly every thing was confusion, the men seizing every thing indiscriminately, and throwing it overboard. These things had previously been placed upon the deck in anticipation of such a catastrophe; but as the vessel, by its rising and falling motion, was constantly breaking the ice, and as no care was taken how or where the things were thrown, I got overboard, calling some of the men to help me, and tried to move what I could away from the ship, so it should not be crushed and lost; and also called out to the men on board to stop throwing things till we could get the things already endangered out of the way; but still much ran under the ship.

"It was a dark night, and I could scarcely see the stuff— whether it was on the ice or in the water. But we worked away three or four hours, when the ice on the starboard side let the ship loose again. We had been tied to the floe of ice by ice-anchors and hawsers, but when the piece on the starboard drifted off she righted from her beam-ends and broke away. I had been on board just before she broke loose, and asked Buddington 'how much water the vessel was making?' and he told me, 'no more than usual.'

"I found that the engineer's statement was a false alarm. The vessel was strong, and no additional leak had been made; but as the ice lifted her up, the little water in the hold was thrown over, and it made a rush, and he thought that a new leak had been sprung. When I found she was making no more water, I went on the ice again to try and save the provisions, if possible. While so engaged, the ice commenced cracking; I told Buddington of it, he meantime calling out to 'get every thing back as far as possible on the ice.' Very shortly after, the ice exploded under our feet, and broke in many places, and *the ship broke away in the darkness, and we lost sight of her in a moment.*

" 'Gone!
But an ice-bound horror
Seemed to cling to air.'

"It was snowing at the time also; it was a terrible night. On the 15th of October it may be said that the Arctic night commences; but in addition to this the wind was blowing strong

"THE SHIP BROKE AWAY IN THE DARKNESS, AND WE LOST SIGHT OF HER IN A MOMENT."

from the south-east; it was snowing and drifting, and was fearfully dark; the wind was exceedingly heavy, and so bad was the snow and sleet that one could not even look to the windward. We did not know who was on the ice or who was on the ship; but I knew some of the children were on the ice, because almost the last thing I had pulled away from the crushing heel of the ship were some musk-ox skins; they were lying across a wide crack in the ice, and as I pulled them toward me to save them, I saw that there were *two or three of Hans's children rolled up in one of the skins;* a slight motion of the ice, and in a moment more they would either have been in the water and drowned in the darkness, or crushed between the ice.

"It was nearly ten o'clock when the ship broke away, and we had been at work since six; the time seemed long, for we were working all the time. Hannah was working, but I did not see Joe or Hans. We worked till we could scarcely stand. They were throwing things constantly over to us till the vessel parted.

"Some of the men were on small pieces of ice. I took the 'little donkey'—a small scow—and went for them; but the scow was almost instantly swamped; then I shoved off one of the whale-boats, and took off what men I could see, and some of the men took the other boat and helped their companions, so that we were all on firm ice at last.

"We did not dare to move about much after that, for we could not see the size of the ice we were on, on account of the storm and darkness. All the rest but myself—the men, women, and children—sought what shelter they could from the storm by wrapping themselves in the musk-ox skins, and so laid down to rest. I alone walk the floe all night.

"Morning came at last; I could then see what had caused the immense pressure on the ship, though I knew she must go adrift when I heard the ice cracking. The floe to which the ship was fastened had been crushed and pressed upon by heavy icebergs, which was the immediate cause of its breaking up. This I could not see last night, but I saw all in the morning.

"Fortunately, we had the two boats on our piece of the floe. This was a nearly circular piece, about four miles in circumference. It was not level, but was full of hillocks, and also ponds, or small lakes, which had been formed by the melting of the ice during the short summer. The ice was of various thicknesses.

Some of the mounds, or hills, were probably thirty feet thick, and the flat parts not more than ten or fifteen. It was very rough; the hillocks were covered with snow; indeed, the surface was all snow from the last storm. Some of the men whom I now found on the ice were those whom I had picked off of the smaller pieces last night in the darkness. I could now see who they were. These men were thirty or forty yards from the main floe, and I pushed off the boat and went for them. Some of the men, too, had taken their shipmates off of small pieces. I do not think any body was lost last night. I think all that are not here are on the ship. I should think they would soon be coming to look for us.

"Those who laid down on the ice were all snowed under—but that helped to keep them warm. Perhaps I should have lain down too, if I had had any thing to lie on; but the others had taken all the skins, and I would not disturb them to ask for one.

"*Oct.* 16. Why does not the *Polaris* come to our rescue? This is the thought that now fills every heart, and has mine ever since the first dawn of light this morning. I scanned the horizon, but could see nothing of the vessel; but I saw a lead of water which led to the land. The gale had abated; it was almost calm. I looked around upon the company with me upon the ice, and then upon the provisions which we had with us. Besides myself there were eighteen persons, namely:

"'Frederick Meyers, meterologist; John Herron, steward; William Jackson, cook.—*Seamen*: J. W. C. Kruger (called Robert); Fred. Jamka; William Lindermann; Fred. Anthing; Gus. Lindquist; Peter Johnson.—*Esquimaux*: Joe; Hannah, Joe's wife; Puney, child; Hans; Merkut or Christiana, Hans's wife; Augustina, Tobias, Succi—children; Charlie Polaris, baby of Hans's.'

"Now, to feed all these, I saw that we had but fourteen cans of pemmican, eleven and a half bags of bread, one can of dried apples, and fourteen hams; and if the ship did not come for us, we might have to support ourselves all winter, or die of starvation. Fortunately, we had the boats. They were across the crack where I had hauled away the musk-ox skins and found the children; we had hauled both the boats on the ice to save them. I had shortly before asked Captain Buddington if he would haul the boats on board; but he had only answered by ordering every thing to be pulled as far back on the ice as possible.

"As soon as I could see to do so, I walked across the floe to see where was the best lead, so that we could get to shore; and in the mean time I ordered the men to get the boats ready, for I was determined to make a start, and try and get to the land, from which I thought we might find the ship, or at least, if we did not find her, that we might meet with Esquimaux to assist us. I thought that perhaps the *Polaris* had been lost in the night, as I could see nothing of her.

"I had called to the crew to rouse up and see to the boats, and at last succeeded in getting them out of the snow, and fairly awake. I told them we must reach the shore; they thought so too, but they seemed very inert, and in no hurry; they were 'tired' and 'hungry' and 'wet' (though I think they could not have been more tired than I, who had been walking the floe all night while they slept); they had had nothing to eat since three o'clock the day before; and so they concluded they must get something to eat first. Nothing could induce them to hurry; while I, all impatience to try and get the boats off, had to wait their leisure. I might have got off myself, but I knew in that case, if the *Polaris* did not come and pick them up, they would all perish in a few days; so I waited and waited. Not satisfied to eat what was at hand, they must even set about cooking. They made a fire out of some wood which they found upon the ice. They had nothing to cook in but some flat tin pans, in which they tried to cook some of the canned meat, and also tried to make some coffee or chocolate. Then some of them insisted on changing their clothing; for several of them had secured their bags of clothing. But every thing has an end, and at last I got started about 9 A.M.; but, as I feared, it was now too late; the leads were closing, and I feared a change of wind which would make it impossible to reach the shore.

"The piece of ice we were on was fast, between heavy icebergs which had grounded, and was therefore stationary. The wind had now hauled to the north-east. I had no means of taking the true bearings, but it was down quartering across the land, and it was bringing the loose ice down fast. But though I feared it was too late, I determined to try. And at last we got the boats off, carrying every thing we could, and intending to come back for what was left; but when we got half-way to the shore, the loose ice which I had seen coming, crowded on our bows so that we could not get through, and we had to haul up on the ice; and soon after I saw the *Polaris!* I was rejoiced indeed, for I thought assistance was at hand.

"She came around a point above us, eight or ten miles distant. We could see water over the ice that had drifted down, and we could see water inshore. I wondered why the *Polaris* did not come and look for us. Thinking, perhaps, that she did not know in which direction to look—though the set of the ice must have told which way it would drift—and though the small ice had stopped us, it was not enough to stop a ship, I did not know what to make of it. But, determined to attract her attention, if possible, I set up the colors which I had with me and a piece of India rubber cloth, and then with my spy-glass watched the vessel. She was under both steam and sail, so I went to work securing every thing, hoping that she would come for us and take us aboard. I could not see any body on deck; they, if there, were not in sight. She kept along down by the land, and then, instead of steering toward us, dropped away behind the land—Littleton Island, I suppose it is. Our signal was dark, and would surely be seen that distance on a white ice-floe. I do not know what to make of this.

"I wanted some poles to help build a house or tent, and I sent some of the men to the other side of the floe to get some; I knew there must be some there belonging to a house I had built of poles in which to store provisions. In going to this portion of the floe they saw the vessel behind the island, and so came back and reported; they said she was 'tied up.' I did not know what to think of it; but I took my spy-glass, and running to a point where they said I could see her, sure enough there she was, *tied up*—at least, all her sails were furled, and there was no smoke from her stack, and she was lying head to the wind. I suppose she was tied up to the bay-ice, which I could see with the glass.

"And now our piece of ice, which had been stationary, commenced drifting; and I did not feel right about the vessel not coming for us. I began to think she did not mean to. I could not think she was disabled, because we had so recently seen her steaming; so I told the men we *must* get to the other side of the floe, and try and reach the land, perhaps lower down than the vessel was, but so that we might eventually reach her. I told them to prepare the boats. I threw away every thing to make them light, except a little provision—enough to last perhaps two or three days.

"I told the men, while they were getting the boats ready, I would run across the ice and see if there was an opportunity to take the water, or where was the best place, so that they would not have to haul the boats uselessly. I ran across as quick as I could. I was very tired, for I had had nothing but some biscuit and a drink of the blood-soup to eat; but I saw there was an opportunity to get through, and that seemed to renew my strength. The small ice did not now appear to be getting in fast enough to prevent our getting across. But in these gales it is astonishing how quickly the ice closes together, and I knew we were liable to be frozen up at any moment; so I hurried back to the boats and told them 'we must start immediately.'

"There was a great deal of murmuring—the men did not seem to realize the crisis at all. They seemed to think more of saving their clothes than their lives. But I seemed to see the whole winter before me. Either, I thought, the *Polaris* is disabled and can not come for us, or else, God knows why, Captain Buddington don't mean to help us; and then there flashed through my mind the remembrance of a scene and a fearful experience which had happened to me before, in which his indifference had nearly cost me my life and those of all my crew. But I believed he thought too much of Puney and the cook to leave us to our fate without an effort. Then the thought came to me, what shall I do with all these people, if God means we are to shift for ourselves, without ship, or shelter, or sufficient food, through the long, cold, dark winter? I knew that sometime the ice would break up; that at last it would break up into small pieces — too small to live upon. From the disposition which some of the men had shown, I knew it would be very difficult to make them do what was needful for their own safety. And then there were all those children and the two women!

"It appeared to me then that if we did not manage to get back to the ship, that it was scarcely possible but that many, if not all of us, would perish before the winter was over; and yet, while all these visions were going through my brain, these men, whose lives I was trying to save, stood muttering and grumbling because I did not want the boats overloaded to get through the pack-ice. They insisted on carrying every thing. They were under no discipline — they had been under none since Captain Hall's death. They loaded one boat full with all sorts of things, much of which was really trash, but which they would carry. We were going to drag the boat across the floe to where we could take the water. I went on, and told the Esquimaux to follow me across the floe. I had not gone more than two hundred yards before a hurricane burst upon me. I nevertheless persevered and got across the ice, and when I got to the lead of water saw that the natives had not followed me! Whether they thought too much of their property, or whether they were afraid of the storm, I do not know; but the cook had followed me, and when he saw they had not come he ran back for them.

"The men still murmured about getting into the boat which they had dragged over so overloaded, but I would have shoved off as long as I had the strength to do it; but when I looked for the oars, there were but three, and there was *no rudder!* I had told them to prepare the boat while I was gone to look for a lead, and this was the way they had done it. I had told them to see that all was right, including sails; but they did not wish to go, and that probably accounts for it. I am afraid we shall all have to suffer much from their obstinacy.

"Perhaps if we had started we could not have reached either land or ship, but it was certainly worth trying. Why they prefer to stay on this floe I can not imagine; but to start with only three oars and no rudder, the wind blowing furiously, and no good, earnest help, was useless. I tried it, but the men were unwilling; and in the crippled condition of the boat it was no wonder that we were blown back like a feather. I was, therefore, compelled to haul the boat back on the ice. The men by this time were really exhausted, and I could not blame them so much for not working with more energy.

"Night was now coming on; our day was lost, and our opportunity with it. We must prepare for another night on the ice.

"We had to leave the boat where she was; we were all too tired to attempt to drag her back. We also left in her the clothing and other things the men had been so anxious to save in the morning.

"I went back toward the centre of the floe, and put up a little canvas tent, and then, eating a little frozen meat and a little ship-bread, I was glad enough to creep in, pull a musk-ox skin over me and get a little rest, drifting in the darkness I knew not whither; for I had had no rest since the night of the 14th—the night before we parted with the ship. All of the afternoon of the 15th I was at work, and all of that night I walked the floe. All the next day I was going and coming across the ice, and laboring with the men and boats, trying to work through the pack; and when night came the ice-floe proved a refreshing bed, where I slept soundly till morning, when I was suddenly awakened by hearing a loud cry from the natives, which made me quickly crawl out from between my wet ox-skins.

"It had snowed during the night; but that was nothing. *The ice had broken!* separating us from the boat which we had left, being unable to haul it the night before. The old house, made of poles, in which there was also six bags of bread, remained on the old floe, and we were left on a very small piece of ice. The Esquimaux, Mr. Meyers, and myself had made our extemporized lodgings on the thickest part of the floe, and when the ice parted we were all on this portion. As soon as I saw the position of affairs, I called the men out, desiring them to go for the boat and bread. It could have been done with safety, for there was no sea running between the broken floe, and they had not separated much at that time; but I could not move them—they were afraid. At least they did not go.

"So we drifted, having one boat on our piece of ice, while one of our boats, part of the provisions, and the house of poles, remained on the main part of the original floe. And so we drift, apparently to the south-west, for I have neither compass nor chronometer with me; my compass is in that other boat, and even my watch is on board of the *Polaris*. *Our* piece of ice is perhaps one hundred and fifty yards across each way.

"*Oct.* 17. Quite a heavy sea is running; piece after piece is broken from our floe. God grant we may have enough left to stand upon! The vessel could now come to us in clear water, if she is in condition either to steam or sail. I told the natives who are with me they must try and catch some seal. Hans was engaged as hunter, servant, and dog-driver; and Joe is one of the best hunters to be found, if there is any thing to catch. If we can only get seal enough, we can live; but without seal we can have no warm food, for we shall have to cook with the blubber-oil, as the natives do. The natives have caught three seals, and could have caught more, but for the thoughtlessness of the men who gathered around and frightened them off; then the weather set in so bad they could do no more; it was thick and heavy. Weather continued bad, but the gale moderated toward the morning of the 18th. When it cleared, I could see the land—about six miles away. I thought it might be the east shore; but, having no compass and no chart, could hardly be sure where we were. 'Young ice,' or new ice, had formed between us and the land; but it was not strong enough to walk upon. I was in hopes it would get firmer, and then we might perhaps get to land.

"One morning—the 21st, I think—Joe was spying around, and saw the end of our abandoned boat on the same floe where we had left it. He called to me, and as soon as I saw it I started off with him to try and recover it. It was about twelve o'clock in the day, and we had not yet had our breakfast. But I was afraid we should not have so good a chance again to get it, and would not wait for any thing, for we could now get across to the old floe from our own piece of ice. Joe and I started, and got it back, with all the things, and also loaded in what bread I could carry. I fortunately had five or six dogs with me. We harnessed them to the boat, they dragging and we pushing over the bad places. We at last got it back safely to the piece of ice we were encamped upon. We saved all. We have now both boats, the natives' kyacks, and are together again.

"*Oct.* 23. We have now given up all hopes of the *Polaris* coming to look for us. All we can do is to wait for the ice to get strong enough for us to get on shore. The worst of it is, we have no sledges; and hauling the loaded boats over the rough ice is likely to injure them, so that they would be unfit for use, should we need to take to them; but it is the only way we can do to get them over to the large floe, which now lies half-way between us and the shore. There is, too, but little time to see to work; all the light we have now is about six hours a day, and not very clear then. On cloudy and stormy days it is dark all the time. But this piece of ice will not do to winter on. So to-day, the ice appearing strong enough, I got the boats loaded, harnessed on the dogs, and started to regain the large floe; succeeded with the first, and then went back for the second. It is fortunate, indeed, that we have the boats. Humanly speaking, they are our salvation, for on an emergency we can use them either for the water or as sledges. Got the second one over safe, and am rejoiced at that; and they do not appear to have received any injury except what can be readily repaired. There are still two kyacks on the small floe. A native will stick to his kyack like a white man to his skin, and Joe and Hans got theirs out of the ship when Captain Buddington ordered them off.

"We had now got all our principal things on the large floe, except a little stuff and these kyacks. I wanted the crew to try and help save them, but could not get them to do any thing toward it. At last Joe started alone, and then two of the men ventured over: one was the negro cook, and the other William Lindermann. One of the kyacks was saved, but the other was lost. These little boats are invaluable to the Esquimaux, who are accustomed to manage them; but no one else can do any thing with them. One might almost as well launch out on an ostrich feather and think to keep afloat, as in these unballasted little seal-skin shells. But I'm glad enough they have got one of them.

"The weather has come on very bad; but, fortunately, we have got our snow-houses built. We have quite an encampment—one hut, or rather a sort of half-hut, for Mr. Meyers and myself; Joe's hut for himself, Hannah, and their adopted daughter, Puney; a hut for the men, a store-hut for our provisions, and a cook-house, all united by arched alley-ways built of snow; one main entrance, and smaller ones branching off to the several apartments, or huts. Hans has built his hut separately, but near by.

"Joe did most of the work of building these huts—he knew best how to do it; but we all assisted. They are made in the regular Esquimau style, and the natives call them *igloos*. The way they go about it is this: the ground is first leveled off, and then one-half of the floor toward the end farthest from the entrance is slightly raised above the other or front half. The raised part is parlor and bedroom, and the front part is workshop and kitchen. The walls and arched roof are composed of square blocks of hard snow, packed hard by the force of the wind. A square of about eighteen inches of thin, compressed snow or ice, or sometimes a piece of animal membrane, is fixed in for a window. The entrance is very low, and is reached through the alley-way, so that one has to almost crawl in. At night, or whenever it storms or is very cold, the entrance is closed up, after the inmates are all in, by a block of snow.

"There is hardly room to turn round in these huts, and an ordinary-sized white man can only just stand up straight in them; it is as much as an Esquimau can do in some of them; but from their form they stand the weather well. A hut is often snowed under, so that it can not be distinguished from a natural hillock; but it can not be blown over; and when there is a sufficiency of oil to burn in the lamps, these kind of huts can be kept warm enough. But from their arched form, and the material of which they are constructed, it can easily be seen that they can not be made spacious enough to properly accommodate a large party of men. The centre of the dome only admits of the upright position being maintained, as from that point the walls slope gradually, until they meet the ground. In the men's hut, for instance, the dais, or raised platform, on which they sleep, just accommodates them, lying like herrings in a box, with no superfluous room in which to turn; and only two or three of them can stand up at a time.

"These huts are only used by the natives in winter. The summer sun is as fatal to them as rain would be if it fell there; but when they begin to thaw and melt, the Esquimaux take to their seal-skin tents for shelter.

"The ordinary lamp in use among the natives is made out of

NATIVE LAMP.

a soft kind of stone, indigenous to the country; it is hollowed out, like a shallow dish, with an inverted edge, on which they place a little moss for wicking, which, when lighted, sucks up the oil from the blubber; and this is all the fire they have in this cold country, either for heating their huts or for cooking. To dry their clothing, they put them in nets suspended over the lamp.

"We, however, did not have even a proper lamp; but we soon contrived one out of an old pemmican can, and having no moss, we cut up a piece of canvas for wicking, and it answered very well for us; but somehow the men could not seem to understand how to use it; they either got the blubber all in a blaze, or else they got it smoking so badly that they were driven out of their hut; and so I am sorry to say that they have begun to break up one of the boats for fuel. This is bad business, but I can not stop them, situated as I am, without any other authority than such as they choose to concede to me. It will not do to thwart them too much, even for their own benefit.

"These boats are not designed to carry more than six or eight men, and yet I foresee that all this company may have yet to get into the one boat to save our lives, for the ice is very treacherous. But they will do as they like.

"I have been taking account of stock. By our successive expeditions, in which we gathered nearly all together which was on the ice when we were first drifted off, I find that we have our two boats (but one is being destroyed) and one kyack, and, thank God, plenty of ammunition and shot.

"Of provisions we have eleven and a half bags of bread, fourteen cans of pemmican, fourteen hams, ten dozen cans of meats and soups, one can of dried apples, and about twenty pounds of chocolate and sugar mixed. The pemmican cans are large, each weighing forty-five pounds; the meats and soups are only one and two pound cans; and the hams are small ones; the dried-apple can is a twenty-two-pounder. Divide that into portions for nineteen people, with a certainty of not getting any thing more for six months (unless we reach the land, or can catch seals to live on), and it is plain we could not exist. And if we have to keep to the floe, it will be April or May before we shall drift to the whaling-grounds.

"We must try once more to get on shore. To-morrow, if the weather permits, I will try and get the house and the lumber where we can have the use of it.

"Have had a talk with Mr. Meyers about the locality of our separation from the *Polaris;* he thinks we were close to Northumberland Island, but I believe it was Littleton Island; he says 'he ought to know, for that he took observations only a day or two before,' and of course he *ought* to be right; but still my impression is that Northumberland Island is larger than the one the *Polaris* steamed behind. I wish I had a chart, or some means of knowing for certain.

"*Oct.* 24, *Morning.* Blowing strong from the north-east, and the snow is drifting; quite cold. Robert and Bill have started for the old house to get two planks to make a sledge to haul the rest of the house over on, and for general use. If it is a good day to-morrow, I hope to get all the lumber and the remains of the canvas from the old place.

"*Afternoon.* The men came back with the planks; they were very hungry—so hungry I was compelled to break the rules, and give them some bread and pemmican to eat.

"We only allow ourselves two meals a day, and Mr. Meyers has made a pair of scales, with which to weigh out each one's portion, so that there should be no jealousy. We use shot for weights. Our allowance is very small—just enough to keep body and soul together; but we must economize, or our little stock will soon give out altogether.

"One bad symptom has appeared: we have only had chocolate prepared for the party four times, and it is *nearly all gone!* Some one has made free with the store-house. It is too cold to set a watch; but it is plain enough to be seen that things have been meddled with.

"The wind is mostly from the E.N.E. Have succeeded in getting a sledge made, and the men have brought in a load of lumber and poles from the old house; no doubt we shall be able to get it all. But our blubber is almost out, and we see no seals; if we do not get some soon we shall be in darkness, and have to eat our frozen food without thawing it—to say nothing of cooking it. We need it, too, very much to melt the fresh-water ice for drink. Fortunately there is enough of this ice in the ponds on this floe, if we can only get the means of melting it.

"Our present daily allowance is eleven ounces for each adult, and half-rations for the children. I was obliged to establish a regular rate, and insist upon its observance, or we should soon have had nothing. There appears to be a good deal of discontent in some quarters, but I fear they will get less before any of us get more. Before this rule was established, some got a great deal more than others. It was hard for some of them to come down to it in consequence; and in fact it has weakened them down; but it is absolutely necessary to be careful of what we have. I am so weak myself that I stagger from sheer want of strength; and, after all, the men bear it as well as could be expected — considering, too, that they do not realize, as I do, the absolute necessity of it.

"Hans has just taken two of the dogs, killed and skinned them, and will eat them. I give each of the natives the same amount of bread, and whatever else we have, as I deal out to myself. But the Esquimaux are, like all semi-civilized people, naturally improvident; while they have, they will eat, and let tomorrow take care of itself. I do not suppose an Esquimau ever voluntarily left off eating before his hunger was fully satisfied, though he knew that the next day, or for many days, he would have nothing. Sailors have some kind of an idea that a ship's company must, under some circumstances, be put on 'short allowance;' but that is an idea you can never beat into the head of a native, and yet of all people they are the most subject to fluctuations of luck—sometimes having abundance, and then reduced to famine; but there is no thrift in them. They will sometimes store away provisions, and build *caches* on their traveling routes; but this is always done when they have more than they can possibly consume at the time — as when they have been fortunate enough to kill a whale or a walrus, and by no possibility can eat it all.

"*Oct.* 26. We lost sight of the sun's disk three days ago—

> "'Miserable we,
> Who here entangled in the gathering ice,
> Take our last look of the descending sun;
> While full of death, and fierce with tenfold frost,
> The long, long night, incumbent o'er our heads,
> Falls horrible.'

"May the great and good God have mercy on us, and send us seals, or I fear we must perish. We are all very weak from having to live on such small allowance, and the entire loss of the sun makes all more or less despondent. But still we do not give up; the men have got another sled-load of poles in to-day; but the ice is very rough, and the light so dim that they can fetch but little at a time. There seems now no chance of reaching the land—we have drifted so far to the west. We are about eight or ten miles off shore. Northumberland Island bears about east from us—should think forty or fifty miles off. Should judge the latitude to be about 77° 30'. Have not drifted any the last three days. The sled has come in with two additional dogs—'Bear' and 'Spike:' these dogs were on the large floe, where the most of our provisions were. I suppose, since we brought the food away, they thought best to follow it. A portion of the sun just showed for a little while to-day—his upper limb about 7' above the horizon.

THE GREAT AUK.

CHAPTER XVII.

A vain Hunt for Seal.—Pemmican.—The Dogs starving.—Blow-holes of the Seal.—Mode of Capture.—Sight Cary Island.—Hans mistaken for a Bear.—Down with Rheumatism.—One Boat used for Fuel.—The Children crying with Hunger.—Joe the best Man.—The Bread walks off.—One square Meal.—Bear and Fox-tracks.—Effects of lax Discipline.—Joe and Hannah.—Our Thanksgiving-dinner.

"*Oct.* 27 comes in with a clear, strong breeze from the northeast. The two natives and Robert have gone to look if they can find any thing more worth bringing away. There are two bags of coal in the old house, or what is left of it, which was thrown over with the rest of the things from the *Polaris;* it is probably the coal Robert wants, as they have not yet learned to use a lamp to cook with. It is so clear to-day that we can see to the west shore. If the ice remains firm, shall still endeavor to find the vessel. At noon to-day the sun showed about a quarter of his diameter above the horizon. We shall soon lose sight of him altogether. Joe and Hans have been out all day hunting for seal, but have found none.

"*Oct.* 28. Wind still blowing clear and strong from the northeast. Have found the dog-tracks, and the natives have secured the dogs, harnessed them to the sled, and taken them with them on the hunt. If they should be so fortunate as to meet a bear, the dogs would keep him at bay till the men could get a shot at him. We are out of seal-meat, and to-day dine on pemmican and bread. The bread, of course, is simply biscuit; it is 'bread' on board ship, and 'biscuit' to landsmen. Some people like this pemmican; it is made of beef cut in thin slices and dried, then either cut up fine, or, as some firms prepare it, ground up and mixed with an equal quantity of fat. I can eat it here; but if I was ashore it would not enter largely into my *cuisine*. It is, however, a very proper compound to take on Arctic voyages, being very 'heat-giving,' which is quite a desideratum in these regions.

"Another day passed and no seal caught. Have no seal-meat left, and very little blubber; must try and save what there is for the lamps; but when hungry natives are around, it disappears very rapidly down their throats. The ice keeps firm, and I am only waiting for the change of the moon, so that we can be sure of light to make another effort to reach the shore.

"In consequence of there being no seals caught, I have nothing to give the dogs to eat. The poor things are almost dead. I can not afford to give them our canned meats. They will have to go.

"*Oct.* 30. Wind lighter, but from the same quarter. Our allowance for the whole company is now two pounds of pemmican,

JOE WATCHING SEAL-HOLE.

six pounds of bread, four pounds of canned meat—twelve pounds in all, to furnish eighteen persons for the day. The natives still continue hunting, but have had no success.

"It is not easy to find the seal in winter; they live principally under the ice, and can only be seen when the ice cracks; an inexperienced person would never catch one. Being warm-blooded animals, they can not remain always under the ice without breathing; and in consequence they make air-holes through the ice and snow, through which to breathe; but at the surface these holes are so small—not more than two and a half inches across—that they are not easily distinguished, especially in the dim and uncertain light which we now have. They are very shy, too, and seem to know when they are watched. A native will sometimes remain watching a seal-hole thirty-six or forty-eight hours before getting a chance to strike, and if the first stroke is not accurate the game is gone forever.

"The natives use barbed spears, and, as the skull of the seal is exceedingly thin, if the blow is well aimed it is sure to penetrate, and the seal can then be held securely until the hole is enlarged sufficiently to pull the body through.

"The natives have come in empty-handed, and report the ice very rough. But I will try it to-morrow, should it not blow a gale of wind.

"*Nov.* 8. I started on the morning of the 1st of November to try and reach the shore. We had loaded the boat with provisions and the most necessary articles, and succeeded in dragging it nearly half-way to the shore on the old piece of floe; then the ice broke, and we were adrift again. Saw Cary Island, about twelve miles to the south-east of us. Since then it has been such thick weather I have seen nothing. We had started very early in the morning, having with us also the dogs and sled, and these were nearly driven into the water before it was discovered. There was a wide crack right across the ice that in the dim light was not discerned; some of the men had crossed over before it cracked, and had a good jump for it to get back.

"Fate, it seems, does not mean that we shall either get back to the *Polaris*, or even reach the shore. To help the matter, bad weather came on, and it has been so bad ever since there has been no possibility of making another attempt. Here we are, and here, it seems, we are doomed to remain.

"We have had to rebuild our huts, and are again sheltered as well as circumstances will permit. I have been sick on my back for the last three days; the exposure and exertion, with insufficient food and clothing, completely prostrated me; but now I am at home (!) again, and have had a little rest, and am better.

"On the 6th, Joe shot a seal, for which I was truly thankful, for our blubber was almost gone. The weather is so bad no one pretends to leave the hut. We are all prisoners.

"On the 10th, Joe and Hans went out hunting. After they had been out some time they got separated, and Joe, after trying his luck alone, made out to get back to the hut before it was quite dark; he fully expected to find that Hans had preceded him, and was much alarmed when he heard that he had not arrived. He persuaded Robert to go back with him to try and find Hans. It seems he had left our floe, thinking there might be better chance for game on another, and had not been able to find his way back. As Joe and Robert were going along, peering through the fast, coming darkness, they saw what they took to be an ice-bear approaching them; they loaded their pistols, and made all ready to give him a warm reception; when, fortunately, the creature coming a few steps nearer, they saw that, instead of a bear, it was poor lost Hans. His fur clothing covered with snow had, perhaps aided a little by their imagination, or their fears, completely deceived them—though, as the ice was very rough, and probably Hans used both hands and feet in climbing over the hummocks, the mistake was not so surprising. They were very glad they had not hurried their fire. The wind is now very strong, and the snow drifting. If Hans had not come in, he would have fared badly such a night as this.

"*Nov.* 13. The men are building a large snow-hut, for what they call a reserve. Peter is sick to-day. The rest all well.

"*Nov.* 15. Change in the weather. It is spring-tide, and the water rises all around our floe; it is beautifully clear, and the moonlight very bright. Our poor dogs are suffering; they got nothing at all to-day. Five have been shot altogether, leaving us only four; I regret it, but it can't be helped. We are now drifting very fast. The men are lining their new hut with canvas. Joe got on a fox-track yesterday, but did not come up with his game; and he also saw three seals, but was not able to secure either of them.

"*Nov.* 19. I am down sick with rheumatism, hardly able to hold a pencil. By the movement of the ice, I judge we are drifting to the southward very fast. The natives tell me that they saw two bear-tracks and five seal-holes; but they brought home nothing. I wish they had better fortune, for we need the fresh meat very much.

"*Nov.* 21. The last few days the weather has been clear and cold; but I have been confined to the hut with heavy cold and rheumatism; but, thank God, I am around again. It has been very difficult for the natives to hunt this month, except the few times the moon shone, on account of the darkness. Some days it was quite impracticable, there being absolutely no light; but to-day, thank God, they have brought in two seals. Without them we should have no fire, one boat being already cut up. We must go without fire or warm food if there are no seal caught. It will never do to touch the other boat; the time must come —if we live to see it—when the boat will be our only means of safety.

"We are living now on as little as the human frame can endure without succumbing; some tremble with weakness when they try to walk. Mr. Meyers suffers much from this cause; he was not well when he came on the ice, and the regimen here has not improved him. He lives with the men now; they are mostly Germans, and so is he, and the affinity of blood draws them together, I suppose. Since he has housed with the men, I have lived in the hut with Joe, Hannah, and Puney. Puney, poor child, is often hungry; indeed, all the children often cry with hunger. We give them all that it is safe to use. I can do no more, however sorry I may feel for them.

"The seals which Joe got to-day will help us very much. In our situation he is the 'best man,' for without him we should get little enough game, I fear. Hans is not so good, though he does well at times; and, as for the rest, they have had no experience. I am the worst off of all, for I have neither gun nor pistol of my own, and can only make a shot by borrowing of Joe. This is a disadvantage in other respects; the men know it; they are all armed, and I am not. After Captain Hall's death, for some reason unknown to me, arms were distributed among the men, perhaps to organize hunting-parties; but, at any rate, while I was looking after the ship's property, the men secured their guns and pistols. Joe has both a shot-gun and a pistol; but he didn't seem to care to give either up, and I will not force him to.

"The men have now moved into their large new hut, and I shall appropriate the other to store provisions. *The bread has disappeared very fast lately: more of this hereafter.* We have only eight bags left. God guide us; He is our only hope.

"We have about three hours' fair light yet on a clear day—like twilight in cloudy weather; but I scarcely know day from night. But, thanks be to God, we are all well, with keen appetites, though scarcely any thing to satisfy our hunger with. *For the first time since separating from the ship I have eaten enough;* but it was of raw, uncooked seal-meat—skin, hair, and all. For the last few days, being sick, I had eaten nothing—scarcely any thing for about a week; and I was so very weak on getting up I found I could hardly stand; and I needed this food very much to give me a little strength. I really need and should have more, to make up for the days I ate nothing; but beyond this one meal, shall not ask for or take it, but will subsist as well as I can on the regular allowance. But this one night I have eaten heartily of seal—yes, and drank its blood, and eaten its blubber, and it will give me strength, I hope. I need strength for many reasons besides my own use.

"We have discovered bear-tracks on our floe, but have not seen the bears. Our four remaining dogs are very thin and poor, and unless we get more food, they will either have to be killed or must starve. It is a great pity, for they would be very useful in bear-hunting.

"*Nov.* 22. Cloudy; wind from the north-east in the morning, shifting to the south-east toward evening. There are plenty of seals around, but it is too dark to shoot; we can't afford to waste ammunition on doubtful shots; besides, a false fire scares away the game, and does no good. Hans made out to get one seal to-day.

"Yesterday and to-day have been very fine; to-day clear and cold, with light north wind. The natives have been out hunting, but got nothing; water all frozen over, and very dark for shooting.

"At midday the stars are visible, even with the moon shining. Saw a fox to-day; he approached the hut, and could have been captured if the men had kept quiet. But they are under no discipline, and have been under none since Captain Hall's death.

"My situation is very unpleasant. I can only advise the men, and have no means of enforcing my authority. But if we live to get to Disco, there they will have to submit, or I shall leave them to shift for themselves. I will not live as I have lived here. But here I am forced to live for the present: there is no escape. It is not altogether their fault either; they were good men, but have been spoiled on board the *Polaris*. For the last year nearly they have been allowed to say, do, and take what they pleased. Such as they were, had they been under good discipline, and left on the ice like we are, I could have saved them; but I don't know how it will be now. But as to that, had there been any discipline we should all have been on the *Polaris* now. And then, too, there appears to be some influence at work upon them now. It is natural, no doubt, that they should put confidence in one of their own blood; but they will probably find out that 'all is not gold that glitters' before they get through this adventure.

"We begin to suffer much with the cold; when the body is ill-fed the cold seems to penetrate to the very marrow. The human system can not repel cold well without a certain quantity of fresh meat, and there is no meat better for the purpose than seal-meat, if we can get enough of it.

"*Nov.* 27. Yesterday and to-day have been very unpleasant; dark and cloudy, with a strong breeze from the south-west. The natives have not attempted hunting; it is entirely too dark to see to shoot. Joe has used the time well, however, in enlarging our hut. I prefer living with him, as both he and his wife, and even the child, can speak English, while in the men's hut I hear nothing but German, which I do not understand; and there are many other annoyances.

"Joe and Hannah have lived for years with and among civilized people. Their native names are *Ebierbing* and *Tookoolito;* they had traveled with Captain Hall on both his previous journeys, and are frequently referred to in his book on 'Arctic Researches.' They had also both been to England, and had been received and entertained by the Queen; they had also lived for some time in Groton, Connecticut. It was, therefore, possible for me to communicate my plans and wishes intelligibly to them, and they to express their ideas to me; while the most of the crew, either could not, or would not, speak any thing but German, with which language I was wholly unacquainted.

"Having to live all the time on such small rations keeps the subject of food constantly before one. It is one of the worst effects of excessively 'short allowance' that it causes the mind constantly to dwell on the matter of eating. While the stomach is gnawing, and its empty sides grinding together with hunger, it is almost impossible to fix the mind clearly, for any length of time, upon any thing else. The scenes that have passed before my eyes during the last weeks were, many of them, worthy of the best efforts of the most accomplished artist, and worthy of description by a poet's pen, but I have not the heart to enjoy or record them; for disgust at the mixed-up way in which I have to live overpowers every other sentiment.

"The spare wood is giving out, and the difficulty of cooking for so many in nothing but flat tin pans is very great; and besides, the lamps, which can only be used with blubber, there is nothing to cook over but a 'stove' made out of an old reflector; but it is something to get the food even thawed, if it can not be cooked.

"To-morrow will be Thanksgiving-day in the States. We shall keep it, too, in our way, thanking God that he has guided us so far in safety, and praying that he will continue to watch over us.

"*Thanksgiving-day, Nov.* 28. It is very dark, and the day comes in cloudy, with a strong breeze from the north-west. I can just see a faint streak of twilight to the south (11 A.M.). The cook has just turned out to prepare his Thanksgiving breakfast for the men — nine in number, including Mr. Meyers, who lives with them. He is cooking with the remains of the wood in the shallow tin pans, and, as it is not easy to cook in such utensils for so many, our company is divided into three messes. Hans cooks for himself and family, and Hannah for herself, Joe, Puney, and I. The natives use and prefer the lamp; it consumes more oil than we can well spare, but there is no help for it — we must have something warm to eat if it can be had.

"We saved the can of dried apples for Thanksgiving, or what was left of them. My breakfast consisted of a small meat-can full of chocolate — it was not a very delicate 'coffee-cup,' but I had used it before; two biscuits, of a size which takes ten to make a pound, with a few dried apples, eaten as they came out of the can. This was the 'thanksgiving' part of the breakfast. To satisfy my hunger — fierce hunger — I was compelled to finish with eating strips of frozen seals' entrails, and lastly seal-skin — hair and all — just warmed over the lamp, and frozen blubber; and frozen blubber tastes sweet to a man as hungry as I was. But I am thankful for what I do get — thankful that it is no worse. If we can only get enough of such food as this we can live, with the aid of our small stores, with economy, until April, and then we must rely on game.

"No doubt many of my friends who read this will exclaim, 'I would rather die than eat such stuff!' You think so, no doubt; but people can't die when they want to; and when one is in full life and vigor, and only suffering from hunger, he don't want to die. Neither would you.

"*Evening.* I have been thinking of home and family all day. I have been away many Thanksgivings before, but always with a sound keel under my feet, some clean, dry, decent clothes to put on, and without a thought of what I should have for dinner; for there was sure to be plenty, and good too. Never did I expect to spend a Thanksgiving without even a plank between me and the waters of Baffin Bay, and making my home with Esquimaux; but I have this to cheer me — that all my loved ones are in safety and comfort, if God has spared their lives; and as they do not know of my perilous situation, they will not have that to mar their enjoyment of the day. I hope they are well and happy. I wonder what they have had for dinner to-day. It is not so hard to guess: a fifteen or sixteen pound turkey, boiled ham, and chicken-pie, with all sorts of fresh and canned vegetables; and celery, with nice white bread; and tea, coffee, and chocolate; then there will be plum-pudding, and three or four kinds of pies, and cheese; and perhaps some good sweet cider — perhaps some currant or raspberry wine; and then there will be plenty of apples, and oranges, and nuts, and raisins; and if the children have been to Sunday-school in the morning, they will have their little treasures, besides all their home presents spread out too. How I wish I could look in upon them! I would not let them know I was here, if I could. How it would spoil their day!

"Well, I set down what I had for my Thanksgiving breakfast; I will give my bill of fare for my dinner also. For the four of us in this hut we had six biscuits, of the size above described; one pound of canned meat, one small can of corn, one small can of mock-turtle soup—each one-pound cans—making altogether a little over three pounds and a half, including the bread, for four persons; and this is an *extra* allowance, because it is 'thanksgiving.' Mixing all the above in one mess together, it was just warmed over the lamp, and our dinner was announced.

"The men had their cans of mock-turtle soup and corn, and whatever there was that was extra just the same; there are no officers' messes aboard the floe. It would have been pleasant and appropriate to have had some general religious service in recognition of our national Thanksgiving; but, perceiving that it could not be unanimous, I did not attempt it. The Germans appreciate Christmas, but are not familiar with our 'Thanksgiving.'

A PERILOUS SITUATION.

CHAPTER XVIII.

Can see the Land.—Hans's Hut.—Nearly dark: two Hours of Twilight.—Economizing Paper.—Northern Lights.—Lying still to save Food.—"All Hair and Tail."—Weighing out Rations by Ounces. Heavy Ice goes with the Current.—The Esquimaux afraid of Cannibalism.—Fox-trap.—Set a Seal-net.—Great Responsibility, but little Authority.—All well, but hungry.—The fear of Death starved and frozen out of me.—The shortest and darkest Day.—Christmas.

"*Nov.* 30. Yesterday it was cloudy, with a westerly wind; to-day it is also cloudy, but almost calm and comparatively bright, with a streak of twilight to the southward at noon. We can see the land, for our eyes have become accustomed to this kind of dim light, and partially adapted to it; as it comes on gradually, of course it does not appear so dark to us as it would to one suddenly dropped down on our floe from the latitude of New York. They would find it perhaps as dark as some of the *shut-up parlors* into which visitors are turned, to stumble about until they can find a seat, while the servant goes to announce them.

"It is a long time—nearly a month—since we lost sight of the land. All hopes of seeing the *Polaris* have also long ago vanished; but the hope of getting to the land is not entirely abandoned. I have been over the floe to the old house to-day after canvas. I called on the men for some of them to go with me. Four responded to the call—the steward, the cook, Peter, and Augustus. I wanted the canvas to line the hut of the native, Hans. He has worked late and early to make the men comfortable, and they have their hut comfortably lined, and the Esquimaux ought to be too, especially as the little children are there; and Hans's wife is continually working for the men, by mending and making for them. More of this hereafter.

"*Dec.* 2. Yesterday, the first day of winter, according to our almanacs—*our* winter commenced on the 12th day of August, when we, in the *Polaris*, were beset in the ice near Cape Frazier. To-day Hans, poor fellow, is sick, and can not hunt; but Joe has been out sealing, in spite of the increasing darkness, but he can find no water. We have a very little light from 11 A.M. until about 1 P.M.—two hours of glimmering light, so that we can just make out to walk over the uneven ice, and then total darkness is on us again. It must be still darker on the *Polaris*, if she is still afloat; for we are farther to the southward than she. If we keep drifting to the south, as I have no doubt we shall, because that is the set of the current, to the S.S.W., the light will increase in proportion to the rate at which we float.

"I do not write every day—*it would take too much paper.* I had some blank note-books in one of the ship's bags. On looking for them a few days ago, found they were all gone. Some of these men seize hold of any thing they can lay hands on and secrete it. But no wonder; they were taught that on board the *Polaris;* they saw so much of pilfering going on there. It would have demoralized worse men than these.

"I can scarcely get an order obeyed if I give one; and if I want any thing done, I try to do it myself, if it is within the compass of one man's strength. One thing, there is not much to do, and hence not much necessity of giving orders or subjecting myself to unwilling obedience.

"*Dec.* 6. Last night there was a fine display of northern lights; the first I have seen since we have been on the floe. There may have been others while we were all asleep. What first attracted my attention was a peculiar dark segment immediately over the horizon; and as I had often seen the aurora borealis spring from just such a beginning, I watched it closely, and soon intensely luminous streamers sprang from it, rising to a height of about thirty degrees; it was a bright and beautiful sight, but not so brilliant as many which I have seen. The wind was light, from the N.W., and I should judge the temperature to be ten or twelve below zero. I have no thermometer, and can not tell exactly.

"*Dec.* 7. Last night, being clear, Mr. Meyers was enabled to take an observation. He has, fortunately, some instruments—a sextant and ice-horizon, and also a star chart; and so he took the declination and right ascension of γ, Cassiopeæ. But he has no nautical almanac to correct his work by, so that he can only approximate our real latitude. He makes it 74° 4′ N. lat., 67° 53′ W. long.; but I do not think it is any thing like that. If we are as far south as that, we have drifted faster than I think we have.

"There is no change in our way of living. We lie still in our

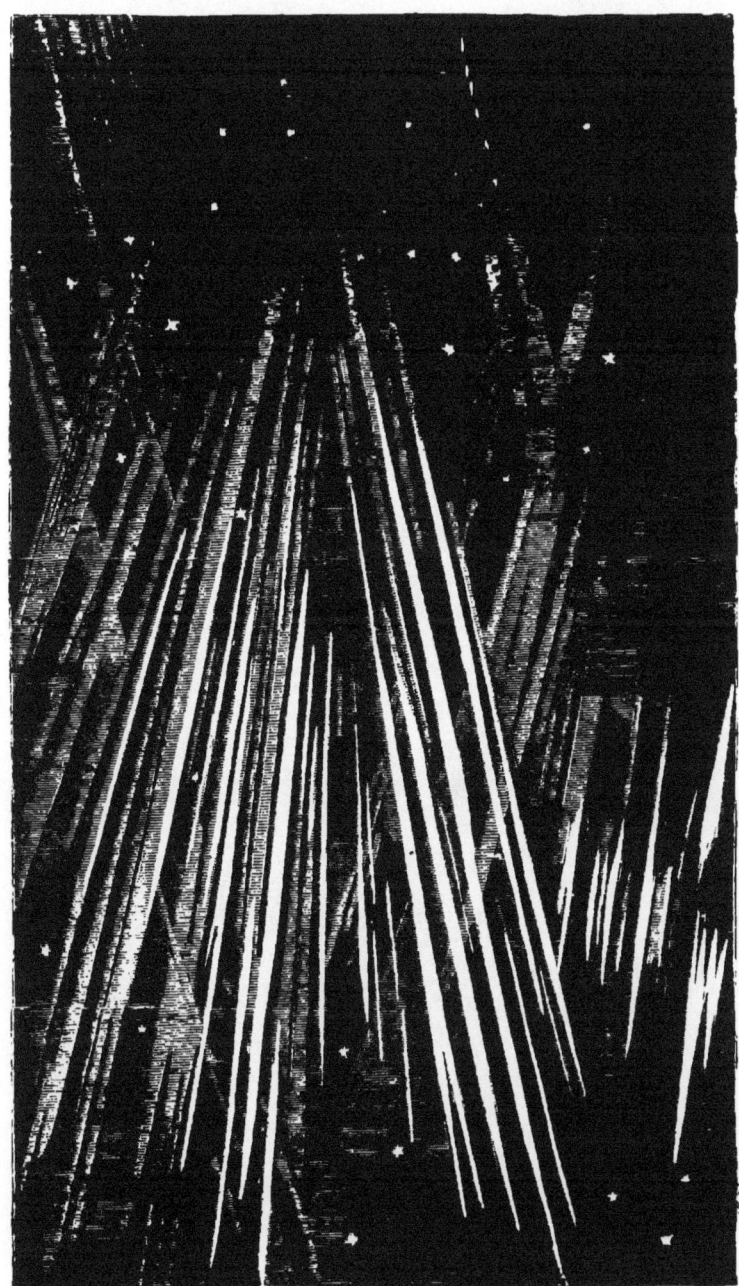

AN AURORA.

snow burrows much of the time, partly because there is nothing to do—it is now too dark to do any thing, if there was—but also because stirring round and exercising makes us hungry, and we can not afford to eat. The stiller we keep and the warmer, the less we can live on; and, moreover, my clothing is very thin and light, quite unfit for exposure in this cold. Being hard at work when the *Polaris* parted from us, and that event being so unexpected to me, I had not on even the usual amount of clothing worn in this climate. The natives have been endeavoring to hunt, but it is in vain.

"The darkness is on us; we must wait for light. The day before yesterday, one of the men—Bill, I think it was—shot a fox; it was a poor, thin creature, with hardly a pound of flesh on its bones—'all hair and tail,' as one of the men said; however, they ate what there was of it, and picked his bones clean.

"Our allowance is now divided out by *ounces*. For a day's rations, we now have six ounces of bread, eight of canned meat, two of ham, and can only allow half of this for the children. These ingredients are mixed with brackish water to season it, and warmed over the lamp or fire; and this is all we have, and it is more than we can really spare from our fast-decreasing store. While the darkness lasts we can not hope to get seals, and bears only come where seals are to be caught; so we need not look for them; and foxes are usually in the trail of bears. I was in hopes, when that poor, thin fox was caught the other day, that a bear might come along, but we have seen none.

"A few days ago it appeared as though we were nearing Cape York; but I am satisfied it is not so. We are surely going to the west side. Mr. Meyers thinks we are going to the east. He judges, I suppose, by the winds being mostly from the northwest; but the ice does not obey the winds—heavy ice, I mean, like this; the loose floating surface-ice often does. But if the currents have not changed their natural course, we *must* go to the S.S.W.

"It would not matter in the least what opinion was entertained as to the course of the floe, only that it makes the men uneasy; thinking that they are approaching the east coast, and nearing the latitude of Disco, where they know there is a large store of provisions left for the expedition. I am afraid they will start off, and endeavor to reach the land on that side; and if they do, it means *death* to some or all of them. None of us, after living on such short allowance for nearly two months, could hope to bear up under the fatigue of walking and dragging a boat-load of things over this rough hummocky ice. The thought has occurred to me whether they might not take the boat, load it with what provisions we have, and leave us and Hans's family without resource; but I will not harbor such thoughts without proof.

"One thing set me on this train of thinking: Joe, who has all along kept his gun and pistol, and did not seem willing even to lend me the latter, has voluntarily brought it to me. He says 'he don't like the look out of the men's eyes.' I know what he fears; *he thinks they will first kill and eat Hans and family, and then he knows Hannah's, Puney's, and his turn would be next.* God forbid that any of this company should be tempted to such a crime! However, I have the pistol now, and it will go hard with any one who harms even the smallest child on this God-made raft. Hannah seems much alarmed.

"Setting aside the crime of cannibalism—for if it is God's will that we should die by starvation, why, let us die like men, not like brutes, tearing each other to pieces—it would be the worst possible policy to kill the poor natives. They are our best, and I may say only, hunters; no white man can catch seal like an Esquimau, who has practiced it all his life. It would indeed be 'killing the goose which lays the golden egg.'

"*Dec.* 12. No change the last five days; the wind still westerly, and very cold—from 21° to 22° below zero. The cold is more piercing and penetrating, from the little heat in our systems, from lack of food. If we had enough to eat, we could do without fire; but no fire, and hardly enough to eat to keep from starving, puts us in such a condition that we can not resist the cold. Probably no one suffers more than I do in this respect, as I have less clothing than any of the others; but Mr. Meyers feels it badly too, for he is not well. The intensity of the cold has frozen over all the cracks, so that no seal can be found, and there is too little light to see them anyway.

"Hans is better. He had fixed up an ice-trap, and yesterday he caught a fox; it was a small white one. If there is any way to catch an animal, these Esquimaux will do it. Now he has made a hole in the ice, and set a seal-net, but, so far, without success. If a seal is not caught soon, we shall be without even a light in our hut; and as there is but a remnant of the boat left, all will have to eat their rations cold. As if, too, all these miserable circumstances were not enough to bear, the men begin to complain, at least the German portion, which is the majority. They do not seem to have self-control or the true courage of endurance. If there were any thing that could be done to relieve us from our uncomfortable position, God knows I would be the first to do it; but at present we are powerless to alter any thing for the better. If they had moved a little sharper, and been willing to abandon their traps the morning after we parted with the *Polaris*, they and we might all have been on board of her now.

"I have since understood that they had heard of the drift of the *Hansa* crew, and the gratuity of one thousand thalers donated by the Government to each man of that party, and that they thought if they should drift likewise they would get double pay from Congress. But little did they realize the difference in the circumstances! The *Hansa* party had ample time to get all they wanted from the vessel—provisions, clothing, fuel, and a house-frame. And then the climate on the east coast of Greenland is moderate in comparison to that of the west. If this did influence them, I fear they will realize, by months of suffering, the sad mistake they have made. However, they are organized now, and appear determined to control. They were masters of the *Polaris*, and want to be masters here. They go swaggering about with their pistols and rifles, presented to each of them after the death of Captain Hall.

"I see the necessity of being very careful, though I shall protect the natives at any cost; any disorder now would be ruinous. I must be wary as well as firm. Situated as we are, there must not be the beginning of quarreling. That would be fatal. They think the natives a burden, particularly Hans and his family, and they would gladly rid themselves of them. Then they think there would be fewer to consume the provisions, and if they moved toward the shore, there would not be the children to lug. With the return of light and game, I hope things will be better, if I can manage to keep all smooth till then. But I must say I never was so tried in my life.

"*Dec.* 16. No material change; still getting on in the old style; wind mostly westerly and very light. To-day there is a strong breeze from the same quarter. It is now noon, and we can see tolerably well to make our way over the jagged ice, but every once in a while some one falls; but it is partly from weakness. We are all what may be called well, yet some complain of pains in the stomach. No wonder!—six ounces of bread, and five ounces of meat per day in this climate! Few could stand it without pain. How long we shall be able to bear it I can not foresee; but we can not use more with any regard to our ultimate safety. If we can weather it till the end of March or the first of April, we can then rely on our guns; but game failing us then, we perish.

"*The fear of death has long ago been starved and frozen out of me;* but if I perish, I hope that some of this company will be saved to tell the truth of the doings on the *Polaris*. Those who have baffled and spoiled this expedition ought not to escape. They can not escape their God!

"*Dec.* 20. To-day seals have been perceived under the ice by Joe and Hans, but they could not get at them. It gives all courage, however, to know that there is life-giving food right under us, only we must wait a little longer for light, so that we can see to shoot them. To-morrow is the shortest day, and then in about three weeks we may hope to see the sun once more.

"*Dec.* 22. We have turned the darkest point of our tedious night, and it is cheering to think that the sun, instead of going away from, is coming toward us, though he is not yet visible. The shortest and darkest day has gone, and I am thankful. Friends at home are now preparing for Christmas, and so are we too. Out of our destitution we have still reserved something with which to keep in remembrance the blessed Christmas-time.

"*Dec.* 23, 24. Strong northerly winds. Both nights there was quite a brilliant aurora; it seems to come timely to lighten up our Christmas-eve. We shall have a slight addition to our rations to-morrow, and a slight change of diet too. All of our hams were used up about a month ago, except one; this we determined to save to celebrate our Christmas. It will be but a small portion for each, but it will be a change, and mark the day. It is not very cold—about zero.

"*Christmas-day!* All the civilized world rejoicing over the anniversary of our Saviour's birth—and well they may; but, though we are out of the civilized world, and in a world of ice, storms, cold, and threatening starvation, we are still trying to rejoice too. We know and feel that God has not forgotten us, that we are his children still, and that he watches over us here, as well as over those who dwell in safety in the cities and in secure country homes. He is trying us by a peculiar providence indeed, but he has not deserted us. We will praise his name forever.

"It is now 12 noon, and the twilight grows a little clearer. I have just finished breakfast. We breakfast late because we only have two meals a day, and the day is better so divided. My Christmas breakfast consisted of four ounces of bread, and two and a half ounces of pemmican warmed over the lamp. Some of the men call this 'soup,' and some call it 'tea.' This is a full ounce over the usual allowance of bread. Even that additional morsel of bread was a treat, and very welcome. Our Christmas dinner was gorgeous. We had each a small piece of frozen ham, two whole biscuits of hard bread, a few mouthfuls of dried apples, and also a few swallows of seals' blood!

"The last of the ham, the last of the apples, and the last of our present supply of seal's blood! So ends our Christmas feast!

PLACING STORES ON THE ICE.

CHAPTER XIX.

Taking account of Stock.—Hope lies to the South.—Eating Seal-skin.—Find it very tough.—How to divide a Seal *a la* Esquimau.—Give the Baby the Eyes.—Different Species of Seal.—New-year's Day, 1873.—Economizing our Lives away.—Just see the Western Shore.—"Plenty at Disco."—Thirty-six below Zero.—Clothing disappears.—A glorious Sound.—"Kyack! Kyack!"—Starvation postponed.—Thoroughly frightened.—Little Tobias sick.—Oh, for a sound-headed Man!—Four ounces for a Meal.—The Sun re-appears after an Absence of eighty-three Days.

"I HAVE been examining what we have left in the store-house, and which I have estimated must last us until April. We have nine cans of pemmican, and six bags of bread. At our present rate of allowance the pemmican will last two months, and the bread three. By that time I hope we shall be where we can get game.

"The natives are wonderfully persevering in hunting, but they can't catch any thing so long as there is no open water. I can just see land to-day, but it is forty or fifty miles off; it is the west coast, and latitude, I should think, about 72°. There is a bright streak of twilight over it. We are gaining fast now on the light; it travels toward us, and we are drifting toward it. I have often thought, while I have been on the floe, how different my feelings would have been if we had been drifting the other way, into the night and north, instead of away from it, as we are. That would have been cheerless, indeed—I may say absolutely hopeless. As it is there is some hope, if we can keep together harmoniously, and are able to get some game. I hope yet to land safely on the coast of Labrador, or, better yet, if we drift safe to the whaling-grounds, we may have the good fortune to be rescued by a whaler.

"I hope we will have the sun by the 20th of January. Our drift, I think, has been about six degrees in two months, and in a south-west direction. There is a strong breeze blowing.

"*Dec.* 26. Yesterday, toward evening, our strong breeze increased to a gale, and is still blowing with a heavy drift of snow; the wind is from the north-north-west. This is our first gale since the latter part of November. It is now two o'clock (afternoon), and the men have not yet turned out to get their pemmican tea. The Esquimaux are off, looking out for their traps. They have two seal-traps set in the ice, and two fox-traps.

"*Dec.* 27. The gale has moderated, but still a strong breeze; wind from the same point. Natives out as usual, looking for seal; and they found the ice broken in many places: they saw two seals, but could not get them. There is still so little light that they can only see to shoot plainly for about two hours in the middle of the day—an hour before and an hour after noon. To-day, even at 12 M., it is dark, being cloudy.

"We are fortunately all well, but very poor in flesh, and on attempting to do any kind of work find ourselves very weak. I think this is the secret of the men keeping so quiet. They are evidently uneasy, and from the talk which goes on in their hut, and which I sometimes hear of, they plan great things; but when they get outside and face the cold, and feel their weakness, they are glad to creep back again to their shelter and such safety and certainty as we have.

"We had, in our hut, saved a few pieces of dried seal-skin for repairing clothing, but Hannah has just cooked some pieces, and we are trying to make a meal of it. The natives have very strong teeth, and can go through almost any thing. I ate some of it, but it made my jaws ache to chew it, it was *so very tough.* We ate up all the refuse of the oil-lamp — tried-out blubber; in fact, we eat any thing we can get that the teeth can masticate, and which will aid in sustaining life until we can get seals to help out our allowance.

"*Dec.* 29. Strong breeze from the west yesterday, and to-day light breeze from east by south. Yesterday Joe and Hans were out sealing; Hans shot one seal yesterday, but lost him. It seemed very stupid, but I suppose he could not help it. If we were getting plenty we should not notice such an accident. Joe also shot a seal to-day, but as it floated away from him, he shouted out as loud as he could call for his kyack, and some of the men carried it over to him. He got in, and was fortunate enough to bag his game; and we have all dined on it this evening.

"When a seal is properly divided, there is but one way to do it. First the 'blanket' is taken off; that is, the skin, which includes the blubber—it is all 'one and inseparable' as it comes from the creature; then it is opened carefully, in such a way as to prevent the blood being lost: it is placed in such a position that the blood will run into the internal cavity; this is then carefully scooped out, and either saved for future use or passed round for each to drink a portion. The liver and heart are considered delicacies, and are divided as equally as may be, so that all get a piece. The brain, too, is a tidbit, and that is either reserved or divided. The eyes are given to the youngest child. Then the flesh is cut up into equal portions, according to the size of the company; with us it was weighed out. Sometimes the person who distributes it cuts it up as fairly as he can, and then, standing with his back to the pieces, another person calls out the names of the company in succession, and each receives his portion, without the distributer being able to show any favoritism. The entrails are usually scraped, and allowed to freeze, and are afterward eaten. The skins are usually saved by the natives for clothing, and also for many other domestic purposes, such as kyacks, oomiaks; the reins and harnesses for dog-sleds, tents; and, in fact, to almost every thing which is worn or used by the Esquimaux, the seal furnishes something. Even the membranous tissues of the body are sometimes stretched and dried, for the purpose of making semi-transparent windows to their huts.

"But we had only two uses for a seal; we ate the whole, except such portions of the blubber as had to be reserved for the lamps. This seal that Joe had shot was but a small one, and when eighteen hungry people had dined there was nothing left but the skin and entrails; we shall eat that too, but not to-night. The blubber derived from this little seal was almost invaluable to us for our lamps; and then its flesh saved so much out of our bread and pemmican; it seems also to put new strength into every body. The blubber will last us for three weeks to warm our food.

"The small Greenland seal (*Phoca vitulina*) is a very pretty creature in the water; its fur is a shiny white, beautifully spotted with obscure dark and black spots on the back and sides, the under part being white; its ordinary weight fifty or sixty pounds. These kinds of seal appear singly or with families, but do not go in shoals, as the 'springing seal' (*Phoca hispida*) does. These appear much more frolicsome than the others, and they play together in the open water very much like the porpoise, except that their movement is more like springing, and not so much rolling as the latter. The large 'hooded' or 'bearded' seal (*Phoca barbata*) is immensely larger than the others, and its movements more ponderous. When assailed it makes a revolution, and goes down head first, like a whale; while the small seal drops backward, tail down, the head disappearing last. The brown seal's fur of commerce is mostly taken from the seals found in Southern or Pacific waters, and is totally unlike the Greenland seal.

"*New-year's Day, Jan.* 1, 1873. 'Happy New-year!' How the sound, or, rather, the thought—for the sound I do not hear—reminds one of friends, and genial faces, and happy groups of young and old! We shall not make many 'New-year's calls' to-day; nor will the ladies of our party have any trouble in ciphering up their 'callers!' Some of our young men, it is true, may be troubled to keep their footing, but it will not be with overmuch wine and revelry. A happy New-year for all the world but us poor, cold, half-starved wretches. It is the coldest day we have had since we have been on the floe—29° below zero. If well fed and clothed, would think nothing of that, but as it is, this bitter wind searches one through and through, letting you know every weak and sore spot in the body.

"*New-year's dinner.* I have dined to-day on about *two feet of frozen entrails* and a little blubber; and I only wish we had plenty even of that, but we have not. In addition to the above, we had a little pemmican tea.

"The natives are out every day hunting, but as constantly fail to find any thing. There is no water, and therefore no seals.

"*Jan.* 3. Joe found three seal-holes to-day, but it was so intensely cold that he could not stay to watch them. Strong breeze, and 23° below zero. The west land can be just picked up in the twilight.

"The men seem quieted down. They have learned at last to use a lamp for cooking; in fact it was absolutely necessary that they should, after the boat was used up, with every bit of spare timber and even the boards they laid on. There was no other resource. Our one remaining boat *must* be preserved at all hazards.

"*Jan.* 7. Been very cold for the last four days; the ice very firm and compact in consequence, and in consequence of that, too, there is no water, and no more seals caught. We have good twilight now about six hours daily. Our pemmican is nearly gone—about eight cans left, and five bags of bread; with economy this could be made to last two months and a half. Well, we have economized our lives almost away already. The question is how we shall hold out.

"The west land can just be seen in the distance—about eighty miles off. We are now in the wide part of Baffin Bay. It still keeps very cold, ranging from 20° to 29° below zero.

"*Jan.* 9. Land still in sight, and just visible in the bright twilight. Ice continues firmly closed, and no water anywhere.

"The provisions are disappearing very fast—faster than the distribution of rations will account for: there must be some leak. I would set a watch if it was possible for us to stand outside in the cold nights, but in our reduced condition of flesh it would be fatal; and my own clothing is too wretchedly thin to think of it. The men are under no control. They keep talking of trying to start overland, and say they are determined to go next month; and if they do, the poor, deluded wretches will go to the east, misled by false advice, thinking they can reach Disco and 'plenty,' when we are all the time, and have been, except perhaps a little while at the start, drifting to the western shore; but they have full faith in their chosen adviser, and will not listen to reason. If they were only risking their own lives it would be bad enough; but, by divided counsels and divided action, the safety of the whole is imperiled, and especially as they are determined to take our only boat. Having burned up one, they now do not hesitate to appropriate the other. I am alone; no one to assist me in case of need. I think even Joe fancies he could reach the shore; but none of them could do it—no more than they could walk that distance over hot coals of fire. If they persist in taking the boat, we shall all have to go too, for what can we do on the ice, when it breaks up, without a boat? We must all inevitably perish. We have no sled, as they have burned the old one and all the pieces of plank out of which one might be made; so we shall have to drag the loaded boat, with the children and women in, besides provisions, over this rough, hummocky ice—and we are all too weak to stand that kind of work now, at least most of us: there are a few strong ones, who have kept themselves so by pilfering the food from our store-house. They are very troublesome men. Much as it goes against me, I must try and conciliate them, and turn them from their purpose. There is some little time yet to operate; for they dare not start in January.

"We are now in lat. 72°, and about in the middle of the strait—Davis Strait. There is more chance of open water on the Green-

CAPTAIN TYSON IN HIS ARCTIC COSTUME.

land or eastern side, but little or no chance of our drifting toward it, either this month or next. To the westward all looks solid ice, so compact there is no chance for a single seal. Could we only have a good strong southerly gale to break up the ice, there would be some hope. Fresh food is what we want, and we can not get it with this condition of the temperature—*thirty-six below zero at noon.* How would those poor, foolish men stand the night without their huts, and exhausted with travel and hauling the boat? But I have no one to co-operate with me. All in the men's hut are Germans but two—Herron, who is English, and the cook, colored.

"*Jan.* 12. The last three days it has been very cold; thermometer standing from 35° to 37° below zero. To-day I thought of looking for some little clothing—a few shirts and drawers, which I had in a bag; there were also a pair of pants, and a vest and several pairs of stockings. This bag was thrown over with the other things on the ice almost by accident, and I had saved them thus far, thinking, whenever we took to traveling over the ice, I should need them still more than while I had the shelter of the hut. But they are not to be found; bag and all is gone. All my little additional store, which I had relied on for traveling, has been stolen. It was very little, but I needed it so much, being but half clad at the best.

"I have thus far wintered without either coat or pants, having only short breeches. I was saving the pants to walk in. What clothing I had on me when we parted from the *Polaris* I have still, not having had them off my back, realizing how much more I should need the change when traveling; and now, when they have made up their minds to travel, without consulting me, they have robbed me of every thing.

"There is no alteration in the ice, and as there has been so much provision stolen we may be compelled to risk the attempt to get to the land in search of water and game. I do not suppose we can hold out; but I fear it will have to be tried if no seal are caught, which at present there is no prospect of; and if we fail, we simply perish. But as things are going now, we shall perish here if we stay. *It will be a struggle for life.*

"*Jan.* 15. Very cold the last few days; day before yesterday the glass (Mr. Meyers's thermometer) stood at 40° below zero. Yesterday it was cloudy and thick, and toward evening a strong gale sprung up from the westward, and is now (noon) blowing very heavy, with a thick snow-drift. Hence we are compelled to keep in our snow burrows. The weather, however, is not so cold; it has moderated considerably since the gale came on. Yesterday it was only $-14°$, and to-day it has ranged from 14° to

−17°. I am greatly in hopes that this gale will open the ice, so that we can get a few seals. We have been very saving of our oil, so we have a little left, enough, perhaps, to warm our food for two or three days.

"*Jan.* 16. The gale has abated; it is calm this morning, but quite thick; the change of weather has, I trust, been favorable for us. The Esquimaux went off early looking for seals, which I hope in God they may find.

"I hear a pleasant sound, because it is a promising one for water; the ice is pushing and grinding, which will surely open cracks. It seems strange to think of watching and waiting with pleasure for your foundations to break beneath you; but such is the case. In our circumstances food is what we most want; with enough of seal-meat we can face all other sorts of danger, but with empty stomachs we are ill prepared to meet additional disaster.

"11.30 A.M. A glorious sound — a life-inspiring shout! I heard the natives calling for their 'kyack!' That means they have found water, and water means seals. I called to the men to help get the kyack; they had not yet turned out, and it seemed to me a long time before there was any response to my call. But I was so impatient that even moments seemed long. We had to carry the kyack about a mile. Found the natives had shot a seal, which we got, and with which we returned in triumph.

"It seems as though God lets one get just to the verge of despair, and then sends some mitigating circumstance to relieve the gloom. But I have never quite despaired yet, and don't mean to while life remains. This seal appears to have come just in time to turn the men from their purpose of traveling. If they can get enough to eat they will be content.

"I ordered the seal to be taken into Joe's hut. As he did the most toward getting food, I thought this was right. One of the men, however, took upon himself to take it into their hut. What with the evil counsel of this man and three or four others, I sometimes fear it will be impossible to save this party of disobedient and lawless men. They have divided the seal to suit themselves, and I hope they are now satisfied; but it does seem hard on the natives, who have hunted day after day, in cold and storm, while these men lay idle on their backs, or sat playing cards in the shelter of their huts, mainly built by these same natives whom they thus wrong. They are anxious to get to Disco, and would peril their lives for the rum which they think they could get there. I should be as glad as they to get to some civilized station, if it was only to get clean clothes; but reason tells me we could not accomplish that journey.

"Our dogs—we have two left—came in somewhat disabled; they appear to have had a skirmish with a bear. A bear would not be a bad addition to our impoverished larder.

"*Jan.* 17. There is a strong breeze from the north; quite cold, with a slight snow-drift. The natives are out as usual, hunting for seal. They only got a small portion of the meat and a little blubber of the one last caught, the men keeping an undue proportion for themselves. This way of managing discourages the natives very much; they labor, and see others consuming the fruits of it. But they dare not say much, for they are afraid of their lives; but I know how they feel, for they have less reserve with me. There does not appear much difference among the men, excepting the English steward, Herron—he I believe to be more conscientious.

"These men brought themselves into this scrape last fall by not obeying orders, when I endeavored to get back to the ship. I have since learned that some of them thought to get notoriety by drifting down on the ice, and persuaded themselves that they would get double pay, which accounts to me now for their inertness then, and their not having the rudder shipped, and only the three oars in the boat; but they find drifting on an ice-floe not so pleasant. The provisions not holding out as they supposed, they are now *thoroughly frightened.*

"I expect to get within forty or fifty miles of the land next month, and hope to save most of the party. Hans's little boy, about six years old, is sick now, but all the rest are well. But to travel one hundred or one hundred and fifty miles, which we must do to reach a settlement, dragging the boat on her bottom, with sleeping-gear, guns, ammunition, provisions, and the children, who are too young to walk, but must be carried over the rough ice—it simply can not be done: it is impossible. In a few days we should have no boat; her bottom would be torn and destroyed. Yet the men appear to hold their determination to unshelter the women and children in the month of February.

If I had even one sound-headed man to assist me, I might possibly prevent them; but situated as I am—powerless against nine —except so far as words have power; and just now they seem insensible to reason.

"There has been, I suspect, an error of sixty or seventy miles in Mr. Meyers's brain as to latitude from the start. It could not have been, I think, Northumberland Island, after all, from which we started. Yet how he could be so far out I don't see; but he has been wrong all the way as to our being near the Greenland coast. But then, as he says, he has made 'observations;' so that ends the matter—for the present. There is probably some fault in the instruments.

"*Evening.* The Esquimaux have returned; they found water and saw some seals through the cracks, but did not get any. It was so very cold standing watching for a chance to shoot, that even they could not bear it, and returned. The glass marks 38° below zero, and with a strong wind blowing, the cold is very piercing.

"It is remarkable how soon men become accustomed to hunger, dirt, and cold. I do not suffer so much as I did with hunger, though I have even less to eat than at first; dirt I have become in a measure accustomed to; and the cold would not conquer me had I a fair allowance of clothing. But as it is, I suffer very much from this cause.

"We are living on less than twelve ounces of food daily; that, if divided in three meals, would be but four ounces to a meal, which is not enough to furnish heat or enable the system to resist this Arctic cold. Joe and Hans say that they have very often suffered before for the want of food, but they have never before been obliged to endure any thing like their present experience. Considering that they are out of the huts so much more than the rest, walking and hunting around, they really ought to have a larger allowance of the food. I would gladly give it them, but it would cause open mutiny among the men; and such harmony as *can* be preserved I am bound to maintain, for the good of all. Notwithstanding all my discomforts, my dark and dirty shelter, my bed of wet and musty musk-ox skins, fireless and cheerless and hungry, without one companion who appreciates the situation, I shall be well content, and even happy, if I can keep this party—worthy and worthless—all together without loss of life until April, when I hope for deliverance.

"*Jan.* 18. Blowing strong from the north-west; thermometer —28°. Esquimaux sealing. We can see a few holes of water in a northerly direction. If the ice is propelled by the wind we should be making considerable drift to the south-east, but the current will prove the stronger, I think, and take us the other way; but the wind may possibly prevail, for we lost sight of the west coast several days ago. If we *are* drifting with the wind, I shall hope to raise the east coast as the sun comes to the horizon. I think he will appear about the 24th—six days more. It will be a blessed sight.

"How our two dogs live I know not. A few days ago Joe discovered where one of them had been off hunting on his own account, and had evidently encountered two bears, indicated by the appearance of the ice, and held them at bay for some time. One of the bears must have hit him, for he came bleeding to the hut. The wound is but a slight scratch, however, and will soon heal.

"*Jan.* 19, A.M. Fair, with light, variable winds. Joe and Hans hunting. Yesterday they found water, and saw a number of seals; but it was blowing heavy, and was very cold. Joe says he tried to shoot, but that he shook so with the cold that he could not hold his gun steady, and that his fingers were so numb that he could not feel the trigger of his gun, and so the seal escaped. The Esquimaux work hard for themselves and for us; but to-day there is little prospect that they can find water. The wind moderated in the night, but it was so cold that all the holes froze up.

"*Afternoon.* Since writing the above, at 9 A.M., the great event has occurred: the sun has re-appeared, after an absence of *eighty-three days*. The very sight gives happiness such as those can not know who see his cheering beams every day. The sun means more than light to us: it means better hunting, better health, relief from despondency; it means *hope* in every sense, for his beams will lead us, God willing, in the path of final rescue. He has come this time earlier than I expected. We must have drifted faster than we have realized. Last year, when we were on board the *Polaris*, the sun was absent one hundred and thirty-five days; but that was in our winter-quarters, in lat. 81° 38′ N.; and we are now so much farther south that the sun comes to us earlier. The sun remained above the horizon about two hours, dipping a little after one o'clock. Mr. Meyers reports, having taken an observation of Polaris (the 'north star'), and of γ (Cassiopeia) at their lower culmination, that our latitude is about 70° 1′ 40″ N., and our longitude 60° 0′ 36″ W. We should probably have seen the sun even yesterday, had it not been for some intervening icebergs of great height, which limit our view in the direction in which he appeared. The sun will now continue to cheer us, and God in his goodness has sent us a seal to-day, to warm and lighten our dreary hut.

GOING THROUGH AN ICEBERG.

CHAPTER XX.

Belated Joe.—Wrong Calculations.—Drift past Disco.—Beauty of the Northern Constellations.—Hans unreliable.—"Where Rum and Tobacco grow."—Forty below Zero.—An impolite Visitor.—One hundred and third Day on the Ice.—Perseverance of the Natives in hunting.—Hans loses a good Dog.—Beautiful Aurora.—The Mercury freezes.—Too cold for the Natives to hunt.—A little Blubber left.—Trust in Providence.—Effects of Refraction.—Relieving Parties on the Ice.—Our Lunch, Seal-skin with the Hair on.—A natural Death.—One hundred and seven Days without seeing printed Words.

"JOE shot two seals to-day, about five miles from the hut, where he found a hole of water, but was only able to land one; the young ice carried the other away. Encouraged, I presume, by the appearance of the sun, he staid out later than usual, and it was very dark—perhaps seemed all the darker from contrast with our two hours of sunlight. I had a light of burning blubber set outside for him to guide him to the hut; for it is very easy to lose one's self on the pack-ice. Hans has already done so once, and it was considerable trouble to find him.

"Joe has got in all right, and I have had a feast of seal-skin—hair and all; also a piece of 'lights,' and my share of pemmican tea. And now to bed, such as it is.

"*Jan.* 20. The natives off before there was any light, sealing. It is now 9 A.M., and I can see very well. What a satisfaction it is to know that we shall now have the sun every day, for a few hours at least, unless it storms. To-day it is fair, with a light north wind.

"11 A.M. The sun is here, and right welcome. Last year, on board the *Polaris*, I was the first to see and greet him after his long absence. This year I was also the first to salute him on his return. I was outside, banking up our snow-hut, and was alone; the men had not turned out yet, not knowing of the pleasure in store for them. I did not expect to see sunlight for four or five days, and was, therefore, almost as much surprised as delighted. We must have drifted rapidly during the last two north-west gales.

"Many people complain at home if they have but one week

of dull, sunless, or rainy weather; how do they imagine they would like to spend three months without a sight of the sun, and then only see him for an hour or two in the day for nearly as long a time preceding his disappearance and following his return.

"I thought yesterday I could see the west coast, but am not quite sure. Mr. Meyers, who is the fountain of all knowledge for his German brethren, places us within a few miles of the land, and that on the east coast. His inexperience in these waters, and the inexactness of the instruments, or his handling of them, has made great trouble in misleading the men, and inspiring them with a confidence of what they can do, which has no basis in fact. When Meyers says 'we are within a few miles of the east coast,' of course they believe him; and some of them think that if they should start now they could get to the coast of Greenland in two days. They little know the labor before them; they would get to their deaths—that's where they'd get.

"I expect we shall drift by, if we are not already passed, Disco; but if the weather permits, I hope we may be able to make Holsteinborg in the boat some time in March, if they will have patience till then. It is not safe to start either this month or next—not with these men. They have been housed all winter, and when in their hut are valiant and brave, and talk of the great things they will do; but let them get out in the cold for a short time and the pluck is all gone. This is the explanation, no doubt, of why they have not as yet attempted to carry out the plan they have so often discussed.

"*Jan.* 21. Clear and cold. I stopped outside as long as I could stand it to-night, admiring the beauty of the stars. The northern constellations seem to me more brilliant here than I have ever noticed them at home. Ursus Major and Minor—if I remember right, these regions are named for the Northern Bear—αρκτος, Orion, Andromeda, Cassiopeia, the Pleiades, and Jupiter, so bright—part of Draco too. What a splendid night it would be for telescopic observations! The air is so clear and pure, there is neither cloud nor fog, nor any visible exhalations from this icy land, or frozen sea, to mar the crystal clearness of the atmosphere; but the cold pinches, and I had to leave the stellar beauties above me, and crawl into my dirty burrow and horrid ox-skins, to keep from freezing. The glass to-day has showed from 35° to —38°.

"The Esquimaux saw seals yesterday, but could not shoot them, or get at them, on account of the young ice. They saw also a great many dovekies, but they had no shot with them to kill small birds. The dovekies are a good sign; with proper shot, we may soon be able to diversify our fare with these birds. I have seen no land to-day either east or west, so the prospect of an early release looks doubtful. If all would be content with the regular allowance, our pemmican and bread would still last two months. Then we must shoot game or starve.

"*Jan.* 22. Yesterday the sun appeared at the horizon at 10.45 A.M. Joe saw two bears, and could have shot them if it had not been for Hans. This Hans acts like a fool sometimes. He is the same Hans who deserted Dr. Kane, and the same who was the cause of Dr. Hayes losing two good men on his expedition. He played the 'pious Moravian' on the good-hearted Kane, and Hayes could not bring the conviction home to him of what he suspected; but I never could read the account of the death of Sontag, and the profit to his own family which Hans made out of it, without feeling a little shaky about him. But he has worked well for us, and is now older, and ballasted with a wife and four children, which may add to his reliability.

"Meyers now states to his small constituency that they are only thirty-eight miles from Disco; so they renew the project of starting eastward as soon as the cold moderates a little, which they expect it will do when the sun is ten or fifteen days high. They have no notion of passing Disco, where they say 'rum and tobacco grows.' No other place will suit them. There they expect to take what they please out of the stores left by the *Congress* for this expedition. They say 'it is all paid for,' and as much theirs as any body's.

"One or two gales from the north or north-west, and we shall certainly be drifted past Disco; for I can see no land on either side, and therefore judge we continue in the middle of the strait, and not far from 70° N.

"The Esquimaux took the two dogs with them to-day, so that if they should have the luck to see another bear they can make sure of him. They started at daylight. It is very cold, —40° *below zero* at 10.15 A.M. The sun makes his appearance a little earlier every day, and continues in sight a little later. Joe is not very well. I hope he will not get down sick, for we depend greatly upon him. Though such a little fellow, he is 'a mighty hunter' in his way.

"*Jan.* 23. Clear; light west wind. No success yesterday with the hunters, but their motto is *nil desperandum.* Off again at daylight this morning; but, from the still cold, no wind to ruffle

HANS, WIFE, AUGUSTINA, AND TOBIAS.

the surface, I fear they will find no water. It is now, 9 A.M., quite light to the east and south-east; but I see no land anywhere, though the *man of wisdom* locates us within a few miles of Disco. This same party caused Hall considerable trouble in his time on board the ship, and he seems bound to do the same for me. Hall put him down, for he had the power; but after that unfortunate death the foreign element had the real control, though the nominal was in other hands. I do not know that he wishes to make me trouble, but his illusions have that effect on the men.

"The provisions are going fast. I know they are stolen, but can not stop it without shedding blood; and I shall avoid that, unless to prevent a crime that I suspect has been more than once contemplated. I hope to linger on until March, when open water and game may be looked for with greater prospect of success.

"4 P.M. Joe has just returned, bringing a small seal. He had found no water; but after watching at a 'blow-hole' for a long time, this fellow came and put up his snout to breathe, and Joe was fortunate enough to spear him. He is now divided up, and I have eaten a little of the raw meat. A seal of this size, divided into eighteen parts, gives only a small piece to each; but, small as it is, we are very glad to get it. This is fine weather, though so cold; thermometer $-34°$. Yet one could enjoy a walk in such an atmosphere if well fed and clothed.

"I was thinking the other evening how strange it would sound to hear a good hearty laugh; but I think there never was a party so destitute of every element of merriment as this. I can not remember ever having seen even a smile on the countenance of any one on this floe, except when Herron came out of his hut and saw the sun shining for the first time. Well, there is little enough to be merry over; but yet, if there had been a more congruous company and less disaffection, it need hardly have been quite so dismal.

"I have just breakfasted on a small piece of seal-meat, one biscuit, and a small pot of seal's-blood soup. No one can tell, except by experience, how much heat this seal's blood furnishes to the body. I am not surprised at the well-fed Esquimaux enduring cold so well.

"I am now compelled to record an event that occurred last evening. Disgraceful though it be, it is part of this story, and must go in. It will also show the animus of some of the men, and is a specimen of what I have had to endure from them. Robert—I forget his other name—entirely unprovoked, entered my hut, and commenced to abuse me in the most disgusting language, even threatening personal violence; but perceiving, though I said but a few words, that I was entirely willing to afford him every facility for trying his skill in that line of business, he did not attempt to put his threat in execution; and finding he could not provoke me to assault him, he shortly subsided and left. I suppose the foolish fellow had probably been boasting of what he could do, and the others had set him on by 'daring him' to do it. However, he walked off feeling a good deal smaller, I think, than when he came in. He came back soon after and offered an apology—such an apology as a man of his character can offer. If he had stood alone, the incident would have had less significance; but it was evident that he had his backers. Meyers, of the Scientific Corps, is their chief counselor; but whether he was knowing to this, or had any hand in it, I can not say. I hope not, for his own credit's sake.

"I know not how this business will end; but, unless there is some change, I fear in a disastrous manner. They are like so many willful children—all wanting to do as they please, and none of them knowing what to do. If four, or perhaps five, of these men were out of the road, the others, I think, would do well enough.

"Since writing the above, the weather has come on thick; light southerly wind, with snow, and temperature moderating.

"4 P.M. The Esquimaux have returned from their day's hunt, bringing a fine seal. He is considerably larger than the one caught yesterday, and will furnish us a fine meal; and, with full stomachs, I hope the men will find themselves in a better frame of mind.

"I have just dined on a small piece of liver—raw, of course, about one yard of seal's entrails, and my pemmican tea, with a little blood and blubber for dessert. And now I shall smoke my pipe—for I have that comfort yet—and then there is nothing for it but to kick my heels together till I get a little warmth in my feet, and then to bed.

"Had I more paper I might write more; but I have to be very saving, and just jot down the events of the day and my thoughts in a sort of short-hand of my own, to be written out, possibly, for my family hereafter, if I ever get where there are stationers and pen and ink. All the writing I do here is with a lead-pencil, which I fortunately had in my pocket. Whether I shall be able even to read it myself, if I ever get ashore, is somewhat doubtful.

"*Jan.* 25. I should like very much to accompany the natives in their hunting excursions, but I can not, for the want of clothing. Like Miss Flora M'Flimsy, 'I have nothing to wear.' I may say that I am almost without clothing—at least, anyway suitable to be exposed all day to a temperature of 30° or 40° below zero. Joe and Hans have deer and dog skin clothing, and even they complain of the cold; and they have a change too, which I have not. They can take theirs off and dry them over the lamp; but I have not been able to change mine for nearly three months. It sickens me to think of it, saturated as it is with all the vile odors of this hut. I have to sleep in them as well; for I should freeze to these wretched skins under which I sleep if I did not.

"To-day is our one hundred and third day on the ice—the most wretched of my life. The monotony is fearfully wearisome; if I could get out and exercise, it would help to relieve the tedium. But, while this severe cold lasts, it is not to be thought of. There has been more grumbling to-day, but I can get no clear idea of the cause. Idleness is probably the root of much foolish talk that goes on; and there is neither authority, nor any object or motive, to induce them to do any thing. But if I was clothed as well as they are, I would at least go out and assist the two natives in bringing home the game, if I could not capture any—which any one could do, except probably spearing the seal, which requires long practice and great skill.

"The two Esquimaux have hunted nobly all the winter, through the darkness, cold, and storm. They saw the necessity of it as well as I. They knew the men could not live without fresh meat, or without fire to warm their food and melt ice for water to drink; so they have worked hard. Joe has been of great service—Hans not. He is no great hunter, and, in fact, is a little foolish; but he has traveled early and late, and has done what his natural abilities permitted. When they catch a seal, I now take it in Joe's hut and have it fairly divided. At one time the men saw fit to lug them off into their hut, and undertook to cut them up. But skinning a frozen seal is not pleasant work, and they soon got tired of it, and are now willing to have it done for them. The two last seals Joe brought into his hut, where I live also. But I took the precaution to ask two of the men to come in and see that it was equally divided, but they would not come. So I told the natives to give them half of the meat, half of the blubber, and half of the skin, and the other half was divided between Hans's family and ours.

"Now it has sometimes happened that when the Esquimaux have been tramping about for hours on the hunt for seals, and at last get one, they are by that time very hungry; and as, when they bring it in to the party, they know they will get no more than those who have been at home all day, they sometimes open the seal, and eat the entrails, kidneys, and heart, and perhaps a piece of the liver; and who could blame them? They must do it to keep life in them. They could not endure to hunt every day without something more, occasionally, than our rations. Yet the men complain of this, and say they do not get their share: so unreasonable and unreasoning are they, or so selfish. When it is caught and skinned for them, they ought to be willing to give the hunters a generous portion.

"It is a beautiful day—no wind; and, on going out, I thought it much warmer than it really was. The cold soon began to penetrate. The glass tells 40° below zero. The natives returned about six o'clock. They have been a long distance to-day, and discovered signs of water to the eastward, and tried to reach it; but it was too far off, and therefore failed. They had no success in getting any game.

"*Jan.* 26. Fair; light north-west wind, and very cold. Joe and Hans have started off early, to try and reach the water they saw to the eastward yesterday, if they can perceive any evidence of its being there to-day.

"I fear we have lost our best bear-dog. Hans, the simpleton— for he seems little better—had the dog out with him yesterday; and instead of keeping his harness on, to have control over his movements, and also the means of bringing him back, when quite a long distance away, took the harness off and let the dog go; and he has not yet found his way back to us. I fear he is lost. Joe had the other dog with him, hoping they might meet a bear. The dogs are very useful in bringing them to bay, giving the hunter time to make aim and shoot. It is now near noon, and the breeze freshens. The thermometer is −42°, the coldest day yet we have had on the floe.

"Last night, at midnight, there was a brilliant aurora. No doubt many interesting natural phenomena occur without our observing them; for we are too wretched, and I am unable, for want of clothing, to stay out long at a time, and I have no means of taking measurements or of recording observations which would have any scientific value. If I had, I would try and endure the cold sufficiently long to make them. But as I came out of our hut at just about 12, low meridian, the heavens appeared to be all on fire; from the south-west to the north-east, from the horizon to the zenith, it was shooting here and there, like flames of fire in a strong wind. The light was at one time almost overpowering. I wish there had been some artist present capable of representing it, or at least who could have given some faint idea of the scene, for which words are totally inadequate. I dare say we miss many such sights, and also others of meteors and shooting-stars; but if this life should last much longer we shall forget that we have brains, and remember only that we have stomachs.

"3 P.M. Joe has returned, bringing a fine large seal. We already have our pemmican tea, made over the lamp; so we thankfully divide the seal, which is such a welcome addition to our meal, and eat a little of the raw meat, and a few mouthfuls of blubber with it, and then have a smoke. But that luxury will not last long; I am on my last plug of tobacco to-day. The *mercury is frozen*, so we know not how cold it is.

"*Jan.* 27. The day-break comes to us now at about 8 A.M. on a clear day; and, that all the daylight may be used, we get our breakfast much earlier than we did. It makes the day seem very long to us who have nothing useful with which to fill up our time, as there are very long intervals between our meals. We do not get game enough to allow ourselves three meals a day. Perhaps that happy time may come, but it is not here yet. The men breakfast about eleven. They do not go out much, it is so very cold. The moon changes to-day or did last night, and there are now full tides; so I hope the ice will open, which will enhance the probability of our getting a supply of seal-meat. I had my seal's-blood soup, a bit of the meat, and one-tenth of a pound of bread. The mercury continues frozen.

"1 P.M. The Esquimaux have returned, the severe weather proving too much for them, inured to it from childhood though they be. If they could have enough of the seal-meat to keep up a proper circulation of the blood, quickening its course through the veins, for which this kind of food is remarkable, they would have gone on, but possibly without success, as the water is of course all frozen over, and the chance of spearing a seal over its breathing-hole seems slight indeed. We have, I am thankful to say, quite a little store of blubber, which will last us, I hope, until the weather moderates, and also some seal-meat left. And I believe God is watching over us, unworthy though we be, and that he will guide us into safety, and where there is abundance. His providence has evidently been over us so far; for when we have been reduced to the greatest extremity, and thought we could endure no longer, then he has sent a seal to give us fresh strength and hope. So I will trust Him for the future who has preserved us so far on this perilous journey.

"*Jan.* 28. Fair; light wind from the south-west. Joe and Hans off again this morning hunting for meat to feed the hungry. Very cold still; —40°.

"I do not see my way clear yet. Can see no land either to the east or west, so we must be far from both shores, and are probably near the middle of the strait, with a slight set to the west. We can not be near the east coast, that is certain, for they have not so low a temperature there in this latitude. They catch whales off the coast there in February, ordinarily at Holsteinborg, and sometimes even at Disco. Yet the 'German Count,' as the men begin to call Mr. Meyers in jest, makes his countrymen believe that we are near to the east shore.

"What convinces me that we are a long way from Disco, which I know so well, is, that Disco is a very high rocky island, which, if we were near it, could certainly be seen. I have been there many times, and know all the coast south of it well. Disco can easily be seen on a clear day eighty miles distant, and I have seen it when one hundred miles off, raised by refraction—not an uncommon phenomenon on the Greenland coast.

"If Meyers had been left on board the *Polaris*, these foreigners would probably have behaved better, for then they would not have had any one to mislead them about our position. His influence is naturally considerable over them, because they think he is educated, and ought to know; and being also their countryman, they probably fancy he takes more interest in their welfare; just as if it was not as much my interest to get to dry land as theirs! But I have sailed these seas too often to be much deceived about our course.

"I know not whether I can keep these men quiet until the

temperature rises. Perhaps it may moderate in March, and then they may yet be saved; but, should they start for the shore in February, they are lost. The sun has not yet much influence. They will find no water to drink, have but little to eat, must sleep unprotected except by their wet ox-skins, if they have the strength even to drag them along; in fact, they must perish. But if they can be induced to hold on until the season is further advanced, many cracks will be found in the ice, and some of them may lead us near the coast, or at least to open water; and in these cracks we shall find plenty of seals, and on them we can live till it is a suitable time to attempt reaching the land. At our present rate of drift, we may even be picked up by some whaler.

"I have relieved parties on the ice. They had not drifted so long, to be sure, nor come so far, nor so many of them; they were all men, too—not a boat-load of women and children—but they were far away from their ships, hungry, and destitute. There were some runaways from the *Ansel Gibbs*, and also another party —I forget the circumstances now—from the brig *Alert*. I have also relieved Captain Hall two or three times on his former voyages; so I hope Providence may send *us* a rescue before it is too late.

"It is now, past 3 P.M., quite light. The mercury is frozen again. It is extremely cold. Joe and Hans have not returned yet. The men are cooking, or, rather, trying to warm, some seal-skin, which serves us all to-day for lunch. We eat it *hair on*, as there is not sufficient heat to scald it off. Boiling water will take it off, but we can't get that. It is very tough. My jaws and head too ache with the exertion made to masticate it. The dogs have the advantage of us there; they will bolt down long strips of it, if they are so well off as to get it, without apparently any chewing at all. They will eat any thing but stone or metal, and make very short work of their harness, or any thing of that kind, which is left in their way.

"6 P.M. The natives have returned; have had no success, and we have now lost our only dog. Joe had him with him to-day. On returning, the poor animal was taken sick and died. I fed him last night on what I was eating myself, seal-skin and pretty well-picked bones; it may be that the bones caused his death, as they swallow such large pieces, or it may be something has happened to him that I do not know of. Well, it is the first and only natural death that has occurred, and that, surely, is wonderful; but it is astonishing what men can endure. It must be that the *hope* keeps us alive, and the poor beasts have not that to sustain them. They feel all their present misery, and can not anticipate relief. It will be a very difficult matter to capture a bear now, without a single dog.

"*Jan.* 29. Foggy, with light east wind. The Esquimaux off, as usual, on the hunt. They do not stop for fog, cold, or wind. They understand the situation they are in, and consequently they are the only ones here I can in any measure rely on. Were it not for 'little Joe,' Esquimau though he be, many, if not all, of this party must have perished before now. He has built our snow-huts, and hunted constantly for us; and the seals he has captured have furnished us not only with the fresh meat so essential to our position, but without the oil from the blubber we could neither have warmed our food nor had any means of melting ice for drink. We survive through God's mercy and Joe's ability as a hunter.

"We are all well but one—Hans's child, Tobias. I can doctor a sailor, but I don't understand what is the matter with this poor little fellow. His stomach is disordered and very much swollen; he has been sick now for some time. He can not eat the pemmican; so he has to live on dry bread, as we have nothing else to give him. The wonder is not that one is sick, but that any are well.

"The mercury is still frozen. The men are seldom outside of their hut now. From the nature of the food we live on, and the small quantity of it, there is no imperative necessity which calls them outside—perhaps not more than once in fourteen days. Oh, it is depressing in the extreme to sit crouched up all day, with nothing to do but try and keep from freezing! Sitting long at a time in a chair is irksome enough, but it is far more wearisome when there is no proper place to sit. No books either, no Bible, no Prayer-book, no magazines or newspapers—not even a *Harper's Weekly*—was saved by any one, though there are almost always more or less of these to be found in a ship's company where there are any reading men. Newspapers I have learned to do without to a great extent, having been at sea so much of my life, where it is impossible to get them; but some sort of reading I always had before. *It is now one hundred and seven days since I have seen printed words!* What a treat a bundle of old papers would be! All the world over, I suppose some people are wasting and destroying what would make others feel rich indeed.

"As it is, the thought of something good to eat is apt to occupy the mind to an extent one would be ashamed of on shipboard or ashore. We even dream of it in our sleep; and no matter what I begin to think about, before long I find, quite involuntarily, as it were, my mind has reverted to the old subject. Some of the ancients, I believe, located the soul in the stomach. I think they must have had some such experience as ours to give them the idea. I miss my coffee and soft bread-and-butter most. Give me domestic bread-and-butter and coffee, and I should feel content until we could better our condition.

"Joe has returned (at 1 P.M.); the weather too thick and cold for him to accomplish any thing. He was, of course, very hungry; so was I. We had two or three yards of frozen seal's entrails left from the last seal, and on that we lunched, eating a little blubber with it. Poor Captain Hall used to say he really liked blubber. I like it a good deal better than *nothing!* To men as hungry as we, almost any thing is sweet; this that we ate was frozen as hard as the ice we are on.

"*Jan.* 30. The change of the moon has not benefited us. There is no opening in the ice; the weather is too calm and cold, —34°. Could we get a heavy southerly gale, it would rapidly break up the ice; but we have not had a strong gale from the south all winter.

ARCTIC HOSPITALITY.

CHAPTER XXI.

A solemn Entry made in the Journal, in View of Death.—More Security on the Ice-floe than on board the *Polaris*.—Eating the Offal of better Days.—Tobias very low.—Anticipations of a Break-up.—Hope.—Joe, Hannah, and little Puney.—" I am *so* hungry."—An interior View of Hans's Hut; his Family.—Talk about reaching the Land.—Inexperience of the Men misleads their Judgment.

"It is as well to look the future fairly in the face, and none of us can tell who will survive to see this business out. Death is liable to come to all men; and especially may one in my situation prepare himself for it at any moment; and therefore, considering the possibility, I wish here to set down a few facts, as well as my own opinion, which, whether I live or die, I sincerely hope will come to light.

* * * * * * * *

"I make the above statement not knowing whether I shall get through this affair with life. I have told Joe and Hannah, should any thing happen to me, to save these books" [this, with other notes, was written on small pocket blank-books.—*Ed.*] "and carry them home. It is very badly written with pencil, in a dark hut, and with very cold fingers; but, so help me God, it is all true.

"My present life is perilous enough; but I can truly say that I have felt more secure sleeping on this floe, notwithstanding the disaffection of some of the men, than I did the last eleven months on board the *Polaris*.

"*Jan.* 31. Fair; light east wind; the natives off hunting very early. They found water yesterday, but got no seals. The weather is much warmer—only 22° below zero this morning. We are evidently drifting westward. I hope to see the land soon; but both east and west there is a heavy mist, which the sun has not power enough to disperse.

"*Afternoon.* It has now come on thick; wind north-east. I have just lunched on seal-skin. This time we have been enabled to cook it, and I discover that it is all the better—quite tender. We not only ate the skin, but drank the greasy water it was boiled in. The time occupied in heating five quarts of water over the lamp is from two to three hours.

"Hannah is now pounding the bread, preparing our pemmican tea. We pound the bread fine, then take brackish ice, or salt-water ice, and melt it in a tin pemmican can over the lamp; then put in the pounded bread and pemmican, and, when all is warm, call it 'tea,' and drink it. It reminds me very much of greasy dish-water; but in this climate a man can eat many things which in a warmer latitude the stomach would revolt at. The offal of better days is not despised by us now. As to dirt, we are permeated with it; and the less I think about it the better I feel, for I know not how it is to be remedied. We can scarcely get water enough melted to serve for drink.

"The temperature this evening is 34° below zero—6 P.M. The Esquimaux have returned again without game. They have been a long distance to the eastward in the direction where they discovered water yesterday, but to-day it was all frozen over. They started at seven this morning, and have but just returned; and they do all this traveling on a few ounces of food daily. It is indeed a hard struggle for life, and the result doubtful.

"We have just had our pemmican tea, and have each taken a few scraps of refuse from the dirty lamp. It all helps to fill up, and keep the blood circulating. Poor little Tobias is very low—nothing but a skeleton; he can eat seal-meat, but steadily rejects pemmican. I wish I knew what to do for him.

"*Feb.* 1. It is blowing very heavy from the north-west; too much wind for any hunting to-day. We keep closely housed in our dens. Should an accident happen to our floe serious enough to turn us out of our burrows, leaving us shelterless in such a storm of wind, with our blood so thin, we should none of us live long.

"We are poorly off indeed to-day; not even a bit of skin or entrails to appease the biting hunger. For the last six or eight days we have had *something* to lunch on—either skin or frozen entrails; to-day we have neither; and now we realize the value of those unsavory morsels, and feel the want of them more and more every hour. So do the most unappreciated 'blessings brighten as they take their flight.'

"This powerful wind will surely carry us past Disco. We ought to be nearly west of the island now. The Germans appear very sad at the thought that their 'promised land' is gone from them forever. They begin to think now that *any* land where they can get something to eat is good enough. Who knows? They may come to their senses yet before we get through. When men refuse to take advice, they must see their error first before they can repent of it.

"The Esquimaux inform me that the cracks in the ice where they have been sealing are not limited to the 'young ice,' but cut clear through the old—which is an intimation that our floe may now split up at any time if the wind continues. As it is, heavy pieces from the edge of the floe have gone; and the huge icebergs which have accompanied us on our long journey are moving rapidly before the wind. Every thing feels its power, but as yet we have not been disturbed, although surrounded by bergs heavy enough, if propelled upon our encampment, to crush us to atoms. We have thus far floated safely, yet it causes reflection: the ice, we know, must break up sometime; and whether we shall survive the catastrophe we can not tell. We are at the mercy of the elements, and can do very little for our own protection.

"Even now the storm rages without, while fierce hunger rages within; and though sometimes overcome with sad thoughts, as I think of family, children, and friends at home, I am not without hope. God, in creating man, gave him hope. What a blessing! Without that we should long since have ceased to make any effort to sustain life. If our life was to be always like these last months, it would not be worth struggling for; but I seem to have a premonition, though it looks so dark just now, that we shall weather it yet. *Hope* whispers, 'You will see your home again. The life-spark is not going to be extinguished yet. You shall yet tell the story of God's deliverance, and of this long trial, and praise him for his mercies.'

"Our old pemmican can, cut in half, which has served us for a cooking-vessel this winter, and in which all our food has been warmed that has not been eaten raw or frozen, got full of holes; but Hannah has managed to doctor it up so that it will hold out a little longer. I dare not look to see *with what* the holes have been stopped. It is dark enough inside of this 'igloo,' as the natives call it, but nevertheless I am compelled to shut my eyes on many occasions. The wind still continues to blow violently.

"Joe and Hannah are sitting in front of the lamp, playing

checkers on an old piece of canvas, the squares being marked out with my pencil. They use buttons for men, as they have nothing better. The natives easily learn any sort of game; some of them can even play a respectable game of chess; and cards they understand as well as the 'heathen Chinee.' Cards go wherever sailors go, and the first lessons that the natives of any uncivilized country get are usually from sailors.

HANNAH AND JOE PLAYING CHECKERS.

"Little Puney, Joe and Hannah's adopted child, a little girl, is sitting wrapped in a musk-ox skin; every few minutes she says to her mother, 'I am *so* hungry!' The children often cry with hunger. It makes my heart ache, but they are obliged to bear it with the rest.

"Still that breeze is blowing from the north-west; but, in spite of its violence, Joe and Hans are going to try and hunt. Poor fellows, they know their situation to be desperate. We have drifted rapidly in this gale, and must, I think, be south of Disco.

"As soon as the weather moderates a little, I shall try and get a false keel on the boat, so as to protect her some when traveling. I saved the keel of the other boat in view of this. It is too cold as yet to work with tools, and we have but little to work with. But we must get ready soon. Unless there is some change, we must move or starve. We must find water, for game will soon be our only reliance; and right glad would I be if I could be sure of getting enough seals to keep the breath of life in eighteen souls—nineteen; for the baby, Charlie Polaris, can not get nourishment if the mother remains unfed.

"On going into Hans's hut the other day to see the sick boy, the miserable group of children made me sad at heart. The mother was trying to pick a few scraps of 'tried-out' blubber out of their lamp, to give to the crying children. Augustina is almost as large as her mother, and is twelve or thirteen years old. She is naturally a fat, heavy-built girl, but she looks peaked enough now. Tobias is in her lap, or partly so, his head resting on her as she sits on the ground, with a skin drawn over her. She seemed to have a little scrap of something she was chewing on, though I did not see that she swallowed any thing. The little girl, Succi, about four years old, was crying—a kind of chronic hunger whine—and I could just see the baby's head in the mother's hood, or capote. The babies have no clothing whatever, and are carried about in this hood, which hangs down the mother's back, like young kangaroos in the maternal pouch, only on the reversed side of the body. All I could do was to encourage them a little. I have nothing that I can give them to make them any more comfortable. I was glad, at least, to see that they had some oil left.

"It is yet very uncertain when we shall have an opportunity of reaching the land. We are at the mercy of the current, and it depends entirely on how and where we drift this month. I do not think the floe will break up till considerably later in the season. We must have gales from the south to disintegrate such a floe as this. There is no telling—the seasons vary so much; but I think, even if we can not make the land, that the floe will hang together several weeks yet, and we must drift toward game before long.

"The Esquimaux have just come in, and found no water. The breeze holds strong, and cuts like a knife, though it is only 16° below zero. The ice is cracked and opened in many places by the pressure of the bergs; but it is so cold that they freeze up again almost instantly. The men scarcely show their heads out of their hut. I think they have at last become convinced that they can not carry out their project.

"We feel the want of seal-meat very much, and fully realize, by its absence, what a benefit we derived from it. In comparison with our present sensations, we were actually comfortable while even a small portion of our last supply remained. Now the cold takes fierce hold of our shrunken frames.

"We have only three bags of bread left—about four hundred pounds in all; a little over five cans of pemmican, weighing forty-five pounds apiece; and this to feed eighteen souls for I know not how long. It is now the 2d of February only. We must count on six or eight weeks of very cold weather yet, and perhaps with but very few seals to help spin out our little store.

"*Feb.* 3. The gale continued, with drifting snow, until yesterday. The snow is very fine and penetrating, and so thick that you can scarcely tell whether it is really snowing, or whether it is simply being drifted about by the wind. The wind, too, has been variable; for whichever way I looked the snow came in my face. To-day the wind is light, from the west, and cloudy, but not snowing. We were all snowed under the other night, and had some little difficulty in digging our way out to air and daylight. This morning the weather is more moderate. The glass shows 15° below zero.

"Dark clouds hang low in the horizon, preventing one from seeing the land, if there is any to be seen. The temperature rising so much, after the strong north-west gale which we have had, give me some hope that the wind has mastered the current, and forced us toward the Greenland coast. It is always warmer on the east than it is on the west side of Davis Strait; but as yet the weather is too thick.

"The Esquimaux are on the hunt once more, and we who stay at home (!) are praying for their success. The men, knowing now that they have lost the long anticipated Disco, appear more reasonable. Before we passed Disco, Meyers and a part of the crew thought they could manage this business, and so wished to control. They have annoyed me very much; and their assurance of soon getting to a land of plenty has been the cause, I fear, of many raids upon the provisions, and of more being consumed than even they would have risked had they not been deceived as to the course of our drift. It seems they were persuaded that they would reach Disco in February, and so I suppose they thought it hard to be put on such low rations. They could not see the necessity of trying to make our little stock last until April, or even March; but now they begin to comprehend that they did not know as much about these seas as they thought they did.

SURROUNDED BY ICEBERGS.

CHAPTER XXII.

Dreary, yet beautiful.—The Formation of Icebergs.—Where and how they grow.—Variety of Form and History.—"The Land of Desolation."—Strength failing.—Travel and Rations.—Unhealthy Influence of mistaken Views.—Managing a Kyack on young Ice.—Secures the Seal.—"Clubbing their Loneliness."—Poor little Puney's Amusement.—Any Thing good to eat that don't poison.—Narwhals, or Sea-unicorns.—A royal Seat.—Hans criticised.—Cleaning House.—"Pounding-day."—Our Carpet.—Lunching by the Yard on Seal's Entrails.—"Oh! give me my Harpoon."—No Clothing fit to hunt in.—Inventory of Wardrobe.—Narwhals useful in carrying off Ball and Ammunition.—Pleasant Sensations in Retrospect.—The Skin of the Nose.—Castles in the Air.—Violent Gale and Snow-storm.—Digging out.—Three Feet square for Exercise.—Dante's Ice-hell.

"This afternoon it has cleared off. The weather is beautiful; the thermometer says only 13° below zero. I look anxiously for the land, but all is ice and icebergs. The ice and the sky is our only 'view.' Dreary, and yet beautiful; when the sun shines on the bergs, and lights up their massive or fantastic forms, sometimes through the crystal pendants or projecting peaks, we see all the prismatic colors as in a rainbow. The forms, too, vary to a surprising extent. Every berg appears to have had an individual history, and presents, in its contour, the effects of battles with wind and water, rain and storm, and rough jostling with its fellows which it has experienced from its birth. I say its birth; for icebergs do not grow in the water, as many imagine, but originate at the foot or outlet of the glaciers which everywhere on the north-western coast of Greenland project themselves into the sea.

"The process seems to be something like this: nearly all the interior of Greenland, or at least a very extensive tract of country, appears to consist of an immense *mer de glace*, or sea of ice; and this throws off a large number of rivers of ice, or glaciers, and these slowly, very slowly, make their way to the coast, often reaching the shore over high rocks. But no matter what is in their way, they push on, and even into the sea; the foot of the glacier, which may often be measured by miles (you may sail along the face of the Great Humboldt Glacier for as many as sixty), projects under the water as well as above it; and when it gets beyond a certain height and depth the tides force themselves under it, and this, combined with its own weight hanging over the precipice, finally forces it off from the parent glacier, and at its disruption it may be said 'an iceberg is born.'

"The berg then sails off, and, like the human race, each one fulfills its own destiny. Some are grounded, perhaps, not many miles from their birthplace; others travel on, and get shored up on a floe like this, and keep it company, as ours have done, for hundreds of leagues; others pursue their solitary and majestic course toward the open sea, and gently melt away their lives in the deep swell of the Atlantic; some, like desperadoes of the highway, make straight for some noble ship, and send her foundering to the bottom, with all her precious freight of human souls. And as they are different in their history, so are they varied in appearance; some being wall-like, solid ramparts, with square, almost perpendicular, faces, impossible to scale, two or three miles long, and half as many broad; others might, at a little distance, be mistaken for a splendid palace, a Turkish mosque, or a Gothic church,

"'Whose spire
Chimes out to the breezes a song,
And glows in the sunset like fire.'

Occasionally a berg gets worn away at the water-line, while the base below the water is intact, and supports an extended surface on a comparatively narrow stem; others are tunneled or arched; in fact, there is no limitation as to form or size. The most beautiful and the most grotesque may sail side by side; one may be a mile square, and the other only forty or fifty feet. Whether large or small, but a small proportion of either is seen; the great mass is always below the water. The proportion varies according to the amount of salt in the water; but a berg never shows more than an eighth or a seventh of its size. But for the terror and the beauty combined, if any one is interested in the birth, life, and death of icebergs, let them read Dr. Hayes's book, called 'The Land of Desolation'—meaning Greenland.

"As I stand, this beautiful morning, and look up on the white waste of desolation around me, with here and there a splendid spectacle of illuminated ice spears, I think of Tennyson's words:

> "'Break, break, break
> On these cold ice blocks, O sea!
> And I would that my tongue could utter
> The thoughts that arise in me.'

But language is too feeble. I give it up.

"It is getting cold again; our one short, pleasant day has gone; and the night comes on drear and cheerless; the mercury has fallen to 26° below zero. Joe and Hans have returned, too, empty-handed. They did *see* cracks in the ice, but they were closed over, and on the young ice they could see plenty of seal-holes, but the ice was too thin to bear them. However, this gives promise of seals as soon as the young ice thickens a little.

"*Feb.* 4. We are again confined to our huts by a strong gale from the north, the snow drifting so much that it penetrates our clothing from head to foot; so, as we all wish to keep as dry as possible, there is very little going out-of-doors. This gale is sending us fast to the southward, but we do not seem to approach either shore, which makes the prospect of eventually saving all this party very doubtful. If I succeed in that with God's help, I shall have something to be thankful for all my days.

"My great fear constantly is that all our bread and pemmican will be consumed before the season is sufficiently advanced to get seals enough to keep the life in all these men, women, and children. We are, I fear, but surely starving, though slowly. The men have but little strength left. They themselves know not how weak they are, as they are doing nothing. I try my own strength, when the weather is not too bad, by walking a little, and it is poor encouragement for attempting a fatiguing journey over the ice. I do not dare to commence traveling with such an enfeebled company this month, when we may expect as cold weather as any we have had. It would be an unjustifiable risk to expose these people, who have been housed in some sort of fashion all winter, to the piercing gales of an unprotected ice-pack, while in their present half-starved condition. One storm would probably be death to the whole party. If we should start, it would be absolutely necessary to increase the rations; they could not move forward and drag the boat on their present allowance, and then, with larger rations, how soon all would be gone! It is the east coast on which every mind is set. I have not the means of taking observations, and so getting our correct position; but I am quite certain we are very much nearer to the west coast than we are to the east; *that* is certainly a long way off. It would take a long time to reach it even by strong men well fed and clothed, and unencumbered by women and children.

"We could have been rescued, this whole storm-beaten, weary party, with one hour's steaming or run under sail. Exhausted it must have been known we were, without shelter or fire; some of us poorly clad, and without food; for in the darkness no one could tell whether we had saved any provisions or not, or even that we had a boat; unless they could see the boats from the *Polaris;* and if they could see them and us, then could also be seen the pack-ice, and all would know that it was too thick for a small boat to get through.*

"If God in his mercy saves this party—for man can do but little here—if a just God will grant us life, and send us seals to sustain us, I care not toward what part of the coast we drift; I know it all about here—north, east, and west. If we can only get to land anywhere, the rest is easy. On the west coast I could find ships, and on the east the Danish and Esquimaux settlements. But the coast of Labrador is a barren, cheerless land.

"*Evening.* The gale still raging without, and snowing, but the thermometer has been as high as $-10°$.

"Mr. Meyers promulgates the statement that the straits in lat. 66° N. are only eighty miles wide! He would find it a long eighty miles' walk over the ice. This man is very troublesome, the more so that I have no chart to show them to the contrary. His interference destroys all discipline. They think they are going to escape easily, and this, of course, makes them less careful and prudent than men ought to be in our precarious position. We are so short of provisions now that I wish to impress each one of the danger of using more than just enough to sustain life until the season is further advanced, and there is a reasonable probability of getting seals.

"*Feb. 5th* comes in with a strong southerly breeze: the natives off hoping to find food. I shall now have the keel of the boat repaired and made ready for traveling as soon as possible. We have been so long without seals that the Esquimaux are very much afraid of starving to death—it upsets them very much. It

---

* See Letter of Mr. R. W. D. Bryan's, in the Appendix.

is not surprising; and sometimes I believe they fear worse: they are afraid of some of the men.

"It is not very cold this morning—only 17° below zero.

"*Evening.* A streak of luck to-day, or, rather, I should say a Providential gift. Joe has brought home a seal. He shot two others, but lost them in the young ice. Hans got the seal after it was shot. It does seem as though when we get to the last gasp a seal just comes in the way to prevent a fatal catastrophe. This seal is a very little fellow; but we shall make a better meal on him than we have had for many days. There will not be hide or hair of him left, or any thing inside or out but bone and the gall; *that* even we throw away. All else is consumed.

HANS GOING FOR A SEAL ON YOUNG ICE.

"It has been pleasant to-day, and during the gale last night there was plenty of water around us; but as soon as the wind died away the new ice made instantly, so that the capture of a seal was scarcely expected. Thermometer —24° to-day.

"*Feb.* 6. Yesterday I helped carry the kyack over to Hans, to help him get our little seal. There was no water, only young ice, and I was curious to see how he would manage. The seal had, unfortunately for himself, just stuck his head through the young ice apparently to gaze at the sun, and its glare, instead of aiding his perception, appears to have dazed him, or charmed him, so that he was less on the alert for enemies than he should have been. We wanted him too badly to respect his sentimental admiration of the great luminary, and Joe took advantage of his rapt attention to put a ball through his head. The body of the seal lay some sixty yards distant from the old ice on which we stood. Hans got in the kyack. I pushed him on the young ice; he then, by sticking his paddle in the ice and by movements of the body, propelled the kyack toward the seal. The ice would not have borne him had he attempted to walk over it; the weight, being extended over a larger surface in the kyack, it bore him. And then, too, had the ice broken, he was safe with the kyack under him. He finally reached the seal, and, making one end of a line fast to its head and the other to the kyack, he turned the latter with the same peculiar movements with which he had propelled it, and got back with the seal safe to the thick ice. When Hans landed with the welcome prize he was perspiring freely, for it was hard work. The natives usually drag the seals behind them over the ice; and each one soon goes the way of all flesh, only reserving the blubber for oil.

"This morning, as usual, the natives renewed their hunt; but it had been blowing strong all through the night, and though it cleared a little in the morning, it soon came up a thick snow-drift, and in half an hour was too thick to see through. When I was out yesterday, I noticed particularly the state of the ice of which our floe and surroundings are composed. It is very rough now, piled and crushed up in every imaginable shape. How long it will hold together is problematical, but I think some time yet. Joe and Hans returned at 3 P.M. No success to-day. Wind now from the north-east; thick and snowing.

"7 P.M. This evening the wind has suddenly hauled to the south; still thick and snowing. Thermometer only −5° this afternoon. I can see but a few yards before me. The ice is in our pan over the fire—that is, our little lamp. It is being melted to make our pemmican tea. Joe and Hannah at their usual game of checkers. When I see people who don't know what else to do resorting to checkers, it always reminds me of what Dickens wrote of a forlorn old couple who tried to consider that it was a social way of spending the evening. He said it was more like 'clubbing their loneliness' than indulging in sociality. It is certainly so here.

"Little Puney, for want of occupation or amusement, keeps digging with a knife into the snow-wall of our hut. She has had her share of seal-meat, and is not quite so hungry to-day;

while I, having used all the paper I can afford to date—for I have to economize my only blank paper as carefully as our bread—sit and bite my whiskers of some months' growth, all of us regardless of the storm raging outside, though it is quite certain that *some* southerly storm, if it finds us later in the season on this floe, will break us up and set us adrift, God knows where.

"*Feb.* 7. The southerly gale did not last long; it abated at midnight, and this morning we have a light breeze from the west. Fine weather, but I see the effects of the gale. Some of our winter friends, the bergs, have changed their positions, and the ice has opened in many places, though the young ice has formed over the cracks.

"I want, if possible, to get seal enough in the next forty days so as to extend our bread and pemmican out until April. We can do this with the allowance now dealt out, *if we lie still*, and do not attempt to travel; we can make it last, perhaps, until the 20th of March. If we get on our feet, it will disappear like chaff before the wind.

"Our last seal we ate partly raw and partly cooked; latterly we have cooked the skin and drank the greasy water. Joe says, 'Any thing is good that don't poison you.' 'Yes,' I respond, 'any thing that will sustain life'—and down it goes; and repulsive though it be, it is astonishing how this warm greasy water, with a little seal's blood in it, stimulates the flagging energies. We save the blood by letting it freeze in hollows made in the ice. The days are not very long yet. Day-break at 8 A.M., but dark at 4 P.M. It is quite cold again; thermometer 26°, and inclined down.

"3 P.M. Esquimaux returned, and we are all rejoicing over another feast of seal-meat. Hans shot one about noon. They also shot two narwhals—Joe one, and Hans one—but could not get them; they both sank. There are now many openings in the ice, and numbers of narwhals going north.

"These narwhals are sometimes called sea-unicorns, or *monodons*, on account of the long horn—six to eight feet long—which projects from the upper jaw. In reality, however, this appendage is not a horn, but an elongated tooth, including the rudiment of a second tooth. This formidable weapon is quite straight, tapering from base to point, and has a spiral twist from left to right; the animal (for it is not a fish) is fifteen or sixteen feet long. On the back, instead of a dorsal fin, there is a low fatty ridge, extending for between two or three feet. The usual color of the skin is dark gray, with darker spots or patches on the sides; some are lighter colored than others. The young of the narwhal has a bluish-gray tinge. The narwhal usually go in schools of ten or twelve, sometimes more; they are migratory, and when they come playing round the whale-ships are regarded as forerunners of the right whale. They feed on mollusks and other soft animals. When attacked, they sometimes fight fiercely. They are harpooned for their oil and ivory, and also, by the Greenlanders, for their flesh. The flesh is highly relished by them. The oil is excellent; the ivory is both hard and white, and takes a good polish. The Danes and natives work it up into many articles of domestic use.

NARWHAL.

"When these seas were first visited by Europeans, they carried home all sorts of fabulous stories about these 'sea-unicorns,' as they always called them; and the Danes relate that the throne seat of the king of Denmark was made out of narwhal ivory. There is nothing so improbable in that. If polished and carved, it would make a very handsome, and certainly a very uncommon, chair.

"We have had some little trouble over our seal this evening. Hans, if he gets a seal — which is very seldom, for he has shot but very few, wishes to appropriate it all to his own and family's use, without considering that he and his family get their daily allowance of bread and pemmican with all the rest. He must not be allowed to have more than an equal share. He is a very thoughtless Esquimau, or selfish; he is not a successful hunter, like Joe, nor has he his sense. He does not know how to build a hut for himself, or, at any rate, he did not do it. Joe built it for him. He could not take care of himself in a country so sparse of game as it is about here.

"I do not wonder at Sontag's (Dr. Hayes's astronomer) freezing to death, or at any thing else happening to him, or to any white man left in the hands of such a miserable creature as Hans has proved. He threatened this evening 'not to hunt any more.' Let him try it. He will go very hungry in that case, for I shall not allow him any thing out of our stores should he persist. He was hired (and will be paid, if we ever get home) for the very purpose of hunting for the expedition. It is no favor on his part.

"*Feb.* 8. The westerly breeze did not last long this morning: we have it now to the southward; very light breeze, however. Morning clear, and fine weather. The ice is still open in many places, the young ice forming fast. Seeing the narwhals so plenty looks very favorable. There must be considerable open water not many miles distant, and that open water should lead to land. The horizon, nearly every day, is dark and heavy to the east; as the sun gets more power we shall see better. Nothing can as yet be done. Hans has come to his senses: he has gone with Joe. Thermometer this morning, at 8 A.M., −37°, but inclined up.

"I think I have not heretofore recorded the fact that since we have had sufficient daylight we have our 'cleaning-house' day; or, rather, more accurately, cleaning-hut day. As may well be imagined, the moisture which arises from cooking, as well as the exhalations of our own lungs, condenses and clings to the inside lining of our hut. It gets so thick at times that it falls on our bedding, making it very uncomfortable. So we have our 'pounding-day,' when we attack the canvas tapestry of our apartment, and beat off the clinging icicles, and all that has frozen to the surface which we can get off. As we beat it from the walls, of course it falls on the carpet. The 'carpet' is a bit of old canvas which is spread over our floor of ice. This 'carpet' is a sight to behold, incrusted with the accumulated drippings of grease, blood, saliva, ice, and dirt of four weary months—all of which can not be removed by our limited means of cleaning, namely, taking it out of the hut, and shaking and beating it. Any civilized being would be astonished, on looking in upon us, that human beings could have lived so long in such a wretched hole. We shake our carpet every day, now that the weather permits.

"4 P.M. The natives have returned; saw but few seals, and could not get those; but they saw plenty of narwhals, and shot five, but did not kill them—they all got away. Some of the men are discouraged now, because they think that the narwhals drive away the seals; but they do not. A strong wind from the southward, and very cold. Thermometer $-20°$

"*Feb.* 9. During the night the wind hauled from south to north; this morning blowing heavy, with a dense snow. It drifts so that we are compelled to keep in our burrows. Our huts are already nearly buried in the drift, and we shall have some digging to do to get out. Should the gale continue much longer, it will send us past Holsteinborg. We are making a rapid drift. Heaven only knows where we shall get to before the weather will permit us to start for the land. It would be folly to start now. We can get nowhere until the water makes: then we must rely on our rifles for a living. That resource failing, *we know our fate!*

"2 P.M. The wind has changed to the north-west, but still blowing heavy; but it is warmer—mercury just beginning to fall again, however. I have just indulged in a lunch of sealskin: two small pieces, with a little cup of the greasy water it was boiled in, Joe, Hannah, and little Puney lunching with me on the same delicacies. To-morrow we anticipate the pleasure of varying the bill of fare with a yard or so of entrails and a piece of lights which is in reserve.

"5.30. Wind moderating; snow stopped drifting. Joe discovered a hole of water close to our floe. He and Hans instantly started with their rifles. On arriving at the water, found narwhals very plenty. Each shot a narwhal, but they both sank.

"When we started on this ice-craft we had with us one of the ship's whaling harpoons, but, like every thing else, it has been destroyed—one of the men cutting it up for a spear. Had I that harpoon with the boat-warp now, it would be comparatively easy to get a narwhal, one or more—as they lie in tempting proximity alongside of the ice. One of them would supply us with meat, and blubber too, for a long time. By throwing the harpoon in one with the line attached, the men holding the line, we could then get across the rough ice, and kill them with a spear; but though they may be killed by a shot, they almost uniformly sink, and are thus lost.

"*Feb.* 10. The gale has been very severe during the night from the north-west, with another heavy snow-drift; it is still strong, though moderating. Joe and Hans are waiting for the snow to cease, when they will be off to the holes of water, if they have not closed over, while I have to stay at home, because I have not clothing fit to face such weather in. In fact, I have scarcely any clothing with me fit for the climate; and that which I have on my back has not been off for four months. All I have is drawers and an old seal-skin pair of breeches, three years old, and very much worn, full of holes; an under and over shirt; a light cotton jumper; and a Russian cap, and all of this as greasy and dirty as it is possible to imagine, after the length of time I have lived in this suit, the sort of work I have done in it, including cutting and handling dead seals and greasy blubber. As for washing face or hands, there has been no means of doing either; and I have been deprived of that luxury for the whole period we have been on the ice, not having even a pocket-handkerchief to use for a towel. I comb my hair with the only comb in the encampment—Hannah's coarse comb—and call it my morning wash.

"1 P.M. We hear the loud call of Joe for 'kyack! kyack!' I rush out, and, mustering up seven of our party—the others are 'sick,' but not too sick to eat—we take the kyack and start toward Joe. I also took with me Joe's long knife and long seal-line, not knowing but he had shot a narwhal where he hoped to secure it. On getting to him, however, found it to be two seals; they were soon hauled out on to the firm ice, and dragged to the huts, where there is at this writing some 'going in' on seal-meat. This time they can all have a good meal. Joe and Hans have gone back after some narwhals, seen by them near by where Joe shot the seals. The weather is quite pleasant this afternoon; only 12° below zero.

"5 P.M. The natives returned, having had no further success.

Joe shot one narwhal; but, like the others, it got off, carrying the ball, or sank. To-day I have fared well, having had some cooked seal-skin and cooked entrails with the soup, and this evening some raw meat and my pemmican tea. I hope there will be no worse fare than that while on the ice. The most I concern myself now about what I eat is, to get something, and sufficient of it, to keep from getting any weaker; I should like to feel that I was gaining strength. I don't know what draft will be made upon our endurance yet, or what I may need my strength for.

"*Feb.* 11. Have heard the narwhals all night 'blowing' in the holes and cracks. As the weather will, I trust, soon permit me to keep out, at least a few hours at a time, I must, if possible, get hold of a rifle, and then I can help do something to support this party. But through February this old dirty clothing will do little to protect me from the strong piercing winds. I am trying to recall the pleasant sensation of putting on clean clothing, and how, while whaling, when I got my feet wet and cold, what a comfort it was to get on a clean pair of stockings or socks—how much warmer they felt than the damp, soiled ones; yet, perhaps I had only worn the discarded ones a few hours, and now I am sitting with four months' dirt clinging to me, and no way of ridding myself of it. Oh, for a good wash and some clean underclothing! I would care nothing for my old torn breeches and my filthy cotton jumper if I could get something clean beneath. Well, perhaps the waters of Davis Strait will yet wash me clean; so I won't grumble too much.

"The thermometer is at 16° below zero; but I do not feel cold at all, having had a good meal of raw seal-meat, a drink of the blood, and some blubber. Nearly all the men have what is called 'the skin of the nose'—skinning, it should be; and Hans's face is much swollen and sore from being frost-bitten.

"I hear nothing lately from the men of Disco and Copenhagen. They have come to a realization of their situation—some of them, at least. Through the long dark winter it was all they could think or talk of. They were 'sure of getting to Disco in February.' *There* there was abundance to eat and drink; they 'could take what they pleased—clothing, liquors, eatables'—and from Disco to Copenhagen, where 'liquor was cheap,' cigars were 'cheap,' every thing was 'cheap,' in fact—or rather in their imaginations. Each was to buy a gold watch, and then they were to go home by steamer, passengers—the 'Government paying their passage' of course—and each a hero in his own estimation. It was a pretty dream, these castles in the air, but somehow the elements did not favor its consummation; and here they are yet, 'clothed and in their right minds,' comparatively speaking. But I know not when or how the delusion may appear again.

"The wind has increased to a gale, and is accompanied with the usual snow-drift.

"2 P.M. Joe and Hans returned without success to-day; they saw many holes in the young ice, where the narwhals had been during the night. But they could not stay out long on account of the weather.

"*Feb.* 12. The gale or hurricane, as it might be called, has been very violent during the night, moderating some, but still a strong gale this morning. We are completely buried in the snow-drift. It will require long digging with our little wooden shovel to get out into daylight. Joe attempted to go out this morning, but failed. Our general passage-way is in a most filthy condition. Joe came back into the hut very indignant, saying, 'They talk about Esquimaux being dirty and stinking, but sailors are worse than Esquimaux.' His indignation was not without reason.

"It is evening once more, after a long and dreary day, blowing and drifting so one could not get out to exercise; and the only space in our little hut in which one can turn round is just about *three feet square!* Time hangs heavy—hungry, cold, and dirty; and the last is the worst. I, at least, feel it wretchedly.

"If a man ever suffered on earth the torments of wretched souls condemned to the 'ice-hell' of the great Italian poet, Dante, I think I have felt it here. Not a countryman of my own on the ice—all foreigners. Not one to talk to or counsel with; a load of responsibility, with an utterly undisciplined set of men; impossible to get an order obeyed or to have any thing preserved which it is possible for them to destroy. They take and do what they please.

CHAPTER XXIII.

Patching up Clothes.—Captain Hall's Rifle.—Cutting Fresh-water Ice for Drink.—Salt-water Ice to season Soup.—Four months' Dirt.—Sun Revelations.—"You are nothing but Bone."—That chronic Snow-drift.—Seal-flipper for Lunch.—Watching a Seal-hole.—Eating his "Jacket."—Dovekies.—The Solace of a Smoke.—Native Mode of cleansing Cooking Utensils.—The West Coast in Sight.—Joe's Valuation of Seals.—Prospects dark and gloomy.—Bill falls Overboard.—Death to the Front.—Evidences of Weakness.—The Natives alarmed.—Washington's Birthday.—A novel Sledge.—The "right Way of the Hair."—Discussions about reaching Shore.

"I was preparing myself for hunting, patching up my old thin clothes as well as I could. Hans had a very nice rifle, which he did not use, preferring his old Danish rifle. Joe has one of the same make. One of these rifles belonged to Captain Hall, the other to Dr. Bessel. On the death of Captain Hall nearly all of his things went to destruction. His rifle is the one Hans had. I charged him to keep it safe until such time as I could use it. On inquiring for it now, I find one of the men has it, and it is broken. It could be used notwithstanding, but he refuses to give it up. They took possession of every thing from the first, and are very insolent and do as they please; and as I am entirely alone, I seen no way to enforce obedience without shedding blood; and should I do that and live, it is easy to see my life would be sworn away should we ever get home. These wretched men will bring ruin on themselves and the whole party yet, I fear.

"*Feb.* 13. Clear and cold in the morning; before 3 P.M. it was blowing a gale once more from the north-west, accompanied by the usual snow-drift. Perhaps Providence is showering this snow upon us with mighty winds, to prevent these foolish men going off to court their own destruction. Fortunately, this morning, while it was calm, I laid in a store of both fresh and salt water ice which can be got from different parts of the floe. The fresh has to be cut from the ponds which formed during the summer by the rain and ice melting from the high hills and hummocks and running into these depressions. The salt ice we melt to season our 'soup,' or 'tea,' and the fresh to drink as water.

"Joe and Hans are still out. I wish they would return, for the drift is so thick I fear they will lose their way.

"February thus far has been a wretched month—gale upon gale; and, in consequence, we must be southing fast. It was 22° this morning, but inclined up.

"4 P.M. Joe and Hans returned safe; had found no water, but had seen many narwhals in the young ice; but had no opportunity to shoot, as they could not get near enough, the ice being too weak to bear them.

"*Feb.* 14. The snow-drift and gale not quite so bad as yesterday. The natives, being well clad, do not mind the weather, and, as we are entirely out of seal-meat, they have started out, despite the drift. We have part of a skin left, which will serve for lunch. Oh, the filth, the utter filth one is compelled to eat in order to appease the fierce hunger, and to secure a little life and warmth to the body! To-morrow night it will be *four months* since we were set adrift on this ice. It is a long time to be starving and suffer as we do, and yet there is no prospect of escaping for a long time to come. There is only one month's provisions left at our present rate of consumption; but we could very easily eat it all in eight days, and not have too much—not have enough.

"Joe took his narwhal line and spear to-day. If the wind was not so strong, I would have gone with him; but through my clothing, or any kind of civilized clothing, this wind would cut like a knife through tender meat.

"12 M. The sun is shining through our little ice-window, made from fresh-water ice. It is about twenty by twenty-six inches in length and breadth. It is the first day that the sun has reached and penetrated our little hut; but there is no blessing without its drawback; though the sun is so welcome, it reveals too plainly our filthy condition. I thought I knew the worst before, but the searching sun has made new revelations.

"The men are actually infected with the spring-fever, and are *cleaning house*. I should think a good cart-load of black, smoky ice was taken from their hut to-day. This morning, too, Puney seemed to be enlightened by the sun. She sat looking at me for some time, and then gravely remarked, '*You are nothing but bone!*' And, indeed, I am not much else.

"4 P.M. That same chronic snow-drift has re-appeared with a strong breeze. I varied my lunch to-day, by compulsion, with a few dirty scraps of refuse blubber from the oil-lamp. At six o'clock I was cheered by the sight of Joe and Hans bringing along one seal. Joe had shot two narwhals, but, as usual, both sunk. They seem principally serviceable in carrying off our ammunition. I had some frozen seal to-night, and, the weather moderating, I feel more like a normal human being. Foxes, too, have re-appeared; three were seen to-day. We have not seen any before for a long time.

"*Feb.* 15. Snow and blow, blow and snow, that is the order again to-day. I was so fully determined to go with the natives on their hunts by the middle of February, that I started despite the drift this morning; but the wind, shortly after increasing, went through my clothing as if it had been gossamer, and struck to the very marrow, if there is any in my bones, which I doubt. My apparel is too poor and thin, and full of holes. I must try and patch up again this evening; for I was soon compelled to return this morning. While I was out I saw no opportunity of shooting either seals or narwhal. The young ice had formed through the night, but too weak for walking on; but I saw plenty of the narwhals, and one or two seals. After I returned Joe and Hans went on, but they got nothing; shot at narwhals with the usual result—a loss to our powder and shot. Joe got adrift, too, on the young ice, and was very near spending a night out, but toward evening succeeded in regaining our floe. The wind subsided in the evening.

"*Feb.* 16. Same sort of weather. Very little water to be seen. Yesterday I ate neither bread nor pemmican. I dare not eat it now every day, if there is a morsel of any thing else to substitute. For breakfast this morning I had a little pemmican tea, with a strip of frozen entrails and a few mouthfuls of blubber. We have a seal-flipper in reserve for lunch.

"I can not resist sometimes giving poor little Puney a part of my scanty rations.

"I have been cleaning house again this morning, shaking carpet, and bringing ice to cook with: the most of such work I do myself. It is here the same as on board the *Polaris*—one man is as good as another, and a little better! I got through my house-work about 10 A.M. The wind was quite moderate, so I started with my rifle (I had at last obtained the loan of one), to see if I could get a seal. Saw signs of water to the north-west about two miles distant. On getting there, found it a very fine place for shooting, providing there had been any thing to shoot. Staid two hours, and saw two or three narwhals, but at too long a distance off, and then the wind came on again and drove me home.

"On arriving at our hut, found Hannah had lunch ready—a piece of seal-flipper, and, what refreshed me very much, a pot of seal-blood soup. It was quite a heavy lunch.

"The natives came in at dark; they had not had any better luck than I; had seen the narwhals and a few seals, but had not got any thing.

"*Feb.* 17. Joe and Hans and myself got off at sunrise this morning; very little water to be found. Thermometer at 20° below zero, but, not being so windy, I stood it better.

"While I was watching a narwhal-hole this morning, one came along; but there was no chance to shoot with any prospect of killing him, so I saved my ammunition. Hans was more fortunate. A seal came to his water-hole, and he firing, the seal quickly lay dead, floating upon the cold water. We soon had him out, and now he is skinned and dressed up. He is quite a small fellow, as all the seals we get are. But he will make a meal, eating him 'jacket' and all. Joe missed two seals to-day. I had his favorite rifle, he taking another; and to this I attribute his failure. There were only four seals seen to-day. At the best, in this weather it is hard work to stand hour after hour, scarcely moving, watching an uninteresting orifice in the ice; and with the temperature from 20° to 30° below zero it is scarcely endurable, clad as I am. I have passed twelve winters in the Arctic regions, and I have been considered 'tough,'* but if we were not in danger of starving, I would not stay outside, in my present habiliments, when there is a strong wind, a moment.

"Joe shot a dovekie, and one of the men shot two of them. These little speckled birds only weigh about four ounces; they have a very plaintive cry, and, as they paddle about in the icy water, do certainly look more 'forlorn,' as Dr. Hayes says, than the strong and voracious gulls or the comfortable eider-ducks; but we are not in a position to indulge the pathos of sentiment, like Dr. Hayes, who, if I remember rightly, was so impressed by

---

* See Captain White's letter in Appendix.

this friendless appearance that he declined to make a 'specimen' of one, though desiring it greatly. In fact, I believe we are beginning to look upon all living things, without a thought of science, only as so much life-sustaining matter.

"*Feb.* 18. The monotonous western gale and snow-drift. There will be no success in hunting, I fear, to-day, unless the gale abates.

"2 P.M. Gale still blowing, but Joe has been off since 11 A.M., hoping to find more dovekies, or a seal, but it is very unpromising weather for either. Hannah, Puney, and I have just had our lunch of seal-skin, and there is a piece over the lamp for Joe when he returns.

"Sitting in the hut (though less dangerous to health and life, than exposure to the sharp winds in a cotton jumper), knocking my feet together to get a little heat into them, is far more wearisome than hard work outside. If I had clothing like Joe, I think I could stand it about as well as he. Could I have foreseen the sort of voyage I was to make, I would have looked out and had at least two things, a warm suit of clothes, and a rifle I could call 'my own.'

"I have about three pipes more of tobacco. When that is gone, I shall feel more lonesome still. It is the only companion I have; and I think the most fastidious lady or the most inveterate 'anti-tobacconist' would hardly object to smoking on an ice-floe. At any rate, I am not afraid of discoloring our curtains: it would be hard to tell what color our canvas hangings are by this time.

"At 4 P.M. Joe returned; saw nothing but one narwhal, though he found considerable water on the north-east side of our floe.

"6 P.M. Have just had some pemmican tea. In the act of drinking it, my lips came in contact with something which I knew did not belong to it, as the pemmican dissolves when warmed. It was not bread, for it felt long and slender; I examined it, and found it to be an innocent piece of seal's entrails, which had doubtless remained in the uncleansed pan. Put it in my mouth again, from whence it descended of course to my stomach. *Nothing is wasted by me!* Hannah had cooked some seal-meat for breakfast, the entrails with the meat, and had neglected the usual cleaning. Do not imagine that our pan is ever washed. It is cleansed as well as may be with the *fingers*, and I doubt not sometimes with the *tongue*, for that is the true Esquimaux fashion. But I can not allow myself to dwell on this subject.

"*Feb.* 19. Clear and cold at 6 A.M. *The west coast in sight!* I think it to be in the vicinity of Cape Seward, and distant thirty-eight or forty miles. If the ice were in condition to-day, I would try to reach the shore. In this latitude I could find Esquimaux, and we could live as they do until June; then we could get to Pond Bay, and find English whalemen. But these are castles in the air; the state of the ice forbids the attempt. We must bide our time.

"It has been a fine, pleasant day. Joe, Hans, and myself off at sunrise to hunt for seals. They were very scarce; saw only one to-day, and Joe shot him, and he is now divided into eighteen portions. It is a small seal, but we shall try to make a meal of him for all hands; each person having with his share one-tenth of a pound of bread. This seal the men took possession of, in one of their freaks, skinned, and divided it as they pleased. Joe was very angry, which was no wonder; he and Hans do all the traveling and labor, and drag the seals home, and it is natural that they should be provoked to have it seized upon in this way. The men are seldom out of their huts, or off their beds before eleven or twelve o'clock. Joe, Hannah, and myself breakfast about day-break, and have but one meal in the day besides that. Joe and Hans are exposed many hours every day to the wind and cold, and it comes very hard that these idle men should take the seals from them. Joe always expected me to divide the rations, and he is willing for the men to have their share of whatever he brings in; but he don't like this way of theirs. When they take it into their heads to do the cutting up, they never say to him or to me, 'Can I do it?' or 'Shall I do it?' but just lay violent hands on it. Joe says, 'I think they ought to be made to pay a hundred dollars apiece for each seal they have taken from me, for their bad conduct.' If seals were to be had for the buying, *I* would gladly give a hundred dollars apiece for them; but money, if we had it, would buy nothing in the ice-pack. Thermometer to-day at 6 A.M. —25°; later, —14°.

"*Feb.* 20. Cloudy; light south-east wind; —4° to —11°. It is quite warm, and evidently plenty of water. Joe, Hans, and myself off early to the openings, looking for seals. Saw no narwhals to-day; and neither Joe nor I saw any seals. Hans says he saw three. Shot eleven dovekies. It is dreadful to think that game is so scarce; winter not over yet, and only two bags of bread and three cans of pemmican left, and no hope of an early escape. It may prove to be June or July before we are relieved from this isolated and dangerous position. It depends on the weather, which varies very much in different seasons.

"It is snowing again this evening. I feel more than usually weak, and, as a natural consequence, somewhat desponding. The boat is undergoing repairs, and will soon be ready, should a favorable change occur; but I do not look for any change for a long time to come, unless it is the change of absolute starvation, which will soon be on us unless game of some sort appears more abundantly.

"*Feb.* 21. The thermometer to-day, for the first time since we have been on the ice, stands above zero. I read, to my astonishment, $+3°$ at 6 A.M. At that hour the natives and myself started on our usual tramp, in hopes of getting some fresh food; but we had no success. I saw one seal and two narwhals in the course of the day, but neither of them where they could be reached. We know that food is under and around us, but day after day passes and the prey escapes us. On returning, lunched, or dined, whatever our second meal may be called, on a dovekie and piece of blubber. At 4 P.M. Joe and Hans returned as light-handed as they went. The prospect looks dark and gloomy — eighteen to feed, and nothing, as it were, to feed upon. The little allowance of pemmican and bread will not keep the breath of life in us much longer, and we have not even a bit of skin or entrails to eat. We have, and are willing still, to eat such stuff; but we can not get even that.

"*Feb.* 22. Thermometer reads $+20°$; cloudy, with light southeast wind — very comfortable weather. We were off, as usual, this morning as soon as it was light. I have watched at one seal-hole since morning, but have seen no living thing. One of the men, Bill, was out to-day, and shot two dovekies, and varied his adventures by falling overboard. He has been out of the hut so little, he is not sure of his feet. He could not swim, but somehow managed to get on to the floe again.

"Our situation is getting desperate — plenty of ammunition, but no game. Every thing is now ready to push for the shore, but how to get there I know not; there seems no feasible way either by boat or foot. The ice is in such a fearfully rough condition, that it could not be traversed with even a light back-load;

and here we have not only our own safety to consider, but women and children to be cared for. If it was only us men, we might risk more; but I can not advise a course which would make the death of these poor creatures almost inevitable.

"The men are frightened; they seem to see Death staring them in the face and saying, 'In a little while you are mine.' Joe is frightened too. He feels that if he and his family were alone on the shore, without this company of men to feed, he could catch game enough for his own use, until it was more abundant; but to catch a living for eighteen discourages him, and indeed it seems impossible, without some great change occurs.

"If we had drifted toward the east, as Meyers thought we were doing, I should consider it far less risk to make the shore; because, if we once got there, we should be almost certain to get assistance from the natives, and game is plenty; but on this side of the strait it is different. I know well what this barren and wretched coast is in winter; but to satisfy all parties, I think I shall have to make the attempt. It is a poor outlook; we shall probably fail, for want of strength, in dragging the boat; and if we should even succeed in making the shore, we should be as likely to starve there as here, for it is most emphatically an ice-bound coast, dreary and devoid of life at this time of the year. If it was spring it would be different.

"To show how little strength the men have, I will relate what occurred this morning. Being about a quarter of a mile off from the men's quarters, I had shot a seal; and, as it was in the water out of my reach, I ascended an elevated hummock with a pinnacled top, and shouted out to the men to bring the kyack. Though weak in other ways, my lungs were sound, and my voice had not, I think, lost any of its power. A kyack is so light that it can easily be picked up and held in the hand; yet several of the men had to take hold to help bring the kyack to me, and were completely tired out and exhausted by the effort. This seemed in a measure to open their eyes as to their unfitness for travel, and discouraged them about setting out. Once before I had seen three or four of them trying to move the large boat, and, though it was then empty, they could not stir it from its place.

"Joe and Hans returned at 4 P.M. They had seen one seal and four narwhals, but got nothing. They, like the men, are anxious to start for the land; so I shall endeavor to do so as soon as the wind abates, which is now (evening) increasing from the south-east.

"There is a double game working around me. I must be on the watch. It is plain to me that the Esquimaux are anxious to get on shore to preserve their own lives from other dangers than scarcity of game. I shall protect them to the utmost extent of my ability.

"*Feb.* 22, *Evening.* To-day, all over the United States, I suppose there have been military parades and rejoicing, and balls and other festivities in the evening, being 'Washington's birthday' anniversary. We might have raised a flag, had I been in spirits to do so. But I forget; there is no one here who knows or cares any thing about Washington—foreigners all.

"*Feb.* 23. This morning there is a strong breeze from the northwest; thick weather and snowing. I have just been in the hut, and had a talk with the men. They have not been inclined to listen to me heretofore, but I hope they will now. I have told them that, when the weather will permit, we will start for the shore, but it would be madness to do so now, dragging boat, ammunition, and sleeping-gear. I told Herron that it might be possible to make a light sleigh out of some of the skins, and put our provisions and clothing in that, which would be easier dragged than the large boat, and that the kyack would answer the purpose of ferrying us across the cracks. He knew that we could not get the boat along over the rough ice with the little strength we have left. But neither he nor any of the others seem to consider that the women and children could scarcely be cared for by this arrangement, nor the great risk we should all be in by abandoning the boat. That is our only anchor of salvation if the ice should suddenly break up, which it may do any time under a strong southerly gale. Separated from our boat, if we did not make the land safely, we should be at the complete mercy of every wave that rolled.

"Either a bear or seal skin will answer the purpose of a sledge without any frame-work whatever, if the dogs can be tackled to it; placing it, of course, *the right way of the hair*, it will run very smoothly over ice or snow, better than any runners, as long as the hair lasts. I have seen it done by the natives, and have tried it myself.

CHAPTER XXIV.

Decide to make the Attempt.—Foiled by successive Snow-storms.—Down to one scant Meal a Day.—Land thirty-five Miles off.—God alone can help us.—Canary-bird Rations.—Bear-tracks.—A Bird Supper.—A Monster Oogjook.—Six or seven hundred Pounds of fresh Meat! Thirty Gallons of Oil!—Oogjook Sausage.—Our Huts resemble Slaughter-houses.—Hands and Faces smeared with Blood.—Content restored.—Taking Observations.—Out of the Weed.—A Present from Joe.—Heat of Esquimaux Huts.—Desponding Thoughts.—"So I sit and dream of Plans for Release."—Terrific Noises portend the breaking up of the Floe.—An unbroken Sea of Ice.—Hans Astray again.—That "Oogjook Liver."—The Steward convinced.—An Ice-quake in the Night.—The Floe breaks twenty yards from the Hut.—Floe shattered into hundreds of Pieces.—Sixty Hours of Ice, Turmoil, and utter Darkness.—The "Floes" become a "Pack."—Storm abates.—Quietly Drifting.—A Choice for Bradford.—Our Domain wearing away.—Twenty Paces only to the Water.—Whistling to charm an Oogjook.—A Relapse into Barbarism.

"*Evening.* It is at last decided that when the ice permits we start, taking the boat with us. We will try to reach a place called by the Esquimaux Shaumeer, a little to the north of Cape Mercy, in lat. 65° N. There is game there, and sometimes Esquimaux. Should we start and fail to get on the land, we must, of course, return to the floe, and continue to drift; and then we shall probably have to take our shot and cut up all the spare balls into slugs, shoot all the dovekies we can—and seal, if there are any—living as best we may until we drift to lat. 62° or 63° N., which will bring us to March, when I hope to find the bladder-nose seal on the ice. Mr. Meyers, on this occasion, agreed with me that it was unsafe to leave the floe at present. If he had worked with me from the first, we should all have been better off. If all fails—well, we perish; and there is one man up north that can go home rejoicing.

"The day has been stormy, and the wind is now north-east, and still snowing. I have had the tent enlarged, and the ammunition divided and put into several bags of ten or twelve pounds each, so that, in case of an accident to a part, some may be saved. It will be bread to us very soon. We are doing every thing that we can to prepare for our contemplated journey. If I start, I shall do my utmost to succeed; but we all need heavier rations than we can afford to use. It has been warm to-day—+26°; but it is growing colder, and beginning to snow.

"*Feb.* 24. The deep fall of snow has rendered it impossible for the most fool-hardy to think of starting to-day; so went off hunting instead. About noon found one little hole, and Joe shot the one seal that has been seen to-day. After a fall of snow it is very difficult to find the blow-holes of the seals, especially without a dog. They will scent them when men can not find them. The breathing-holes are very small, and scarcely distinguishable in the snow.

"I have had a long talk with the men again this evening. I have explained to them that I hope soon to get to the ground of the bladder-nose seal, which in March come on to the ice, not far from where we are now, to breed. This seal is sometimes called the 'hooded,' or 'bearded,' seal. I have told them—what they know as well as I—how little bread and pemmican we have left; and, that we may not find ourselves perfectly destitute, if we are a little later than I expect reaching the new sealing-ground, that we ought to live on still less than we have been doing. It seemed hard to ask them to live on one *short* meal a day, but they have consented to come down to that allowance. I think the snow to-day was a great damper to them in regard to our projected journey. God knows we have been living on little enough, but we must try and have a morsel saved to last until April.

"The land is visible about thirty-five or forty miles off. But it is simply impossible, in the condition of the snow and ice, to get even as far as that, carrying sufficient with us to sustain the company; and at least three or four of the children must be carried also, as they are too young to walk. The women, I suppose, could walk almost as well as the men, if they were not encumbered with these young children. And without carrying some night-gear to shelter us, we should freeze. If it was April, I would go for the land, carrying nothing but guns and ammunition.

"*Feb.* 27. Clear and cold. The mercury has gone down again to 24° below zero. Such a set of skeletons as we would have had a poor chance camping out such a night without the shelter of our huts. We may be thankful for the snow which prevented our starting. God alone can help us. We seem placed where we can neither move one way nor the other to help ourselves.

The wind is very strong, but I went with the natives to hunt at daylight. Saw a few seals, narwhals, and dovekies, but the young ice would not allow us to get near enough to shoot.

"We are now on our allowance of one meal a day; and all that is allowed for the whole company of eighteen—for we don't count the baby in the matter of eating—is five and a half pounds of bread, and four pounds of pemmican. Well, a man can be trained to live on the rations of a canary; but I do not like the training.

"*Feb.* 28. Cold, and strong breeze from the north-west; thermometer reads 28° below zero, and I can see no water anywhere this morning. Joe and Hans, however, thought they would go and look as usual.

"Yesterday I forgot to mention that a large bear had passed close to our huts, knocking down a spear and gun. We can tell the size and weight of a bear by the tracks of his feet on snow; they are so heavy they sink deep. Having no dogs to give the alarm, he escaped us. A bear would have been a splendid addition to our present limited allowance. The hunters gladdened our eyes with thirty-seven dovekies. We take two apiece, and having these, use no bread. We cooked them, eating every thing but the feathers. The children had one dovekie apiece.

"*March* 1. Clear, and very cold. Thus March begins with winter weather. We had not been so gluttonous as to eat all our share of dovekies at once, and so had part for our breakfast this morning. We had cleaned the birds, and then replaced every thing but the gall, and then we had the water in which they were cooked to drink. Joe, Hans, I, and two of the men went out with our rifles this morning. Altogether there were sixty-six shot. This is a good deal better than nothing, for it saves our bread, though the flesh is not heat-giving, like seal-meat. We all returned at 6 P.M., glad that the day had brought us something to eat. This morning when we went off it was 34° below zero; this evening it is 25° below.

"*March* 2. We were off to the cracks again this A.M., hunting for either seal or dovekies. The men say after they have had their breakfast they will come too. I shall try hard to get a narwhal—one narwhal would set us up in meat for a long time, compared with all the dovekies that we can shoot.

"5 P.M. I did not get my narwhal, but Joe has shot a monster oogjook—a large kind of seal—the largest I have ever seen. It took all hands to drag him to the huts. Peter fairly danced and sang for joy. No one who has not been in a similar position to ours can tell the feeling of relief which his capture produced. How we rejoiced over the death of this oogjook it would be impossible to describe. It was, indeed, a great deliverance to those who had been reduced to one meal of a few ounces a day.

"Hannah had but two small pieces of blubber left, enough for the lamp for two days; the men had but little, and Hans had only enough for one day. And now, just on the verge of absolute destitution, comes along this monstrous oogjook, the only one of the seal species seen to-day; and the fellow, I have no doubt, weighs six or seven hundred pounds, and will furnish, I should think, thirty gallons of oil. Truly we are rich indeed. Praise the Lord for all his mercies! A few dovekies were also shot, but the oogjook is the joy of our eyes; and dovekies, so prized only yesterday, are scarcely regarded to-day.

"*March* 3. We eat no bread or pemmican now—oogjook is the only dish; and it does me good to see the men able once more to satisfy their appetites. And they are bound to do so—they are cooking and eating night and day. We have had oogjook sausages for breakfast, the skins being stuffed with blubber, and with this some of the meat boiled. Our 'civilized food,' as we call our bread and pemmican, is being kept for a 'rainy day.'

"The men, after such long fasting, can not restrain their appetites, and some of them have eaten until they are sick. But one can not find fault with them, knowing that they have been living on nine ounces of food a day. When first killed, the warm blood of the seal is scooped up in tin cans, and is relished like new milk. The mammary glands of the female seal, especially when distended with the lacteal fluid, is a very delicate morsel. Our glorious oogjook proved, on measurement, seven feet nine inches in length from head to tail, excluding the latter; adding the hind flipper, he measured fully nine feet. What a godsend!

"Our huts now look like slaughter-houses. Meat, blood, entrails—dirt all over every thing. Our hands and faces are smeared with blood, and one coming among us now would take us for carnivorous animals just let loose upon their prey. We have plenty now for a few days, at least; both oogjook and dovekies, I hope, for a week to come.

"And yet—and yet—now we have enough of this for a little while, we discover that it is hard living—'nothing but meat, blubber, entrails and blood.' Like the Israelites with their quail, some murmur at the monotony of the diet. Well, it *is* monotonous, very; but it will give us strength. I thank God for it. A few such will save this party, at least from starving. The men are all feeling more contented; and they see now that, if they had started when they wished to for the shore, they would all probably have perished.

"We are now approaching the Cumberland Gulf—my old whaling-ground. Should the weather prove favorable, I shall have no hesitancy about trying to get clear of the floe; for there, finding ships, we should end our misery. But should we be drifted past the gulf, why then we can try Hudson Strait, and, getting on Resolution Island, could safely wait there for Hudson Bay vessels or American whalers, who now go there every year. But of course we should be obliged to depend on our rifles until succor arrived.

"The day has passed; cold, windy, and drifting snow. We have dovekies for supper this evening. With two meals a day, which we have had since the capture of the oogjook, it takes sixty-six dovekies per diem for the party—not very full meals, about eight ounces. A well-kept dog receives more, and many a one would reject food that I have had to eat this winter.

"*March* 4. At sunrise mercury down to 30° below zero. Started on the hunt, but found no water, and returned at noon. Joe and Hans did not come up till 4 P.M.; had seen one seal. One of the men shot four dovekies. It is quite clear to the westward, but I can see no land. There is no more thought nor talk of the east coast; that is seen to have been all a delusion, inspired by the desire to have it so. What some people wish they soon believe.

"Meyers, I believe, has given up taking observations, his countrymen having lost all confidence in him since finding how his prophecies have failed. If they had been in this state of mind four months ago, we should have had at least one month's more provisions on hand than we have now.

"*March* 5. Wind from north-west, and cold, accompanied by the ever-attendant snow-drift. Unless the wind moderates, there can be no hunting to-day. The weather is indeed very bad, blowing continually, and intensely cold at night—mercury going from 35° to 40° below zero, rising perhaps at noon to 25° or 20° below. We must be drifting southward very fast. I think that we are in lat. 65° N. This wind continuing will soon carry us to the coast of Labrador.

"A great many persons in this world of ours cry out against the use of tobacco in any form; but should they ever have occasion to live as we do, which God forbid, they would find out the comfort of it—how consoling it is to a cold and hungry man. I have been out of the weed about twelve days, and I feel the deprivation very much. Perhaps I may get used to it—in fact, I must; but it is an additional hardship, not easy to get reconciled to. Had I any thing pleasant to occupy my mind, or any thing really to do—any active employment, which was of any possible use to any body—I might more easily forget it; but when the weather forbids hunting, and one must sit still in a cold ice-hut for twelve hours at a time, it is an immense solace to have one's tobacco hold out. If any one don't believe this, let him try it. Joe has just given me out of his limited store two pipefuls, for which I am truly grateful. The men—the most of them—have got tobacco yet.

"That happened in this way: after we found ourselves left on the floe, I was looking in every direction for some possible chance of escape and return to the ship, about which several of them seemed very indifferent, at the right time to have availed ourselves of the open water; and while I was thus on the lookout for release, they were looking out for the contents of the bags which had been thrown on the ice. So, when all hope of escape from the floe was lost, I looked round too, to see what there was, and then found that all the good clothing and nearly all the tobacco had disappeared, and I was left with nothing. All my things were left on board the *Polaris*.

"We are now getting some meat ready for lunch. Joe is separating the pieces with a hammer, they are frozen so hard; though the pan in which it was placed has been setting within one foot of our lamp-fire for the last thirty hours, which conveys some idea of the coldness of the hut in which we are now spending our days and nights. Our inside lining is white with frost. I have just tried to eat a piece of this frozen meat, but it was too hard, and I gave it up. I have seen written descriptions of the heat of an Esquimaux hut, but I have never felt it. It keeps off the wind, but it is cold—cold. Ours is not crowded sufficiently, I suppose, and we have only one lamp giving out a flame as large as an ordinary gas-jet.

"*Evening.* The gale has been very severe through the day, and still continues to blow with great violence. Thermometer —32°, and the snow-drift very heavy. We dined on part of the oogjook's head this evening; it was very tough, but with it and the addition of a pot of blood we contented ourselves.

"Oh what a wretched life to live! Sometimes I feel almost tempted to end my misery at once, but thoughts of the divine restriction hold me back. Had our Maker left us free to choose —had not

> "'The Everlasting......fixed his canon
> 'Gainst self-slaughter,'

I think there would soon be one wretched being less in this world. God alone knows what we suffer; no pen can describe it—at least I can not, but I can *feel* it.

"*March* 6. This gale has been the most severe of any that we have had while on the ice, and it still continues. We are again completely buried in the snow-drift, and can not get out of the outer passage-way until it abates; and then in time we can dig out with our little shovel.

"Had it not been for the providential supply of that oogjook, we should now be in a still more deplorable condition, living on a few ounces of bread and pemmican, without oil enough even to warm it. But the meat we have left of it yet will last us for eight or ten days to come, and there is blubber for a month; and by then I hope to be where there are more large seals—in lat. 62° N.

"In March and April these large bladder-nose, or hooded, seals are usually found in great numbers on the ice; and if we drift there in the right season for them, all danger of starvation will be at an end; but about that time other dangers will assail us. We can not foresee what will occur when this floe breaks up: it will be miraculous if the whole of this party can be saved when that happens, and come it must. There is no hope of escape while these whirling winds and snows continue, and when they cease a southerly gale may break us up at any time.

"We are drifting southward very fast, and I am sincerely glad of it, for if we can make the shore at all, it would be much better that we do so on the coast of Labrador. Could we even get ashore where we are now, it would be summer, or quite late in the spring, before we could hope to get to the whaling-ships—it might even be July or August. But should we make a rapid drift to the coast of Labrador, there would be some hope of finding a 'fishing station' by May, and at any of these we could get temporary relief; and from there we could reach St. Johns, and thence easily get to my own happy land.

"So I sit and dream of modes of escape, and all the minutiæ of travel, and the management of the party—if they would consent to be managed: thoughts such as these fill my mind. One day one mode seems feasible or possible, then all the difficulties present themselves before me, and the result looks disastrous. Then I turn over in my mind some other plan, and all the while I know that, plan as wisely as I can, all may come to naught; for we are entirely at the mercy of the elements. A gale and a heavy sea may totally prevent my carrying out any one of these schemes for release. God's will be done. He alone can lead us out of this perilous condition, and 'set our feet upon a rock,' if he sees fit.

"The day has passed, and a bitter cold day it has been. No one has been out but Joe, who cut his way out, but was driven back in a few minutes, with a frozen face. The gale still continues, though it has somewhat moderated. We are closely housed, and the lamp burning continually; but all things in the hut, except the human beings, are frozen as hard as the ice we are on. Many of the Esquimaux huts have two or three large lamps, which combined give out considerable heat—very unlike ours.

"*March* 7. The weather is better this morning, but there is a good strong breeze still, and some snow drifting yet. But the sun is shining brightly, and I hope for better weather before night. It is a relief to see the sun once more, and be able to put one's head out-of-doors. For some days past, and all last night, the ice has been cracking and snapping under us, sounding like distant thunder. This betokens the breaking up of the floe; but it may freeze together again. It received a severe shaking last night several times. I think the noise and commotion is caused partly by loose pieces of ice getting underneath our floe, and rolling along until they come to an opening, when they come grinding up, and rise to the surface. These noises startled me from my sleep; several times I thought our ice was breaking in fragments. I begin to have some idea of how people in earthquake countries must feel when the ground is trembling and shaking beneath their feet, especially in a dark night, when one can not see a foot before him, and knows not which way lies danger and which safety—if, indeed, there is safety anywhere. It is impossible to describe, so that, without the actual experience, the sounds of breaking ice floes and bergs can be realized. No two sounds appear alike, except the repetition of the grinding and explosive, which are a horrid sort of refrain. But somewhere I have read words like these, which partially give an idea of their variety and fearfulness:

> "'Hark! a dull crash, a howling, ravenous yell,
>   Opening full symphony of ghastly sound;
> Jarring, yet blunt, as if the dismal hell
>   Lent its strange anguish from the rent profound.
> Through all its scale the horrid discord ran;
> Now mocked the beast—now took the groan of man.'

And even this does not begin to convey an idea of the overwhelming *power* of these pushing and grinding masses. Their force and human helplessness compared, makes one realize that there are yet elements in nature which man's ingenuity can never control.

"Joe and Hans went out to try and find water, and found one little hole; but all they shot was two dovekies. Wind still strong from the north-west; thermometer $-24°$.

"*March* 8. Fair; light west wind. I have roamed some ten miles over the floe to-day, back and forth, in search of water, going in every direction, and necessarily retracing my steps many times; but I find only a solid, unbroken sea of ice—not one little hole where a seal might pop his head up and receive the welcome of a bullet. Joe has also been looking, but found none. Hans may have been more fortunate: he has not yet returned, and it is 5 P.M.

"Our cheerful day is short-lived; there is another storm brewing; the sun looks pale and sick. It is now near 6 P.M., and Hans has not returned. Joe fears he is lost, and is getting ready to go out and look for him. This Hans is a plague.

"Hans just come in all right. He says he has been a long distance to the north-west, but could find no water; he had shot at one seal in a crack, but lost him; had seen several others, but got nothing.

"*March* 9. Cloudy; latter part of the day strong breeze from the north-east, and snowing. Joe, Hans, and myself went off early looking for something to eat. I discovered a fresh bear-track, and followed it for some distance, but did not see him; and not feeling well to-day, I returned home. Our exclusive meat diet does not agree with me. In fact, it is surprising that any of us keep as well as we do. Hans's little boy is the only one who has been really sick for any length of time, though Meyers has not been well; he is not down sick. Now and then one of the men will complain, but they soon get better.

"Joe and Hans returned at 5 P.M., Joe bringing a small seal. Water, I am glad to state, is making again. Thermometer in the morning, −20°; at noon, 11°.

"*March* 10. We have another storm on us. These storms seem endless. Snowing again, and drifting. There will be no hunting to-day. Another long day of misery. We have passed through many such, and I have survived; but I can not reconcile myself to this life, or this way of living. Captain Hall learned to enjoy it, but I can not. What filth and dirt I am compelled to eat! But if this was necessary to accomplish any valuable purpose, I could bear it as well as any one—except, perhaps, Captain Hall. Since it has come light, I can see it more plainly, and it is *horrid!* But I must endure, and say nothing.

"I have a little incident to note about the oogjook and the men.

"When this great creature was shot, the men were, naturally enough, all happy, thinking of the feast they were going to enjoy. Now, in the common seal, as in most animals, the liver is considered a delicacy, and the men were always anxious for their share; but I knew the liver of the oogjook, like that of the bear, to be poisonous, and told the men they must not eat it, and that the liver of the full-grown animal, like the one we had, was especially dangerous. What they thought I don't know; perhaps they suspected it was a rare tidbit, which they were being deprived of—many sailors have that sort of chronic jealousy. But perhaps it will be better to give the conversation about it, as it occurred. It will show what sort of persons I have to deal with.

"'You must not eat the liver, steward.'

"'Why?'

"'Because it is poisonous.'

"'Oh, d—n the odds; we'll eat it, won't we, Fred?'

"'Well, you can do as you please. I give you fair warning that it will make you sick.'

"They took the liver, thinking, no doubt, that they had deprived our hut of a great luxury, and I heard nothing concerning it for some time. It had been stormy, and it was no uncommon thing for the men to keep within their hut. Not a word about the liver until yesterday, when the steward came in to see me. We have none of us washed while on the ice, and of course we are all very dirty; but I saw, through the dirt of five months' accumulation, that the steward looked sick, and I saw some peculiar white spots on his face; so I asked,

"'What is the matter, steward?'

"'Oh, captain, that oogjook liver played the d—l with me.'

"'Well, you know I told you not to eat it.'

"'That's so; and I'll bet I eat no more of it, or bear's liver either, unless—yes, we might get a *young* bear, and then, perhaps, the liver would be good; but no, I'll be d—d if I trust it. No more liver for me.'

"Finally, I found out that they had eaten that oogjook's liver a week ago, and that the most of them have been sick ever since; but the only one to acknowledge that the liver was the cause of the sickness was Herron. Yet the skin is all coming off their faces, hands, and breasts.

"*March* 11. Last night was one of great anxiety. The gale raged fiercely through the day; about 5 P.M. the ice began a great uproar; our heavy floe commenced working, cracking, with a constant succession of dismal noises, mingled with sharp reports and resounding concussions, and these noises seemed to have their centre immediately under our huts. These sounds commingling with the raging storm, the crushing and grinding resulting from the heavy pressure of the bergs, and heavy ice around us, gave us good reason for alarm. Blowing a gale, with a thick swirl of snow so that one could scarcely see their hand before them, and knowing not but each succeeding moment would bring our snow

BREAKING UP OF ICE-RAFT.

tenements tumbling about our ears, we had got every thing ready to catch and run—but where to? That was the question.

"About nine o'clock, hearing a heavy explosive, and then grinding sound, Joe and I felt our way down in the darkness some twenty yards from the entrance to the hut, and there found the floe had broken. The sides of the severed pieces swaying back and forth, then rushing upon each other and grinding their sides with all the force which the sea and the gale could give them, caused the alarming noises I had heard. We crept back and watched through the night, but nothing more serious occurred.

"The gale still blows this morning, and there is some sea under the ice. Should the ice break up still farther, and should we be obliged to abandon our little snow-burrows, or be actually turned out of them by farther disruption, it would be hard upon the party, with such weather prevailing. . But a kind and merciful God has thus far guided and protected us, and will, I trust, yet deliver us.

"*March* 12. Another twenty-four hours of care, watching, anxiety, and great peril. The gale has been terrible. Yesterday evening, our large floe-piece, on which we have lived all winter, was suddenly shattered into hundreds of pieces, leaving us on a piece about seventy-five by one hundred yards. We passed a dreadful night, expecting every moment that our little piece would follow the fate of the larger, and be broken into yet smaller fragments. But, thank God, it still holds together.

"When I selected the place for erecting the huts, I picked out what seemed to be the thickest and most solid spot, which was not far from the centre; and if it is thick enough, it may be able to endure the shock of riding among these loosened bergs and other fragments, without further disruption; but it is all uncertain; and I almost fear it can not hold together, after the heavy thumping it has already received, and which it still must bear with such a heavy sea as is now running. Most fortunately, our boat remains uninjured.

"The morning of the 12th came at last, and with it the wind moderated. For sixty hours, amidst this fearful turmoil of the elements, with our foundations breaking up beneath our feet, we could not see ten yards around us. But at last the wind has abated, the snow has ceased to fall, and the terrible drift stopped. We can now look around and see the position we are in.

In a vessel, after such a storm as this, the first work, with returning light, would be to clear the decks and set about repairing damages. But how shall we repair our shattered ice-craft? We can look around and take account of loss and damage, but can do nothing toward making it more sea-worthy.

"We see a great change in the condition of the ice; the 'floes' have become a 'pack,' and great blocks of ice, of all sizes and shapes, are piled and jammed together in every imaginable position. On my last extended walk before this storm, the floes had appeared to extend for many miles; they are now all broken up like ours, and the pieces heaped over each other in most admired disorder.

"With the return of moderate weather we recommenced shooting. Seals are scarce, but, there being open water around us and between the cracks, we can now shoot all we see. To-day Joe shot two, Hans one, and I one. So we have four seals this evening.

"*March* 13. It is again blowing strong from the north-east. The weather is much warmer, and I hope the winter is broken. Mr. Meyers took an observation yesterday, and makes the latitude 64° 32', which would place us directly east of Cumberland Gulf. I have no sextant, and no means of accurately ascertaining our position.

"Our reduced piece of ice is now quietly drifting along, and we feel safer; we are surrounded by icebergs which have drifted with us all winter. If Bradford was here now, he might have his choice of bergs to paint. I know well I am not in a condition of mind or body to appreciate the scene surrounding us to-day; but I realize, nevertheless, that an artist would, provided he was on board of some safer craft than this self-navigated fragment of floe.

"Notwithstanding the exciting and dangerous events which we have just passed through, we are all well—which I consider really astonishing. Even Hans's little boy, Tobias, is around again.

"*March* 14. Yesterday and last night it was blowing heavy from the north again; this morning it suddenly ceased, and the day has been fine, with a light south wind. Joe and myself were out looking for seals before any one else was up. But our domain is wearing away at the edges. We can stand in our own hut door and shoot seals now, for our piece is so reduced that it is only twenty paces to the water!

"Soon after sunrise I espied a large oogjook. Joe was at a distance; and not having had so much practice as he, and fearing I might not kill it with my inferior rifle, I beckoned to Joe to come along with his 'Springfield.' In the mean time, to keep the creature from slipping away, I commenced whistling. Seals are really attentive to such sounds, whatever some writers say to the contrary; if they hear music, singing, or whistling, or even a pleasantly-intoned voice, they will keep still and listen. So I whistled away until Joe crept along to within shooting distance, and killed my oogjook. He has also killed three seals to-day, but one sank and was lost.

"The thermometer to-day has been up above zero; in the evening $+6°$. Our latitude now is said to be $64° 19'$ N., which would make thirteen miles' drift southward in the last forty-eight hours.

"*March* 15. A strong breeze from the west, but clear and pleasant. No seals to-day. I think the strong wind unfavorable for seal-hunting. No doubt they scent the hunters when the wind sets toward them; and when it snows, with wind, and is otherwise bad weather, they appear to keep out of sight—don't like it, perhaps, any better than we do to be out in a snow-drift. Thermometer at sunrise $-2°$; at noon rose to $+10°$.

"I am looking around our snow-hut this evening, and can not describe how nasty and dirty it is. I know it is impossible to be really clean living as we do, but one would hardly think that any one could relapse into quite such horrid practices who had ever lived among civilized people, as Hannah has done for years. But among the Americans Hannah learned one thing that has been of no benefit to her, and which has added many annoyances to our inevitable discomfort the past winter. She observed among the white folks that it was the custom for men to support their wives, instead of using them as slaves, as her own people do in their natural condition; and, in order to be as much like a white woman as possible, she has positively declined to do—has certainly omitted to do—many things which would have made this hut more tolerable. Perhaps, however, should a new-comer see *me*, he might criticise my enforced habits, as I am now criticising hers.

"*March* 16. No luck to-day. Before sunrise Hans shot one seal, but lost him in the ice, which has been moving rapidly all day. No other seal has been seen. Cloudy this morning, with strong north-west breeze.

"*Afternoon.* Clear; wind has shifted to the north-east. This afternoon I saw several narwhals; put three balls into one, but he carried them all off. The harpoon is the only thing for a narwhal, and that we have not got. Had one, but it was made over into a spear by one of the men. A spear supplements a harpoon remarkably well, but it is no substitute for one. Thermometer did not get above zero to-day, $-5°$."

"*March* 17. Cloudy, and wind to the east; snowing.

AN ESQUIMAU PILOT.

CHAPTER XXV.

A Bear prospecting for a Meal.—The Ice in an Uproar.—Seven Seals in one Day.—Spring by Date.—The "Bladder-noses" appear.—Off Hudson Strait.—A Bear comes too close.—A lucky Shot in the Dark.—Description of *Ursus maritimus.*—Milk in the young Seal.—Fools of Fortune.—We take to the Boat.—Rig Washboards.—A desperate Struggle to keep Afloat.—Alternate between Boat and Floe.—Striving to gain the west Shore.—Dead-weights.—Ice splits.—Joe's Hut carried off.—Rebuild it.—Ice splits again, and destroys Joe's new Hut.—Standing ready for a Jump.—Our Breakfast goes down into the Sea.—No Blubber for our Lamps.—The Ice splits once more, separating Mr. Meyers from the Party.—We stand helpless, looking at each other.—Meyers unable to manage the Boat.—Joe and Hans go to his Relief.—All of us but two follow.—Springing from Piece to Piece of the Ice.—Meyers rescued.—He is badly frozen.—Mishaps in the Water.—High Sea running.—Washed out of our Tent by the Sea.—Women and Children stowed in the Boat.—Not a dry Place to stand on.—Ice recloses.—Sea subsides.—Land Birds appear.—No Seal.—Very Hungry.

"No seal-meat to-day. This morning I discovered a bear, about five o'clock; had quite an exciting chase after him, but he got away. Both Joe and Hans fired at him, but missed. The bears have been within twenty paces of our hut through the night. Their tracks are everywhere around us, but we have not yet succeeded in getting one. These bears are almost as much water animals as the seals. I have seen them swimming among the loose ice a hundred miles from any land.

"At meridian our latitude was 63° 47' N., showing a drift of thirty-two miles in three days.

"*March* 18. No game to-day, and nothing seen but two or three narwhals, which it appears impossible to get. The weather is quite cold again, $-15°$.

"*March* 19. The ice is commencing another uproar, crushing and grinding—berg against berg, and bergs against the pack-ice, and the separate portions of the pack crowding and pushing as if each separate block was determined to get to the front. Colder than yesterday by one degree.

"*March* 20. These March winds have been very cold. We have another north-wester. Though the ice is continually opening, it freezes over again in a few hours strong enough to bear.

The ice opens more or less every day now, keeping us constantly on the lookout for the safety of our huts. Have seen a few seals to-day. Hans fired at several, and got one little fellow. Joe also fired several times, but the wind seemed to carry the bullet from its course. The wind is both strong and cold, and unfavorable for shooting.

"February and March have been two dreadful months, blowing and snowing almost continuously, but, with all the bad weather, we have been mercifully provided with enough meat during March. Had it not been for the oogjooks I know not how we could have subsisted, for our bread and pemmican would then have had to be eaten, and, on the smallest allowance, would have been gone by the 1st of April. But now we eat nothing but meat, and we consume every part of these creatures, except such as is dangerous to health and life. The diet is not agreeable, but it is strengthening.

"*March* 21. Clear and cold. Strong breeze from the northwest. Joe and Hans have been sealing, and have had a fortunate day. The hole where they found the seal was a mile off. They traveled over the newly cemented ice, and succeeded in shooting, Joe six, and Hans one—seven seals! Our stock is increasing fast; we have enough meat now to last through March, and I do not fear for April as to the matter of game, as we are approaching still better hunting-ground. But what a gory appearance our little hut presents—a perfect shamble! The blankets of the creatures are, of course, mainly saved for oil; and when we eat enough, it takes two of these small seals to supply the whole company.

"*March* 22. An agreeable variety in the weather; it is both clear and pleasant, with a light wind from the west. Joe, Hans, and myself went off early to the sealing-holes. Considerable young ice had formed through the night, and Joe shot two seals —only one day's fare if there is no restriction put upon the men. These Germans are tremendous eaters and outrageous grumblers. They seem to be possessed with the idea that they can improve every thing—as they did the useful harpoon into a useless spear, and, in consequence, nearly every rifle we had upon the ice but Joe's, which they could not get hold of, has been ruined by their tinkering. They must work away at every thing, and never stop till it is rendered useless.

"The sun entered the first point of Aries yesterday, and is now on his upward course. Spring is here, according to the astronomers, and the weather shows that it is at least approaching. The thermometer has marked 10° to 15° above zero. Oh how I wish that two months more were passed! This is a dreadful life, and we have been a long time in it—over five months now; but we 'still live.'

"*March* 23. Our promised spring appears to have deserted us again. It is blowing strong and cold from the north, and the ice appears to be frozen together again everywhere within seeing distance. There has been no hunting to-day.

"*March* 24. We started about 8 A.M. for a hole of water, which Joe discerned in the distance to the eastward. Saw a few seals; Joe shot one. We also discovered bear-tracks in the vicinity of our huts; we see them now frequently. It has been cloudy to-day, with strong breeze from the north-west. Thermometer varying slightly in the vicinity of zero.

"*March* 25. Off hunting; got two seals. Ice remains the same, and no water within a mile. I went over to the water to-day, but rheumatism compelled me to return at noon. By observation to-day our latitude is 61° 59' N.—the cold strong breeze still blowing from the north-west. We are down now where I expected to find the large hooded seal, or, as we call them, 'bladder-noses.' The weather is so very cold, I think it prevents them showing themselves on the ice. A very few of these large seals, and there would be no more risk of starving.

"*March* 26. The bladder-noses are here! I thought they could not be far off. Shot nine large ones to-day, and saved four—five of them sank. Joe shot three, and Hans one. Thank God, we have now meat enough for eighteen or twenty days. Saw one whale to-day.

"*March* 27. Our whole company feel cheered and encouraged, knowing we have now got to the promised seal-grounds, where plenty can be obtained; and our ammunition holds out well. One of the men, Fred, got a bad, but not dangerous, cut in the thigh; it was an accident, and I think will soon heal up.

"We are now in the strong tides off the mouth of Hudson Strait; but we can see no land. The ice is on the move, but without any present signs of disruption.

"*March* 28. We have got a bear at last! Shortly after dark

last evening, we heard a noise outside of our hut. I had just taken off my boots, preparing for rest. Joe, too, was about retiring, but on hearing the noise thought it was the ice breaking up, and that he would go out and see what the situation was. He was not gone more than ten seconds before he came back, pale and frightened, exclaiming, 'There is a bear close to my kyack!' The kyack was within ten feet of the entrance to the hut. Joe's rifle, and also mine, were outside—mine lying close to the kyack—Joe's was inside of it; but Joe had his pistol in the hut. Putting on my boots, we crept cautiously out, and, getting to the outer entrance, could hear the bear distinctly eating. There were several seal-skins and a good deal of blubber lying around in all directions. Some of the skins we were drying for clothing, and some were yet green. Getting outside, we could plainly see his bearship. He had now hauled some of the skins and blubber about thirty feet from the kyack, and was eating away, having a good feast. Joe crept into the sailors' hut to alarm them. While he was gone, I crept stealthily to my rifle, but in taking it I knocked down a shot-gun standing by. The bear heard it, but my rifle was already on him; he growled, I pulled the trigger, but the gun did not go; pulled the second and third time—it did not go; but I did, for the bear now came for me. Getting in the hut, I put another cartridge in, and put two reserves in my vest-pocket, and crept out again, getting a position where I could see the animal, although it was what might be called quite dark. He saw me, too, and again faced me; but this time, to my joy and his sorrow, the rifle-ball went straight to its mark—the heart I aimed for. Joe now came out of the men's hut, and cracked both a rifle and pistol at him. The bear ran about two rods, and fell dead. On skinning him in the morning, I found that the ball had entered the left-shoulder, passed through the heart, and out at the other side—a lucky shot in the dark!

"This bear will at least give us a change of diet, if it is still meat. He is a fine large animal, and every part good but the liver. The meat tastes more like pork than any thing we have had to eat for a long time.

"It may be thought strange by those who have never lived in this climate in an igloo, that we should leave our guns outside of the hut, instead of keeping them by us; but if brought in they would soon be spoiled, because the exhalations from the lungs condense in this atmosphere, and form moisture, which settles on every thing, and would spoil fire-arms, unless carefully cased, and we have no casings.

"This bear was what is called by the whalers the 'sea bear' (*Ursus maritimus*), and it is almost amphibious, as it swims quite as well as it walks, only I suppose it could not live entirely in the water; and it might live exclusively on land if it could get sufficient food. It is a modification of the common Arctic bear, and necessity makes it seek its food, which is principally seals, either upon the ice or in the water, as opportunity offers.

"*March* 30. Night before last the wind sprung up strong from the north-west. Yesterday it increased to a gale. Huge bergs—and I do not in the least exaggerate when I say hundreds in number—were plowing their way through the ice: there was quite a heavy swell under the ice, and the broad bases of these bergs are sunk many fathoms deep in the water. The floe-ice had refrozen mostly together again, after the break-up in the middle of March, and was now once more in fragments. The gale continued heavy through the night of the 29th, keeping us on the lookout for the safety of our piece. It is still blowing heavy, with considerable swell. In the night I felt a great thump, as if a hammer a mile wide had hit us, and getting out to see what was the cause, found we had drifted foul of a large berg, and the collision had produced the sensation I have described. Well, we thumped a while on the berg, and I did not know but we should go to pieces and founder; but after finally we cleared it, and sailed on, apparently without serious injury to our brittle craft.

"This morning it is snowing again, with heavy drift. We can see but a short distance before us. We are somewhere off the mouth of Hudson Strait, but how far from shore I have no means of ascertaining. Our little ice-craft is plowing its way through the sea without other guide than the Great Being above.

"6 P.M. Still blowing strongly, but little snow drifting. This afternoon saw two "bladder-noses" floating on the ice; got the boat launched, and went for them. The male escaped to the water, but we got the female and her little young one. Hans, later in the day, shot another young one. When the young of the seal can be secured without shooting, it is customary to press them to death by putting the foot down heavily upon them, as by this means not only all the blood is saved, but the milk in the stomach; and among the Esquimaux this milk is highly relished. The men put some of the milk in their blood-soup. These bladder-noses, when attacked, often show considerable fight, if approached with spears or clubs. But they can do nothing against bullets but get out of the way.

"Our piece of ice is gradually wearing away; last night there was a heavy sea, water all round us, and scarcely any ice to be seen; but it may close again. Latitude at noon reported 59° 41' N.

"*April* 1. We have been the 'fools of fortune' now for five months and a half. Our piece of ice is now entirely detached from the main pack, which is to the west of us, and which would be safer than this little bit we are on, and so we have determined to take to the boat and try and regain it. To do this we must abandon all our store of meat, and we have sufficient now to last us for a month, and many other things. Among the most valuable, much of the ammunition will have to be left, on account of its weight—all the powder being put up in metallic cartridges, for preservation against damp and other accidents.

"We got launched, and made some twenty miles west, but were very nearly swamped, for, notwithstanding all we had abandoned, we were still excessively overloaded, what with nineteen persons and the heavy sleeping-gear. When it is considered that the boat was only intended for six or eight men, and that we had to carry twelve men, two women, and five children, with our tent, and with absolutely necessary wrapping of skins for protection from the weather, it is not surprising that we did not make much headway. We were so crowded that I could scarcely move my arms sufficiently to handle the yoke-ropes without knocking over some child—and these children frightened and crying about all the time. Having got about twenty miles, we were compelled to hold up on the first piece of good ice we could find. It was with much difficulty that through these changes I preserved Captain Hall's writing-desk from destruction; some of the men were bound to have Joe throw it overboard, but I positively forbade it, as it was all we had belonging to our late commander.

"On this ice we spread what few skins we had, set up our tent, and ate our little ration of dry bread and pemmican. Hans and his family had the boat for sleeping-quarters.

"On the morning of the 2d we started again, still pushing to the west; but the wind, with snow-squalls, was against us, being from the quarter to which we were steering, and we made but little progress; what we made was S.S.W. Hauled up on another piece of ice, and encamped.

"*April* 3. Spent part of the day repairing the boat, and fitting her up with wash-boards of canvas, to keep the water from dashing over the sides. Seals are so plenty around us now that I do not hear any more croaking about the want of meat. We can get all we want as long as our ammunition holds out. After rigging our boat up, started again, heading to the west.

"*April* 4. After a desperate struggle, we have at last regained the 'pack,' and are now encamped. The sun showed itself at noon, but we are again blessed with a heavy wind from the north and snow-squalls. Our tent is not as good a protection from the wind as the snow-huts. Joe, with a little help, can build a hut in an hour, if the right kind of snow-blocks can be procured. If we were on land we could find stones to help make them of. Mr. Meyers has saved his instruments, and gives us the latitude of our new home as 56° 47′ N.

"We are now on a heavy piece of ice, and I hope out of immediate danger: it looks compact to the westward, but there is no ice to be trusted at this time of the year. We have had a hard battle to reach it, however, and we are all pretty well tired out.

"I did not make any conversation with either Meyers or the men about abandoning the small floe; for the time had come when it was absolutely necessary to do so. I told them in the evening that if the wind abated through the night we must leave in the morning. Some objected to go back into the pack-ice, but wanted to take to the water in the boat. Had I consented to that, most would probably have been lost in the first gale; for we should have had to throw overboard every thing, sleeping-gear, even guns and ammunition; and some of the men, by their expressions, seemed to intimate that they would not have hesitated to throw over the women and children to save their own lives. Then, also, we should have had no water to drink, nor any opportunity to catch game, and, getting once thoroughly wet, our clothes would have frozen on us in the night, and we probably have frozen too, as it is still very cold.

"When we finally got into the boat to try and reach the pack-ice, some again insisted, instead of sailing west, on getting out to seaward, by trying to work south in the boat, which was laden very heavy, and was, of course, low in the water, with nineteen souls aboard, ammunition, guns, skins, and several hundred pounds of seal-meat; and, consequently, the sea began to break over us, and the men became frightened, and some of them exclaimed that 'the boat was sinking.' Of course, I wished to reach the pack without losing any thing more than was absolutely necessary, for we really had nothing to spare; but the boat took water so badly that I saw we must sacrifice every thing, and so the seal-meat was thrown over (the loss of which nearly caused our ruin), with many other things we sadly needed; but the boat had to be lightened, and so I set the example of throwing away some things I prized most highly, that the men might be induced to rid themselves of 'dead-weights;' and after all was done, the boat was still overloaded fearfully; but, turning to the west, by careful management we reached the pack as I have narrated, through great peril and much loss, but with all our company saved.

"*April* 5. Blowing a gale from the north-east, and a fearful sea running. Two pieces broke from our floe at five o'clock this morning. We had to haul all our things farther back toward the centre. Soon after another piece broke off, carrying Joe's hut with it. Fortunately, the snapping and cracking of the ice gave some warning, so that they had time to escape, and also to throw out and save some few things. No telling where it will split next. It has been a dreadful day—the more so that we can do nothing to help ourselves. If there was any thing to be done, it would relieve the mind of much anxious watching. If the ice breaks up much more, we must break up with it. We shall set a watch to-night. Joe has rebuilt his hut, or rather built another. This sort of real estate is getting to be 'very uncertain property.'

"*April* 6. Blowing a gale, very severe, from the north-west. We are still on the same piece of ice, for the reason that we can not get off—the sea is too rough. We are at the mercy of the elements. Joe lost another hut to-day. The ice, with a great roar, split across the floe, cutting Joe's hut right in two.

"We have such a small foothold left that we can not lie down to-night. We have put our things in the boat, and are standing by for a jump.

"*April* 7. Wind still blowing a gale, with a fearful sea running. At six o'clock this morning, while we were getting a morsel of food, the ice split right under our tent! We were just able to scramble out, but our breakfast went down into the sea. We very nearly lost our boat — and that would be equivalent to losing ourselves.

"Of course, while this storm and commotion has been raging around us we could not shoot any seals, and so are obliged to starve again for a time, hoping and praying that it may not be for long. The worst of our present dearth of seals is that we have no blubber to feed the lamp, so that we can not even melt a piece of ice for water. We have, therefore, no water to drink. Every thing looks very gloomy again. All we can do is to set a watch, and be prepared for any emergency. We have set the tent up again, as we held on to that and saved it. Half of the men have got in under it to get a little rest, while the others walk around it outside. This is a very exciting period. If one attempts to rest the body, there is no rest for the mind. One and another will spring up from their sleep, and make a wild dash forward, as if avoiding some sudden danger. What little sleep I get is disturbed and unrefreshing. I wonder how long we can fight through this sort of thing.

"*April* 8. Worse and worse! Last night at twelve, midnight, the ice worked again right between the tent and the boat, which were close together—so close that a man could not walk between them. Just there the ice split, separating the boat and tent, and with the boat was the kyack and Mr. Meyers, who was on the ice beyond the boat. We stood helpless, looking at each other.

"The weather as usual, blowing, snowing, and very cold, with a heavy sea running, the ice breaking, crushing, and overlapping. A sight grand indeed, but most fearful in our position—the helpless victims of this elemental rage.

"Meyers can manage neither the boat nor the kyack—the boat is too heavy, the kyack of no use to any one unaccustomed to its management. Should he get in it, he would be capsized in an instant. So he cast the kyack adrift, hoping it would come to us, and that Joe or Hans could get it and come for him, and bring him a line, or assist him some way. Unfortunately, the kyack drifted to the leeward. However, Joe and Hans took their paddles and ice-spear and went for it, springing from one piece of ice to another, and so they worked over. It looks like dangerous business. We may never see them again. But all the rest of us will be lost without the boat, so they are as well off as we. They are lost unless God returns them. After an hour's struggle through what little light there is, we can just make out that they have reached the boat, which is now half a mile off. There they appear to be helpless.

"It is getting too dark to see the end; it is colder, and the ice is closing around us. We can do nothing more to-night. It is calmer, and I must venture to lie down somewhere and get a little rest, to prepare for the next battle with ice and storm.

"Daylight at last! We see them now with the boat, but they can do nothing with her. The kyack is about the same distance away in another direction. They have not strength to manage the big boat. We must venture off and try to get to them. We may as well be crushed in the ice as remain here without a boat. So I determine to try and get to them. Taking a stick in my hand, to help balance and support myself on the shifting ice-cakes, I make a start, and Kruger follows me. We jump or step, as the case may be, from one slippery wave-washed piece of ice to another—a few steps level, and then a piece higher or lower, so that we have to spring up or down. Sometimes the pieces are almost close together; then we have a good jump to reach the next, and so we go, leaping along like so many goats. On arriving where the boat was, we found our combined strength—Mr. Meyers, well, he was too used up to have any—Joe, Hans, Kruger, and myself —could not stir it. I called over to the other men, and two others got over in the way we had, and still our strength was insufficient. At last all came over but two, who were afraid to venture, and after a long struggle we got her safe back to camp again, bringing Mr. Meyers with us. Both he and Frederick Jamka fell in the water, but were pulled out again. Luckily for them, there were two or three dry suits among the men, so that they could change. We are all more or less wet, and Mr. Meyers badly frozen.

"We have taken our tent down once more, and pitched it nearer to the centre of our little piece of ice, and the boat is alongside, so that we feel comparatively safe once more. Joe has built another hut alongside the tent, and we have breakfasted on a few morsels of pemmican and bread. We have also set a watch to observe the movements of the ice, and the remainder of the men are lying down to get some sleep, of which we are all much in need. Where we are the wind is west-north-west, but outside of the 'pack' there is no wind.

"*April* 9. Things have remained quiet the last twelve hours. During the night the wind was north-west; now blowing a north-east gale outside of the 'pack.' The sun shone for a few minutes—about long enough to take an observation: lat. 55° 51', approximates to that. The sea is running very high again, and threatening to wash us off every moment. The ice is much slacker, and the water, like a hungry beast, creeps nearer. Things look very bad. We are in the hands of God; he alone knows how this night will end.

"*Evening*. The sea washed us out of our tent and the natives from their hut, and we got every thing into the boat once more, ready for a start; but I fear she can never live in such a sea. The sun set clear in a golden light, which has cheered us up with the hope of better weather. The women and children now stay in the boat for safety. The ice may split so suddenly that there would not be time to get them in if they were scattered about. The baby is kept in its mother's hood, but the rest have to be picked up and handled every time there is a change of position on the ice; but we have got thus far without losing any of them.

"The sea keeps washing over, so that there is not a dry place to stand upon, nor a piece of fresh-water ice to eat. We have suffered badly with thirst. The sea has swept over all, and filled all the little depressions where we could sometimes find fresh-water ice with sea-water.

"10 P.M. The ice closing around us fast. The wind and sea going down.

"12 *o'clock, Midnight.* Things look so quiet, and the ice is so well closed, that we have risked setting up the tent once more, and intend to try and get some sleep, for we are quite worn out.

"*April* 10. Last night it was quite calm. To-day it is cloudy and very warm. The ice is closed around, and we are prisoners still.

"The other morning Mr. Meyers found that his toes were frozen—no doubt from his exposure on the ice without shelter the day he was separated from us. He is not very strong at the best, and his fall in the water has not improved his condition.

"*April* 11. Calm and cloudy. We can not, I think, be far from shore. We have seen a fox, some ravens, and other land birds. The ice is still closed around us—nothing but ice to be seen. We have two large bergs almost on top of us; but, fortunately, there is no movement of the ice, or a portion of these overhanging bergs might fall upon and crush us. It is at present calm and still.

"*April* 12. Light wind from the south-east; nearly calm at times. Have seen some seals, but can not get them. Are very hungry, and are likely to remain so. The sun is shining for the first time in a good many days, and the weather is very pleasant. Got an observation to-day: lat. 55° 35′ N.

OOMIAK, OR WOMAN'S BOAT.

CHAPTER XXVI.

Easter-Sunday.—Flashes of Divinity.—Meyers's Suffering from want of Food.—Men very Weak.—Fearful Thoughts.—A timely Relief.—Land once more in Sight.—Flocks of Ducks.—Grotesque Misery.—A Statue of Famine.—A desolating Wave. —A Foretaste of worse.—Manning the Boat in a new Fashion.—A Battery of Ice-blocks.—All Night "standing by" the Boat.—A fearful Struggle for Life.—Worse off than St. Paul.—Daylight at last.—Launched once more.—Watch and Watch.—The Sport and Jest of the Elements.—Lack of Food.—Half drowned, cold, and hungry.—Eat dried Skin saved for Clothing.—A Bear! a Bear!—Anxious Moments.—Poor Polar! God has sent us Food.—Recuperating on Bear-meat. —A crippled, overloaded Boat.—A Battle of the Bergs.—Shooting young Bladder-noses.—Hoping for Relief.

"*April* 13. I think this must be Easter-Sunday in civilized lands. Surely we have had more than a forty days' fast. May we have a glorious resurrection to peace and safety ere long!

"The ice opened again last night, but closed in the morning. It remained open but a few hours, slackening a little to-day. But we can neither travel over it nor use the boat; we can do nothing with it; we might as well be without volition. Our fate is not in our own hands.

"Last night, as I sat solitary, thinking over our desperate situation, the northern lights appeared in great splendor. I watched while they lasted, and there seemed to be something like the promise accompanying the first rainbow in their brilliant flashes. The auroras seem to me always like a sudden flashing out of the Divinity: a sort of reminder that God has not left off the active operations of his will. This, with my impression that it must be Easter-Sunday, has thrown a ray of hope over our otherwise desolate outlook.

"Saw some seals to-day, but the ice being in such a condition, we can not secure any. We should be very glad now of some of the meat we were obliged to abandon. Our latitude is 55° 23′, approximate.

"*April* 14. Wind light, from the north. The pack still close. No chance of shifting our position for a better yet. See seals almost every day, but can not get them. We can neither go through the ice nor over it in its present condition. The weather is fine and the sea calm, or, rather, I should say, the *ice* is calm, for I see no water anywhere. Lat. 55° 13' N.

"Our small piece of ice is wearing away very fast, and our provisions nearly finished. Things look very dark, starvation very near. Poor Meyers looks wretchedly; the loss of food tells on him worse than on the rest. He looks very weak. I have much sympathy for him, notwithstanding the trouble he has caused me. I trust in God to bring us all through. It does not seem possible that we should have been preserved through so many perils, and such long-continued suffering, only to perish at last.

"*April* 15. Nearly calm; very light north wind. The ice still the same. No change except that it was much colder—8° to 10° below zero. Snow is falling very thick, but without wind. Stopped snowing, and sun shining as bright as ever again—a spring 'spurt' of snow. This would be splendid weather to travel now; but we are stayed and can not stir. Meyers looks very bad. Hunger and cold show their worst effects on him. Some of the men have dangerous looks; this hunger is disturbing their brains. I can not but fear that they contemplate crime. After what we have gone through, I hope this company may be preserved from any fatal wrong. We can and we must bear what God sends without crime. This party must not disgrace humanity by cannibalism.

"*April* 16. One more day got over without a catastrophe. The ice is still the same. Some of the men's heads and faces are much swollen, but from what cause I can not discover. I know scurvy when I see it, and it is not that. We keep an hour-watch now through the night. The men are too weak to keep up long together. Some one has been at the pemmican. This is not the first time. I know the men; there are three of them. They have been the three principal pilferers of the party. One of them was caught at it on the 7th of this month. I should not blame them much for taking food, but of course all the others will have less in consequence. We have but a few days' provisions left. We came down still lower on our allowance this morning. Rather weakening work, but it must be done to save life in the end. The idea that cannibalism can be contemplated by any human being troubles me very much.

"*April* 17. Light breeze; S.S.W. The ice the same; no opening yet. Lat. 54° 27'.

"*April* 18. Very light breeze from the north.

"11 A.M. Joe spied a small hole of water about half a mile off. He took his gun, and ventured over the loose ice. Joe is very small and light, and can go where an American can not. He had no sooner reached the spot than we heard the welcome sound of his rifle. He had shot a seal, and called loudly for the kyack, for the water was making rapidly. It took an hour to get the kyack there—an hour of intense anxiety, for we were afraid the seal would float away; but at last, with trouble and risk, it was accomplished, and a nice-sized seal, enough for three meals, rewarded our exertions. We shall have to eat it raw, but we are thankful to get that. It will save us from starving, perhaps worse.

"The water is making quite a lead, and this morning at daylight the joyful sight of land greeted our eyes. It bore to the south-west. We saw it very plainly in the morning, but the weather has become so thick since that we have lost sight of it for the present. It is as if God had just raised the curtain of mist and showed us the promised land to encourage us and keep us from despair. The seal, too, has put new life in us. We have only a few pounds of bread and pemmican left—enough for tonight. The lead has closed up again, but the 'pash' seems to slacken.

"We had visitors to-day—a raven, some other land birds, and a large flock of ducks. I should think there was a hundred and fifty. I wish we could shoot some of them for a meal or two; but they keep off a mile or more. We have eaten up every scrap of that seal, every thing but the gall.

"Poor Meyers, he is tall and very thin. He has on his hands a monstrous pair of deer-skin gloves, ever so much too large for him. It looked quite pitiable, though almost grotesquely amusing (if the case had not been so serious), to see him striving to gather up some bones, once abandoned, to pick at again for a scrap of meat. The gloves were so large, and his hands so cold, he could not feel when he had got hold of any thing; and as he would raise himself up, almost toppling over with weakness, he found time and again that he had grasped *nothing*. If Doré had wanted a model subject to stand for Famine, he might have drawn

Meyers at that moment and made a success. He was the most wretched-looking object I ever saw.

"*Evening.* It looks very threatening; breezing up from the north-east, and the swell increasing.

"*April* 20. Blowing strong from the north-east. There is a very heavy swell under the ice.

"At 9 P.M., while resting in our tent, we were alarmed by hearing an outcry from the watch; and almost at the same moment a heavy sea swept across our piece, carrying away every thing on it that was loose. This was but a foretaste of what was to follow; immediately we began shipping sea after sea, one after another, with only from five to ten minutes interval between each. Finally came a tremendous wave, carrying away our tent, skins, most all of our bed-clothing, and leaving us destitute. Only a few things were saved, which we had managed to get into the boat; the women and children were already in the boat, or the little ones would certainly have been swept into watery graves. All we could do now, under this new flood of disaster, was to try and save *the boat.* So all hands were called to man the boat in a new fashion—namely, to hold on to it with might and main, to prevent its being washed away. Fortunately, we had preserved our boat warp, and had also another strong line, made out of strips of oogjook-skin, and with these we secured the boat, as well as we were able, to projecting vertical points of ice; but having no grapnels or ice-anchors, these fastenings were frequently unloosed and broken, and the boat could not for one moment be trusted to their hold. All our additional strength was needed, and we had to brace ourselves and hold on with all the strength we had.

"As soon as possible I got the boat, with the assistance of the men, over to that edge of our ice where the seas first struck; for I knew if she remained toward the farther edge the gathered momentum of the waves as they rushed over the ice would more than master us, and the boat would go. It was well this precaution was taken, for, as it was, we were nearly carried off, boat and all, many times during this dreadful night. The heaviest seas came at intervals of fifteen or twenty minutes, and between these others that would have been thought very powerful if worse had not followed.

"There we stood all night long, from 9 P.M. to 7 A.M., enduring what I should say few, if any, have ever gone through with and lived. Every little while one of these tremendous seas would come and lift the boat up bodily, and us with it, and carry it and us forward on the ice almost to the extreme opposite edge of our piece; and several times the boat got partly over, and was only hauled back by the superhuman strength which a knowledge of the desperate condition its loss would reduce us to gave us. Had the water been clear, it would have been hard enough. But the sea was full of loose ice, rolling about in blocks of all shapes and sizes, and with almost every sea would come an avalanche of these, striking us on our legs and bodies, and bowling us off our feet like so many pins in a bowling-alley. Some of these blocks were only a foot or two square; others were as large as an ordinary bureau, and others larger; in fact, all sorts and sizes. We all were black-and-blue with bruises for many a day after.

"After each wave had spent its strength, sometimes near the farther edge and sometimes on it, we had then, whenever the boat had got unmoored, to push and pull and drag it back to its former position, and stand ready, bracing ourselves for the next sea, and the battery of the loose ice which we knew would accompany it. And so we stood, hour after hour, the sea as strong as ever, but we weakening from the fatigue, so that before morning we had to make Hannah and Hans's wife get out and help hold on too. I do not think Mr. Meyers had any strength from the first to assist in holding back the boat, but that by clinging to it he simply kept himself from being washed away; but this was a time in which all did their best, for on the preservation of the boat we knew that our lives depended. If we had but 'four anchors,' as St. Paul describes in the account of his shipwreck, we could have 'awaited the day' with better hope; but 'when neither sun nor stars appeared, and no small tempest lay on us, all hope that we should be saved was then taken away'—nearly all. That was the greatest fight for life we had yet had. Had it not been for the strength imparted to us by the last Providential gift of seal-meat, it does not seem possible that our strength would have sufficed for the night; and how we held out I know not. God must have given us the strength for the occasion. For twelve hours there was scarcely a sound uttered, save and except the crying of the children and my orders to 'hold on,' 'bear down,' 'put on all your weight,' and the responsive 'ay, ay, sir,' which for once came readily enough.

"Daylight came at last, and I thankfully perceived a piece of ice riding quite easy, near to us, and I made up my mind that we must reach it. The sea was fearfully rough, and the men hesitated, thinking the boat could not live in such a heavy sea. But I knew that the piece of ice we were on was still more unsafe, and I told them they must risk it, and to 'launch away!' And away she went, the women and children being all snugly stowed in first; and the rest all succeeded in getting in safely but the cook, who went overboard, but, managing to cling hold of the gunwale of the boat, was dragged in and saved. Working carefully along, we succeeded in reaching the piece without other accident; and having eaten a morsel of food, we laid down on our new bit of floe, in our wet clothes, to rest. And we are all to-day well and sound, except the bruises we received from the blows and falls.

"*April* 21. There are no dry clothes for any one to put on, for every sea washed over us, and there is not much sun to-day and but little drying in the air. We have taken off all we can spare to try and dry our clothes.

"The men are now divided into two watches, and part sleep in the boat as best they can, stowing themselves here and there in all sorts of positions. The ice around us is very pashy and thick: we can not force the boat through it, and so must wait for a change. The sun showed himself just long enough to take an observation. Lat. 53° 57'.

"*April* 22. The weather very bad again last night; snow-squalls, sleet, and rain; raining until twelve noon. The ice is closing around us. What we want most now is food. We begin to feel, more than at first, the exhausting effects of our over-strained efforts on the night of the 19th–20th.

"Now, as I recall the details, it seems as if we were through the whole of that night the sport and jest of the elements. They played with us and our boat as if we were shuttlecocks. Man can never believe, nor pen describe, the scene we passed through, nor can I myself believe that any other party have weathered such a night and lived. Surely we are saved by the will of God alone, and I suppose for some good purpose of his own. The more I think of it, the more I wonder that we were not all washed into the sea together, and ground up in the raging and crushing ice. Yet here we are, children and all, even the baby, sound and well—except the bruises. Half-drowned we are, and cold enough in our wet clothes, without shelter, and not sun enough to dry us even on the outside. We have nothing to eat; every thing is finished and gone. The prospect looks bad enough; but we can not have been saved through *such* a night to be starved now. God will send us some food.

"*Afternoon.* If something does not come along soon, I do not know what will become of us. Fearful thoughts career through my brain as I look at these eighteen souls with not a mouthful to eat. Meyers is actually starving. He can not last long in this state. Joe has been off on the soft ice a little way, but can not see any thing. We ate same dried skin this morning that was tanned and saved for clothing, and which we had thrown into the boat when the storm first came on—tough, and difficult to sever with the teeth. Joe ventured off for the fourth time, and, after looking a while from the top of a hummock, saw a bear coming slowly toward us! Joe returned as fast as possible for his gun, all hope and anxiety lest the creature should turn another way. All the party were ordered to lie down (in imitation of seals), and keep perfectly still, while Joe climbed to the top and Hans secreted himself behind the hummock, both with their rifles ready. It was a period of intense, anxious excitement. Food seemed within our reach now, but it might yet escape. The bear came slowly on, thinking, undoubtedly, that we were seals, and expecting to make a good dinner upon us. A few steps more, and he was within range of the rifles; both fired, killing him instantly. We arose with a shout. The dread uncertainty was over. We all rushed to the spot, and bending on a line, dragged him, in grateful triumph, over broken ice to camp —'camp' meaning now our boat and the point of ice where we 'most do congregate.' Poor Polar! he meant to dine on us, but we shall dine on him. God has sent us food.

"The blood of the bear was exceedingly acceptable; for though we had more water than enough on the outside, we had nothing to drink, and were very thirsty. This bear was farther to the south than Arctic bears usually come. His stomach was empty, and he was quite thin; but his flesh was all the better for that. When permeated with fat, it is gross feeding, and very strong. We had no hope of seeing a bear in this latitude.

"*April* 23. Wind east-north-east, and later in the day north-north-east, where I hope it will remain. The weather is still disagreeably full of rain, squally and cloudy. We are now living entirely on raw bear-meat. Every thing wet still, but looking for brighter days. This can not last long at this time of the year; but we are still surrounded by this miserable pash, and can not get free. All well. Mr. Meyers recuperated since refreshed by the bear-meat.

"*April* 24. Wind still north-north-east, sometimes backing to the north; raining all last night, and still continuing; every thing wet through for several days now; no possible means of drying.

"Saw a large flock of ducks this morning, and another later in the day. Can not be far from land, of which we get glimpses now and then, when the falling weather holds up a little, and then, again, we seem to be driven from the coast.

"There was a fine lead of water last night, and I thought we were going to have a chance to take to the boat and get to shore, but it soon closed up again. Another lead to-day, but farther off.

"*April* 25. Wind increased to a gale last night from the north-east; raining all night and all day. If it was not for the bear-meat we should be chilled to death—that keeps some heat in us; but it is not equal to seal-meat for that, though it is tender and good. Now and then, for variety, we have a snow-squall. We launched our boat this morning about five o'clock, determined to try and get to land, though the attempt was dangerous in the extreme; for the boat was badly damaged, with her struggle on the ice and other hard usage. She was scratched and patched, but we have no means now of putting her in repair. It seemed like putting to sea in a cracked bowl. But what were we to do?

"The piece of ice we were on had wasted away so much that we knew it could never ride out the gale. The danger was very great either way. The light, overladen, damaged boat looked as if she would founder; but the ice certainly would before long, if not founder, be broken up into pash, affording us not even a foothold.

"So, with this crippled, overloaded boat, we start, the wind blowing a gale, and a fearful sea running, full of small ice as sharp as knives. But thank God we came safe through it, and, after eight hours' fruitless labor at the oars—for we made no westing —hauled up on a piece of floe, and prepared to camp for the night. It snowed all night and this forenoon; it stopped snowing in the afternoon. We see plenty of water some distance off, but can not get to it. Can take no observation, the sun being absent, and know not how far we have drifted, the weather being too thick for me to recognize the coast. We are all well.

"*April* 28. A gale of wind has sprung up from the westward, and a heavy sea is running. Water again washing over our little bit of floe. Had to stand 'all ready' by the boat again all night, Not quite so bad as the other night, but had snow-squalls all the time, and the following forenoon. The ice seeming unsafe from the effects of the gale, we again launched our boat at daylight, but could get nowhere for the small ice, a heavy sea, and a head wind blowing a gale right in our teeth. Had to haul up on a piece of ice, after an hour's exhausting but useless effort. Laid down and had a few hours' sleep on the ice.

"3.10 P.M. Threatened by some heavy bergs to be smashed to pieces. These bergs were having quite a battle among themselves, and bearing all the time right for us. The gale has set every thing that can float moving—a grand and awful sight. The sounds accompanying these collisions are frightful, combined with the roar of the waves, and the actual danger to such frail supports as either our bit of floe or slender boat. Seeing they were coming too near, I called the watch, and launched the boat to try and get out of the way of these approaching hostile bergs. We left our floe at one o'clock in the afternoon, the ice very slack, and more water than I have seen for a long time.

"Joe shot three young bladder-nose seals as we were coming along, and, not being very large, we took them into the boat. Hope soon to see whalers.

CHAPTER XXVII.

A joyful Sight!—A Steamer in View.—Lost again.—She disappears.—Once more we seek Rest upon a small Piece of Ice.—The Hope of Rescue keeps us awake.—Another Steamer.—We hoist our Colors, muster our Fire-arms, fire, and shout.—She does not see us.—She falls off.—Re-appears.—Gone again.—Still another Steamer.—Deliverance can not be far off.—Another Night on the Ice.—Hans catches a Baby Seal.—"There's a Steamer!"—Very Foggy, and we fear to lose her.—Hans goes for her in his Kyack.—She approaches.—We are saved!—All safe on board the *Tigress.*—Amusing Questions.—A good Smoke and a glorious Breakfast.—Once more able "to wash and be clean."—Boarded by Captain De Lane, of the *Walrus.*—Meyers slowly recovering.—A severe Gale.—Six hundred Seals killed.—Captain Bartlett heading for St. Johns.—The Esquimaux Children the "Lions."—Awaiting the Tailor.—Going Home in the United States Steamer *Frolic.*

"4.30. A JOYFUL sight—*a steamer* right ahead and bearing north of us! We hoisted our colors, and pulled toward her. She is a sealer, going south-west, and apparently working through the ice. For a few moments what joy thrilled our breasts—the sight of relief so near! But we have lost it! She did not see us, and we could not get to her; evening came down on us, and she was lost to sight.

"We boarded, instead of the hoped-for steamer, a small piece of ice, and once more hauled up our boat and made our camp. The night is calm and clear. A new moon, and the stars shining brightly—the first we have seen for a week. The sea is quiet too, and we can rest in peace; for, though one steamer has passed us, we feel now that we may soon see another—that help can not be far off. We take the blubber of the seals, and build fires on the floe, so that if a steamer or any vessel approaches us in the night she will see us.

"We are divided into two watches, of four hours each. We had a good pull this afternoon, and made some westing. The hope of relief keeps us even more wakeful than does the fear of danger. To see the prospect of rescue so near, though it was quickly withdrawn, has set every nerve thrilling with hope.

"*April* 29. Morning fine and calm; the water quiet. All on the lookout for steamers, except those who had 'turned in,' as we still call it. Sighted a steamer about eight miles off. Called

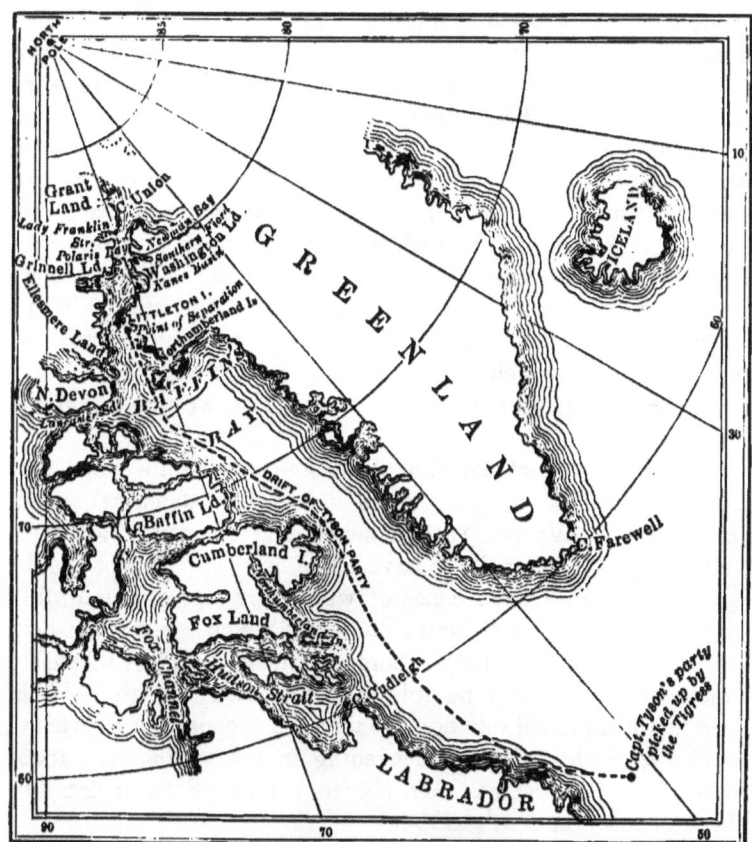

ICE-DRIFT OF THE TYSON PARTY.

the watch, launched the boat, and made for her. After an hour's pull, gained on her a good deal; but they did not see us. Another hour, and we are beset in the ice, and can get no farther.

"Landed on a small piece of ice, and hoisted our colors; then, getting on the highest part of the ice, we mustered our rifles and pistols, and all fired together, hoping by this means to attract their attention. The combined effort made a considerable report. We fired three rounds, and heard a response of three shots; at the same time the steamer headed toward us. Now we feel sure that the time of our deliverance has come.

"We shout, involuntarily almost, but they are too far off yet to hear voices. Presently the steamer changes her course, and heads south, then north again, then west; we do not know what to make of it. We watch, but she does not get materially nearer. So she keeps on all day, as though she was trying to work through the ice, and could not force her way.

"Strange! I should think any sailing ship, much more a steamer, could get through with ease. We repeated our experiment of firing—fired several rounds, but she came no nearer, being then four or five miles off. All day we watched, making every effort within our means to attract attention. Whether they saw us or not we do not know, but late in the afternoon she steamed away, going to the south-west; and reluctantly we abandoned the hope which had upheld us through the day. For a while she was lost to sight, but in the evening we saw her again, but farther off.

"While looking at her, though no longer with the hope that she had seen us or would reach us, another steamer hove in sight; so we have two sealers near—one on each side of us. And though as yet neither have made any sign (except the firing in the morning, the cause of which now appears doubtful), yet we are beginning to count the hours which we can not help hoping will bring us help. Some of these sealers will surely come by us, or we may be able to work down to them. What if we had abandoned our boat, as the men proposed in February!

"*Sunset.* Sighted land this evening in the south-west, about thirty-five miles distant. Mr. Meyers thinks we are in lat. 49°. We are not so far south as that.

"Hans caught a baby seal to-day, the smallest I have seen this season. Our latitude, approximate at noon to-day, 53° 0' 5" N.

"*April* 30. The last day of April, and the last, I hope, of our long trial.

"*Evening.* At 5 A.M., as I was lying in the boat, it being my watch below, but which had just expired, the watch on the lookout espied a steamer coming through the fog, and the first I heard was a loud cry, 'There's a steamer! there's a steamer!' On hearing the outcry, I sprang up as if endued with new life, ordered all the guns to be fired, and set up a loud, simultaneous shout; also ordered the colors set on the boat's mast, and held them erect, fearing that, like the others, she might not see or hear us, though much nearer than the others had been.

"I also started Hans off with his kyack, which he had himself

THE RESCUE.

proposed to do, to intercept her, if possible, as it was very foggy, and I feared every moment that we should lose sight of her; but, to my great joy and relief, the steamer's head was soon turned toward us. But Hans kept on, and paddled up to the vessel, singing out, in his broken English, the unmeaning words, 'American steamer;' meaning to tell them that an American steamer had been lost, and he tried to tell them where we came from; but they did not understand him. We were not more than a quarter of a mile off when we first sighted her. In a few minutes she was alongside of our piece of ice.

"On her approach, and as they slowed down, I took off my old Russian cap, which I had worn all winter, and, waving it over my head, gave them three cheers, in which all the men most heartily joined. It was instantly returned by a hundred men, who covered her top-gallant-mast, forecastle, and fore-rigging. We then gave three more and a 'tiger,' which was appropriate, surely, as she proved to be the sealer *Tigress*—a barkentine of Conception Bay, Newfoundland.

"Two or three of their small seal-boats were instantly lowered. We, however, now that relief was certain, threw every thing from our own boat, and in a minute's time she was in the water, while the boats of the *Tigress* came on, and the crews got on our bit of ice and peeped curiously into the dirty pans we had used over the oil-fires. We had been making soup out of the blood and entrails of the last little seal which Hans had shot. They soon saw enough to convince them that we were in sore need. No words were required to make *that* plain.

"Taking the women and children in their boats, we tumbled into our own, and were soon alongside of the *Tigress*. We left all we had behind, and our all was simply a few battered smoky tin pans and the *débris* of our last seal. It had already become offal in our eyes, though we had often been glad enough to get such fare.

"On stepping on board, I was at once surrounded by a curious lot of people — I mean men filled with curiosity to know our story, and all asking questions of me and the men. I told them who I was, and where we were from. But when they asked me, 'How long have you been on the ice?' and I answered, 'Since the 15th of last October,' they were so astonished that they fairly looked blank with wonder.

"One of the party, looking at me with open-eyed surprise, exclaimed,

"'*And was you on it night and day?*'

"The peculiar expression and tone, with the absurdity of the question, was too much for my politeness. I laughed in spite of myself, and my long unexercised risibles thrilled with an unwonted sensation.

"At this time the captain came along and invited me down into the cabin. I then told him that there was another officer in the party—Mr. Meyers, of the Scientific Department—and he then invited him also to the cabin.

"We had been sitting talking of our 'wonderful,' or, as he called it, 'miraculous' escape, some half an hour. I was very hungry, having eaten nothing since the night before, and I wanted a smoke *so* much; but I saw no signs of either food or tobacco. So I finally asked him if he would give me a pipe and some tobacco.

"He said he 'did not smoke.'

"However, I soon procured both from one of his officers, and had a good long smoke—the first I had had since Joe gave me the two pipefuls, one of those dreary days in our snow-hut. In course of time breakfast came along—codfish, potatoes, hard bread, and coffee!

"Never in my life did I enjoy a meal like that; plain as it was, I shall never forget that codfish and potatoes. No subsequent meal can ever eclipse this to my taste, so long habituated to raw meat, with all its uncleanly accessories.

"*On board the Tigress, May* 1. Captain Bartlett has all his boats down this morning, sealing. Numbers of seals are to be seen lying on the ice. We see also two other steamers not far off engaged in the same business. Joe has joined in with them, and is in all his glory. Captain Bartlett spoke one of these steamers last night, so that, should they arrive home before us, the captain will telegraph the news of our rescue home. God bless the good and kind Captain Bartlett! He is very kind indeed; so are all the ship's company.

"How strange it seems to lie down at night in these snug quarters, and feel that I have no more care, no responsibility. To be *once more clean*—what a comfort!

"We were picked up in latitude 53° 35' N. I have learned

that the steamer we saw on the 29th ult. was the *Eagle,* belonging to St. Johns, Captain Jackman. Captain Bartlett says he could not have seen us, or he would have come for us, or, if he could not, he would have stood by, or in some way tried to save us; that he was noted for his humanity, and had more than once received medals for saving life in these waters. I am glad to know this.

"*May* 2. There is a strong breeze from the north-east this morning. Many seals in sight on the ice, but on the approach of the steamer they instantly take to the water. Three more steamers are now in sight.

"Captain Delane, of the steamer *Walrus,* came on board of the *Tigress* to-day. He was as much surprised as the others had been on hearing my statements. As he is likely to return home before us, he will probably telegraph the news home this evening. Wish I could get a telegram from home to-night.

"It is blowing a gale from the north-east, and snowing; but it is so comfortable in this snug little cabin, that it is almost pleasant to know that there is a storm outside, and that we are sheltered from it—that it may rage without, but can not reach us.

"Mr. Meyers is slowly recovering. He could not have lasted much longer on the ice.

"The boat which has carried us so far, and has served us for store-house and home on the floe-bits, I have made a present to the captain's son—a fine young man here on board. When obliged to leave the floe, I had her fitted up with canvas washboards, to keep out the water. These the men, true to their nature, have commenced destroying. So, to save her from further mutilation, I have given her away; but she is badly worn, and of little use to any body.

"No one, unless they have been deprived of civilized food and cooking as long as I have, can begin to imagine how good a cup of coffee, with bread and *butter,* tastes. When I look back at what we have passed through, I fairly shudder at the recollection.

"*May* 3. The gale has been very violent through the night, and still continues. The good steamer thumps bravely against the ice. Captain Bartlett this morning steamed to the westward, to escape the swell. He says it is the heaviest gale he has experienced this season. Could we have outlived it had we remained exposed?

"*Evening.* The gale continues with unabated violence. The captain has been steaming to the south-west all day, to get clear of the swell which is running under the ice.

"*May* 4. The sun shines brightly this morning, and the wind has hauled to the west, blowing strong and quite cold. Captain Bartlett tells me that he has not experienced any thing like it through March or April. We are now fast in the ice, under which there is a slight swell running. Our latitude at noon was 53° 27′ N. How we would have fared on the ice throughout this long, cold gale, I know not. It is the general opinion on board that we should have perished, being so near the ocean. But He that guided us so far was still all-powerful to save.

"There is no steamer in sight to-day. Captain Bartlett thinks of returning home soon. I hope he will, for I feel sadly worn.

"*May* 5. We had a light fall of snow this morning. But in the afternoon it cleared off, with a light southerly wind. The ice remains close and compact around us—the ocean not far off. Can see four steamers in that direction now. Nearly all the ship's company are off, with gun and gaff-hook, after seals. Saw some four or five miles to the east.

"I had the pleasure last evening of attending divine service. It was a pleasure to me to see the rough men attend so respectfully to the good old captain while he read the Episcopal service. The boys and most of the men kneel nightly in prayer. The Episcopal service has this advantage, that it can be used with sincerity and devotion by men in circumstances where, perhaps, no one would feel competent to lead their companions in extempore exercises. And there can always be found in that Prayer-book a prayer which will suit the circumstances. To any one who has been deprived of united religious exercises for sixteen or eighteen months, it is indeed refreshing to hear the grand old prayers of the Church read by lips that you know are sincere and true.

"*May* 6. It is blowing strong from the north; snow-squalls and cold weather. The crew of the *Tigress* did not get on board till midnight, but they had killed about six hundred seals. The Newfoundland sealers have learned of late years to stow seal-skin cargoes better than they used to. I have heard old sealers tell of having to abandon their ships from the oil making in the hold, and rising so as to flood not only the forecastle, but the cabin; for oil will work through any thing.

"After stripping the seals, instead of trying the oil out immediately, they used to put the skins containing all the blubber loose in the hold, and sometimes, if they met with continued rough weather and storms, the skins would shift and roll about, so as to work out the oil, and the oil, of course, being lighter than the skins, would rise and work through as described. Cargoes would be spoiled, and sometimes the vessel too. But after a while they learned better. It is now their custom, if try-works are not set up, to lay the skins compactly, and secure them from shifting by stakes and beams properly fixed. The seal-fishery is of great value to the United States, and ought to be encouraged. Seal-oil is excellent for light-house lamps.

"This morning the captain forced his vessel to where the dead seals were, but only succeeded in getting between two and three hundred out of the six which his men had killed. The greater part had been taken by another steamer belonging to St. Johns, Newfoundland.

"I regret to record that the captain's son injured his hand by the premature discharge of a Remington rifle—one that was brought on board by the Esquimaux.

"*May* 7. Captain Bartlett has concluded to go home. We have been going south since morning. It is cloudy, with a strong breeze from the north, and snowing. Several of our men are complaining; two are down sick; Joe and Hannah are also ill. Both Mr. Meyers and myself are troubled with swollen feet and ankles. Mr. Meyers's hands are frozen, and need attention.

"*May* 8. The breeze is still strong from the north. Cloudy this morning. Sighted Fogo Island, one hundred and thirty miles from St. Johns.

"Owing to a defect in one of the boilers, we have been running under canvas. Should the weather prove favorable, we shall get to St. Johns to-morrow. We have all been troubled, since coming on board, with colds, swollen feet, sore-throats, and rheumatism; but my appetite is good, and I somewhat astonish our good captain by my able performance at table.

"Captain Bartlett altered his mind during the night, and, instead of going to St. Johns, he put in to Conception Bay, some thirty-five miles north of St. Johns. He will stay here until Monday, landing the boats and sealing-gear and various things from his vessel, preparing her to be hauled out at St. Johns for repairs. The particular port or harbor where we landed is called Bay Roberts.

"*May* 8. While dining on shore with the captain to-day, we were visited by the American consul of Harbor Grace. We gave him the particulars of our journey on the ice, to telegraph to Mr. Molloy, our consul at St. Johns.

"I have furnished the crew with quarters on shore with Mr. J. Kelpam, who is very kind to them, and has taken them out riding to see the country—those who are able to go. They are all more or less complaining, and several of them sick. They will remain on shore till the vessel is ready to start. The consul here furnished me with sixteen dollars to divide among the crew.

"*May* 10. Fine weather. Captain Bartlett is busy lightening his vessel, preparing her for hauling on to the ways. I have received many kind invitations to visit the shore, but not feeling well, and not being suitably clothed, they are of course declined.

"*Sunday, May* 11. A bright, beautiful day. Pressing invitations to dine on shore are sent to me from many quarters; all of which I decline, for reasons named above. Besides, I feel little inclined to mix in general company at present. But I appreciate the good feeling which dictates the invitations.

"Captain Bartlett's wife sent me a basket of apples this morning; quite a present here, and very acceptable. Apples must be very scarce here at this time of the year. I have been jotting down a few dates, and making *mems*. by which I may be able to reconstruct my lost journal in case the *Polaris* is abandoned.

"*May* 12. It is a splendid morning. 6 A.M., I am waiting patiently for the captain and crew to come on board, and get under way for St. Johns. I wish once more to get clean clothes on me, to get properly shaved, and thoroughly cleaned. Many persons from Harbor Grace were on board yesterday, to see the men who had drifted fifteen hundred miles on the ice.

"The Esquimaux and their families, and all of the men, are troubled with heavy colds, swollen feet and legs, since coming on board. It is the scurvy coming out.

"8.35 A.M. We are under way. Several lady passengers are on board, on their way to St. Johns. We have a southerly wind and fine weather. Arrived at St. Johns at 8 P.M. Crowds of people on the wharf, to see the waifs who have drifted so far on the ice. The Esquimaux children attract much attention, espe-

UNIV. OF
CALIFORNIA

THE COMPANY WHO WERE ON THE ICE-DRIFT WITH CAPTAIN TYSON.

cially the baby, Charlie Polaris. Collections of money were constantly being made by successive visitors to the *Tigress*, for the benefit of these little ones. The American consul, Mr. Molloy, was soon on board, and appears ready to do any thing and every thing for our comfort.

"The men are already (8.30 A.M.) ashore, and I hope provided for. Mr. Meyers and I remain on board, awaiting the tailor to make us presentable.

"*May* 13. Cloudy; wind easterly. The harbor of St. Johns is full of drifting ice and bergs in sight. This afternoon received a telegraphic dispatch from Messrs. Harper & Brothers, requesting a photographic group of our party for the *Weekly*. Have also received a dispatch from Mr. Robeson, Secretary of the Navy, ordering me to take charge of the men and of the Esquimaux on their passage home.

"*May* 14. The consul has furnished all necessary funds for new outfits, etc., for the men; and whatever Mr. Meyers and myself required to draw on him for was promptly responded to.

"Many ladies call at the hotel to see Joe, Hannah, and the child, who are stopping with us—some knew of them before, through Captain Hall's book. Hans and family are stopping in a house opposite. Many go to see them also; and almost every one asks, 'How she took care of the baby on the ice?'

"It is little care Esquimaux babies get. They are pulled out of the hood for nursing, and not much else; the only washing they get is such as a cat administers to a kitten; and, while in the hood, they have no clothes on.

"I have been obliged to prohibit visitors to the Esquimaux. Most of the children are sick, partly from the effects of cakes and candies given them by visitors—things they are not used to, and can not bear all at once in profusion, after such different diet. They all need rest and quiet.

"*May* 16. Many visitors to-day. The governor's wife, Mr. and Mrs. Oliphant, and many other distinguished gentlemen and ladies, came to see us.

"Received the intelligence that the Secretary of the Navy has ordered the United States steamship *Frolic* to come for us to St. Johns, and convey us direct to Washington.

"I shall always remember the kindness of Captain Bartlett and the people of St. Johns. And thus ends our strange, eventful history."

CHAPTER XXVIII.

THE SEARCH FOR THE POLARIS, AND THE SURVIVORS OF THE EXPEDITION.

The News of the Rescue.—Captain Tyson and Party arrive at Washington.—Board of Inquiry organized.—Testimony given as to lax Discipline.—The *Juniata*, Commander Braine, dispatched, with Coal and Stores, to Disco.—Captain James Buddington, Ice-pilot.—Captain Braine's Interview with Inspector Karrup Smith, of North Greenland.—*Juniata* at Upernavik.—Small Steam-launch.—*Little Juniata* essays to cross Melville Bay.—Repelled by the Ice.—President Grant in Council with Members of the National Academy of Sciences.—Purchase of the *Tigress*.—Description of the Vessel.—Necessary Alterations.—List of Officers.—Captain Tyson Acting Lieutenant and Ice-pilot.—A Reporter to the *New York Herald* ships as ordinary Seaman.—Esquimau Joe ships as Interpreter.—Several Seamen belonging to the Ice-floe Company ship in the *Tigress*.—Extra Equipments.

WHEN the first news of the rescue of the weary waifs of the ice flotilla was flashed over the wires from Newfoundland, a thrill of mingled astonishment and incredulity swept through the community; but to those who put faith in the first telegram sorrow for the death of Captain Hall was added to the first emotion of surprise. It seemed almost incredible that nineteen persons, including women and small children, with a babe of two months old, could possibly survive a journey of six months through the darkness and cold of an Arctic winter. The telegram was brief, and read thus:

St. Johns, Newfoundland, May 9, 1873.

The English whaling-ship *Walrus* has just arrived, and reports that the steamer *Tigress* picked up on the ice at Grady Harbor, Labrador, on the 30th of April last, fifteen of the crew and five of the Esquimaux of the steamer *Polaris*, of the Arctic expedition. Captain Hall died last summer. The *Tigress* is hourly expected at St. Johns.

Even after the general outline of the story had been given, with sufficient details to convince any reasonable person that it could not be a fabrication, Arctic experts were found who pronounced the story 'impossible' and 'ridiculous.' So wonderful was the preservation, and so fearful the difficulties to be overcome, that some of those who knew those regions best were the last to be convinced of the truth.

Immediately on receipt of the news, the Secretary of the Navy dispatched the United States steamship *Frolic*, Commander Schoonmaker, to the port of St. Johns, with orders to bring all the survivors to Washington.

On the 27th of May, Captain Tyson, with the whole of the rescued party, went on board the *Frolic*, which sailed on the next day, arriving at Washington on the afternoon of the 5th of June. In his official report, Commander Schoonmaker speaks particularly of the favorable impression produced upon him by Captain Tyson.

At once a board of inquiry was organized on board of the United States steamship *Tallapoosa*, composed of Commodore William Reynolds, the senior officer of the Navy Department, Professor Spencer F. Baird, of the National Academy of Sciences, and Captain H. W. Howgate, of the Signal Service Corps, and presided over by the Hon. Secretary of the Navy, George M. Robeson, and an examination of all the adults of the rescued party took place (except Hans's wife, who can not speak English). As the painful details were revealed in the plain, unvarnished tale of Captain Tyson and the rest, incredulity as to the main facts was no longer possible, while still the wonder grew at the miraculous preservation of the party.

On the examination it was learned that, when the vessel was separated from the floe, there was left on board of the *Polaris* fourteen persons:

Captain Sidney O. Buddington, sailing-master; Dr. Emil Bessel, chief of the Scientific Corps; R. W. D. Bryan, astronomer and chaplain; H. C. Chester, first mate; William Morton, second mate; Emil Schuman, chief engineer; A. A. Odell, second engineer; W. F. Campbell, fireman; John W. Booth and N. J. Coffin, carpenters; Jos. B. Mauch, Herman Sieman, Henry Hobby, and Noah Hays, seamen.

The *Polaris* was also reported to be in a leaking condition. Six persons, including Captain Tyson and Mr. Meyers, testified to the drinking habits of Captain Buddington, and all the rest to the lack of discipline in the vessel under him, and several to the fact of his having expressed himself "relieved," and having "a stone taken off his heart," by the event of Captain Hall's death.

Others threw a doubt over the cause of the commander's death, which Dr. Bessel had pronounced apoplexy. Six days were occupied in the examination of these parties, and the result was that the Secretary of the Navy decided to send out immediately a relieving party to search for and bring back the remnant of the *Polaris* expedition. While negotiations were pending for a suitable vessel to send on the search, the head of the Naval Department utilized the intervening time by sending forward the United States steamship *Juniata*, Commander Braine, to form a dépôt of supplies on the coast of Greenland, with orders there to await the coming of the relief party.

THE JUNIATA.

The *Juniata* took with them a small steam-launch about thirty-two feet in length, for the use of parties penetrating the fiords and small inlets along the coast, Commander Braine and his whole party entering with commendable zeal upon the duty of making preliminary search, so far as their means, inexperience, and limited orders permitted. On board of the *Juniata* was Captain James Buddington, an uncle of Sidney O. Buddington, the missing captain of the *Polaris;* the former sailed in the capacity of ice-pilot to the *Juniata*. This vessel was a screw-propeller of some eight hundred tons; and on account of her size, which exposed her to great risks on the imperfectly charted coast of Greenland, and the fact that she was not built to contend with Arctic ice-packs, she was ordered to remain at Disco, or, at the farthest, Upernavik, whence it was hoped news might be received of the missing party.

The *Juniata* arrived at St. Johns, Newfoundland, early in July, making the run from New York in five days and eighteen hours.

While lying in harbor there, some extra iron sheathing was put on her bows to strengthen her, as she would necessarily meet more or less ice, even in getting to Disco. The steam-launch was also partially iron-clad.

The *Juniata* left St. Johns on the 11th of July, and reached Holsteinborg on the 21st, having touched at Fiscanaes and Sukkertoppen. At Holsteinborg dogs for sledges and seal-skins for clothing were purchased for the use of the searching party expected to arrive. From Holsteinborg the *Juniata* went to Disco, and from thence to Upernavik, which she reached on the 31st of July. Here Commander Braine learned from the inspector, Mr. Karrup Smith, that he had in his possession certain records of Captain Hall's (referred to in Captain Tyson's journal), relating to his search for Sir John Franklin. Inspector Smith stated that Captain Hall had fears that he might not return, and wished these records preserved. Considering Commander Braine a proper custodian, Mr. Smith transferred the records to his care.

While the *Juniata* remained at anchor at Upernavik, the steam-launch, the *Little Juniata*, was brought into use. She was provisioned and coaled for two months, and a party of eight—all volunteers, under Lieutenant De Long, and the ice-pilot—started off on a searching trip on the 2d of August, reaching Tossac the next day; and, without special event, made the Duck Islands on the evening of the 4th. Pushing on to Wilcox Point, they encountered some pack-ice, and their fuel getting short, they worked up, under sail, across Melville Bay, until within sight of Cape York, when first a heavy fog and then a severe gale compelled their return, and on the 11th of August they were back at Tossac. Considering that nearly all of this little party were new to Arctic scenes, and inexperienced in the peculiar dangers and difficulties of navigation in the ice-beset waters of Melville Bay and vicinity, we can not forbear our meed of praise for the courage and perseverance displayed under such novel circumstances.

From the first inception of the *Polaris* expedition, President Grant had taken much interest in all that related to it, and after the close of the official examination of the ice-floe party by the Board of Inquiry, the President held a consultation with the Secretary of the Navy, and subsequently a conference was held with President Henry, Professor Spencer F. Baird, Professor Hilgarde, of the Coast Survey Office, and Professor Newcomb, of the Naval Observatory, for the purpose of devising a more thorough search than the *Juniata* could make for the rescue of the remainder of the *Polaris* party. All the gentlemen above named were members of the National Academy of Sciences, which had furnished instructions to the Scientific Corps of the *Polaris*. The result of this consultation was, that the President authorized the Secretary of the Navy to purchase the steam-sealer *Tigress*, and to fit her up for a thorough search in the Arctic seas for Captain Buddington and his companions.

The *Tigress* was bought for sixty thousand dollars, with the privilege granted her owners of repurchasing her from the United States at the reduced price of forty thousand. She was a Canadian vessel, built at Quebec in 1871, expressly for the sealing trade, and was rated at three hundred and fifty tons, though her carrying capacity was something over that. It was the supposed strength of the vessel, and her peculiar adaptation to the Arctic regions, which induced the Government to purchase her, instead of employing one of the vessels lying idle in the Navy Yards. Her build differs from ordinary vessels principally in her keel lines and deck sheer; her bow makes a very acute angle; her cutwater is flat, and widens gradually below her water-line. In other words, she flares more than is usual, which enables her to rise upon the floe-ice, breaking it through by sheer weight. As originally built, she was very strong. Her sides forward, for a distance of twenty feet, were over three feet thick, and for twelve feet she was incased in half-inch iron; her whole frame was iron-braced, and covered with a sheathing of wood; while still further to strengthen her against the force of the ice-packs, there was fixed inside at her water-line a number of heavy beams running from rib to rib. In most respects she was as good a vessel as could be found ready-made for the purpose intended; but, unfortunately, her boilers were made for soft coal, and consequently her flue-room was too contracted to burn anthracite; and as the *Juniata* had carried nothing but hard coal to deposit at Disco for the use of the *Tigress*, this arrangement had to be altered. After a short trial-trip the vessel was brought to the Brooklyn Navy Yard, where the necessary alterations were made. Her cabin was enlarged, and two deck-houses constructed, to accommodate the necessary number of officers, and some other changes were introduced. Her sailing-rig is that of

THE TIGRESS.

a barkentine, and she makes her best time close to the wind. She has two engines, direct-acting, compound high and low pressure, of twelve hundred horse-power. Her propeller is set down very low, so that she can work under the ice.

After all the necessary alterations and repairs were completed, another short trial-trip was made to the compass station near Sandy Hook, for the purpose of correcting the deviation in her compasses. She had also an entire set of new rigging, sails, and topmasts, and steam pipes had been adjusted for the purpose of heating the vessel, as it was uncertain whether she might not have to winter in the north.

The *Tigress* carried a complement of eleven officers and forty-two men, including petty officers and crew, as follows:

OFFICERS.

Commander—James A. Greer, commanding.
Lieutenant-commander—Henry C. White, executive officer.
Lieutenant—George F. Wilkins, navigator.
Lieutenants—Robert M. Berry and U. S. Sebree.
Acting Lieutenant—George E. Tyson, ice-master.
Acting Master—Elisha J. Chipman, assistant ice-master.
Engineers—George W. Melville, first assistant (chief engineer); William A. Mintzer, second assistant.
Assistant Pay-master—George E. Boughman.
Acting Assistant Surgeon—J. W. Elston.

PETTY OFFICERS AND CREW.

Frank Y. Commagere,* yeoman; John P. Britton, master-at-arms and captain of the hold; W. E. Bullock, apothecary; Joseph Brewin, boatswain's-mate; William Sheriff, chief quartermaster; Samuel Randall and George Gray, quartermasters; Henry Clifford, captain of the forecastle; Thomas Hovington, captain of the foretop; Gustavus W. Lindquist, captain of the maintop; Charles Cooper, captain of the mizzentop; George R. Willis, captain of the afterguard; William Boyer, carpenter's mate; William Hurley, cabin cook; John P. Wallace, steerage steward; Richard Davis, ship's cook; John M'Intyre, David M. Howells, and John M'Ewen, machinists; Christopher T. White, James Horan, Samuel Slater, Richard Brenen, Daniel Lynch, and Patrick Devaney, first-class firemen; Frederick Howlett, Jeremiah Murphy, Edward Jokish, John W. Smith, S. W. Harding, William Lindermann, J. W. C. Kruger, and Joe Ebierbing (Esquimau Joe), seamen.

---

* Mr. Commagere was one of those energetic correspondents of the *New York Herald* who are ever ready to do and dare in any field where reportorial honors are to be won. Finding there was no other way to secure a passage, he shipped as ordinary seaman, and was very considerately appointed to the position of "yeoman" by Commander Greer. Mr. Commagere was much esteemed on board, and added not a little to enliven, by his intelligence and humor, the short though stormy voyage of the *Tigress.*

If the *Polaris* had been found in a sea-worthy condition, it was arranged that she should be put in the charge of Lieutenant Commander White to bring to Washington.

Captain Tyson (*pro tem.* lieutenant) went on the *Tigress* as ice-pilot, and Mr. Chipman, of New London, was his assistant.

The Esquimau, Joe, also went on board the *Tigress* as interpreter, that he might act in the very possible contingency of seeking information from the natives on either coast of Davis Strait. Including the crew, there were fifty-three persons on board.

Joe had sent his wife Hannah down to Wiscasset, Maine, but she has spent most of the summer with their adopted child, Puney, in Groton, Massachusetts; and Joe intends to spend the remainder of his life in the United States. He owns a little house and a bit of land near New London, Connecticut, where he can make a living by fishing. All the other Esquimaux, and Hans and family, were taken on board, to be returned to Disco, Greenland, which they preferred to remaining in this country, as they found it uncomfortably warm.

It will be seen by Captain Tyson's journal that the fact of having only thirteen seamen, with such a large number of superior and petty officers, was the cause of much suffering and unusual hardship to the dozen men who had to do the main work of the ship. In consequence of being thus short-handed, there could only be two watches formed. The cruise proving a stormy one, involved much extra work. It bore very hard on the crew.

In regard to stores, the *Tigress* was supplied with a large variety of food, and, with what the *Juniata* took for her use, an ample store for two years. Pork was substituted, to a considerable amount, in place of the regular navy beef, as the former article is considered better adapted to an Arctic climate; and this was complemented by large quantities of antiscorbutics, in the shape of canned potatoes, onions, tomatoes, pickles, and fresh meats.

Some unwonted articles also appeared in her equipment. Among these were three tripod-derricks, iron-shod and spiked. These derricks were for the support of large "cross-cut" or "gang" saws, several of which were taken for the purpose of cutting through heavy ice. Like some other patent articles, the practical value was thought to be scarcely equal to the ordinary mode of sawing by one man-power.

Sledges built on the Esquimau plan, with modern and Caucasian improvements, were taken for the use of searching parties; the bottoms of these sledges were made of slats of very tough wood, curved at the ends, and about five feet in length; the runners were sixteen feet long, and nearly a foot deep, and were made of spruce planking; and their whole length shod with whalebone, except the tips, which were iron-bound. The seats and runners were bound tightly together with leathern thongs inserted through holes drilled in the wood. A large load could be carried on one of these sledges — four or five persons, with food for several days; but it needed a strong team of dogs, a dozen or more, to carry such a sledge fully loaded over the ice.

The officers on board the *Tigress* did not expect to make any scientific observations. Their orders were to "find the *Polaris*, and relieve her remaining company" if they could; and every thing else was to be subservient to that. Yet time was found to make some very interesting geological and mineralogical collections, particularly in the neighborhood of Ivgitut, where large works are in operation for the excavation of the kryolite of commerce, and also of mica, which exists in that locality in large quantities.

Three of the seamen who had come down on the ice-floe formed part of the company: these were Lindquist, Kruger, and Lindermann. Some of the others who agreed to go back failed to appear when the *Tigress* sailed. While she lay at the Navy Yard, a constant concourse of visitors, anxious to see the sealer which had rescued Captain Tyson and company, were constantly ebbing and flowing through the gates, and overrunning the vessel, to the serious embarrassment of the workmen.

CHAPTER XXIX.

The *Tigress*, Commander James A. Greer, sets sail.—Enthusiasm at her Departure.—Hans and Family as Passengers. — "Knowledge is Power."—Arrive at Tessuisak.—Governor Jansen.—*Tigress* proceeds North.—Approach Northumberland Island.—Not the place of Separation.—Make Littleton Island.—Excitement on Board on hearing Human Voices.—Encampment of the *Polaris* Survivors found.—Commander Greer's Success.—Esquimaux in Possession of the deserted House.—Captain Tyson's Advice to seek the Whalers.

THE *Tigress* cast off her lines about six o'clock on Monday evening, July 14, and, amidst the cheers of the thousands assembled to witness her departure, steamed away through the East River and toward the Sound, saluted on all sides by the shrill whistles of passing steamers, who recognized her, and knew the service she was detailed to perform. Her departure created far greater interest and attention than did the sailing of the *Polaris* two years earlier.

"The Hell Gate pilot gave place to the Sound pilot, and the latter left them at Pollock Rip, off Cape Cod, from whence many sent letters home, and then turned their faces bravely to the north. All possible speed was made with sail and steam; the *Tigress* proving an excellent sea-boat going before the wind, but not so comfortable head to it, under which course she was slow, and given to rolling and pitching.

Mr. and Mrs. Hans Christian, with their interesting progeny, did not commend themselves to the dainty olfactories of the naval officers, and they were quartered in the forward deck-house, having for companions the surgeon, pay-master, and a *Herald* correspondent. All the children had fatted up since their sojourn in Brooklyn, and had assumed their normal condition of rotundity. Many persons had given this family articles of clothing for the children, and no matter what the garment was, it was immediately put to use; consequently, though the thermometer was ranging, just before the *Tigress* sailed, between the eighties and nineties, the children, baby and all, were enveloped in numberless dresses, sacques, and shawls wonderful to behold. Mrs.

Hans appeared to have no idea of keeping any thing in reserve. It is no wonder they found it warm!

Joe had shipped as interpreter, but was put to work on board as a seamen, while Hans was a gentleman passenger; but as it happened, the latter could speak but little English, or understand it, so Joe took advantage of his superior knowledge, and led the simple-minded Hans to become his drudge, by gravely informing him that both were equally expected to work; thus exemplifying the old adage, that "knowledge is power," even to an Esquimau.

The *Tigress* had very favorable weather until nearing Cape Race, when the inevitable fog settled down over the ship and her company; and on the afternoon of July 22 they had a narrow escape from running afoul of a large iceberg, which was fortunately revealed on their starboard bow by a sudden lifting of the fog. Early the next morning they dropped anchor at St. Johns.

Under date of August 25, Commander Greer, of the *Tigress*, reported to the Secretary of the Navy that, having met the *Juniata* at Upernavik on August 11, and shipped a supply of coal, he had taken on board a Danish pilot and sailed for Tessuisak, and that, while forging slowly ahead, the engine had caught on the centre; that he had let go his anchor, which failed to bring the vessel up, and she ran, without great force, on a smooth rock, but was backed off in a few minutes without injury.

At this place Governor Jansen came on board and acted as pilot, clearing them from the islands which abound in that vicinity. We may as well here explain that the term governor in Greenland is not the highest title, but is better expressed by the Danish term "colonibestyrere," or *steerer of the colony*—that is, the head man, or chief, of the settlement. This "Governor" Jansen had been with Hayes, and had proved himself a valuable and faithful companion and assistant; he knew the waters thoroughly, and soon set the *Tigress* on her northward course.

On the same night, at two o'clock (August 12), the *Tigress* met the steam-launch *Little Juniata;* but they had seen nothing of the *Polaris* or Captain Buddington's party. Near Cape York the *Tigress* encountered the heavy pack-ice which had repelled Lieutenant De Long; and they were prevented by it from getting close to the shore, but went near enough to have observed any flag or signals, if such had been displayed. A bright lookout

GOVERNOR JANSEN AND FAMILY.

was kept up all the time. Clearing the pack, they skirted the eastern shore, examining North Star Bay on their course, and on August 14 found themselves opposite the Esquimau settlement of Netlik.

They were now approaching Northumberland Island, the generally supposed location of the *Polaris* at the time of her separation from the ice-floe, though Captain Tyson had always suspected it to have been Littleton Island, as he states in his journal on the ice-floe, "supposing" it only to be Northumberland, as he stated in his testimony, because Mr. Meyers, who had the means of accurately knowing, asserted positively that it was. Of course he and others of his party who were on board were keenly on the alert to detect any familiar objects; but as they came within range, it being full daylight too, they found the scene quite unfamiliar. It was evidently not the place; and as they were quite certain they had not passed it—it being in August light through the whole twenty-four hours—they knew they must sail farther north to find it.

They sailed by capes Parry and Alexander, looking sharply around Hartstene Bay not only for Buddington and his companions, but also for the missing locality of the separation. At last Littleton Island, with its lesser companion, M'Gary Island, came in sight, and, with a simultaneous shout of recognition by all on board who had parted from the *Polaris*, this was declared to be the spot where the separation took place. All was now excitement as one and another pointed out the familiar rocks and other points in the view, indelibly impressed upon their memories.

Here the *Tigress* hovered around, in hopes of seeing some signs of the missing ship and men. Between 9 and 10 P.M. a boat was lowered for the shore with Lieutenant White, Captain Tyson, and other officers, and then the distant sound of human voices appeared to come from the shore, and the wildest excitement prevailed; when, Commander Greer ordering "Silence!" the sounds were distinctly recognized as human speech, and in a few moments more the commander was heard to exclaim, from his elevated position on the bridge, "I see their house—two tents; and human figures are on the main-land near Littleton Island!"

No one on board doubted for a moment that these human figures were Buddington and his party, and each one felt that their mission was nearly accomplished; but those in the boat soon discovered the mistake. The "human figures" were not the lost men of the *Polaris*, but native Esquimaux, whose language was unintelligible to all the officers except Captain Tyson. Some of them wore the clothing of civilized men. Captain Tyson obtained some facts from them when the boat returned to the ship, and Esquimau "Joe" was then taken back with them to the shore to act as interpreter, for fuller information. He confirmed what had been learned by Captain Tyson, that Captain Buddington had abandoned the *Polaris* on the day after she was separated from the floe; that his party had built a house on the main-land, where they had wintered; had fitted it up with berths, or bunks, for sleeping, fourteen in number (showing that none of the party had died); and had also furnished it with a stove, table, chairs, and other articles taken from the *Polaris*; that during the winter the party had built and rigged two sail-boats, with wood and canvas taken from the ship, and that "about the time when the ducks begin to hatch" Captain Buddington and the whole party had sailed southward in these boats.

The chief man among these Esquimaux also said that Captain Buddington had *made him a present of the Polaris*, but that soon after the former left the vessel broke loose from the ice in a gale of wind, and, after drifting about one mile and a half toward the passage between Littleton Island and the main-land, had foundered. The chief told in sorrowful accents how he had watched her sinking—sinking, until down she went, out of his sight forever, to his great sorrow and loss.

POLARIS CAMP, 1872-'73.

Commander Greer, in his report, states that exactly "one month and four hours after leaving New York," he had found and visited the winter-camp of the *Polaris* crew. The Esquimaux were in possession of the deserted quarters, and also had two tents made out of canvas belonging to the *Polaris*.

The camp was found to be situated in lat. 78° 23' N., and in long. 73° 46' W. On entering, a scene of disorder and willful destruction presented itself; articles of furniture, instruments, books, the stove, medical stores, and an ample stock of provisions, were scattered about in the utmost confusion. How much of this was the work of the retreating party, and how much of the Esquimaux present, it was not easy to determine; but its condition showed at least that no pains had been taken to seal up or preserve in any way the records, books, or scientific instruments. The most diligent search failed to reveal any writing which indicated the time of their breaking up, or what route they meant to pursue. One expressive article was found, namely, a log-book, out of which was torn all reference to the death of Captain Hall.

Commander Greer took possession of all the manuscripts, the log-book, the medical stores, and remains of instruments, and whatever else was of any use or value, either intrinsically or as relics, and then returned on board the *Tigress*.

As there was no further object in going north, the *Tigress* was once more headed for Tossac, on the supposition that news of the *Polaris* survivors might have reached that settlement. On their southward course a continual and close watch was kept along shore and in every direction for the *Polaris's* boats, but nothing was seen of them. Nor, when the *Tigress* touched at Tossac, was any news obtained.

Passing down to Upernavik, the *Tigress* dropped anchor there, for the purpose of overhauling and repairing her machinery; but could still hear no news of Buddington's party. Commander Greer, therefore, decided to keep to the south as far as Goodhavn, where intelligence was more likely to be obtained. Leaving Upernavik at 2 P.M. of the 23d of August, the *Tigress* arrived at Goodhavn at 2 A.M. of the 25th. Here it was ascertained that the whaler *Arctic*, of Dundee, and the *Aseek*, with seven others, had gone northward, all expecting to sight Cape York. Hence it seemed certain that, if Captain Buddington had kept to the east coast, he had already been rescued by some of these vessels.

But the duty of the *Tigress* could not be considered fulfilled while there was no absolute information obtained as to the safety of the missing men. Captain Tyson's familiarity with the habits of the whalers convinced him that the vessels would at that time of the year be found on the west coast, and he so explained to Commander Greer, who then determined to strike over to the west to try and intercept the whalers, who invariably take that course in working down from the "north water." At the time of writing his dispatch he had one hundred and fifty-five tons of coal aboard, and while that lasted he expressed his intention to continue the search, unless sooner receiving positive intelligence of the parties he had gone to seek.

Official reports are usually dry reading. With all due respect, therefore, for Commander Greer's official communications, we prefer to give Captain Tyson's journal of the daily incidents of this cruise; and in it we are sure that the intelligent reader will perceive many valuable suggestions, intermingled with incidents of an amusing character, though these were not infrequently imminently verging toward tragic catastrophes.

CHAPTER XXX.

CAPTAIN TYSON'S CRUISE IN THE TIGRESS.

Captain Tyson's Journal on board the *Tigress.* — "Too late."—Fire training on board.—*Mal de mer.*—A tall Story.—Angling for Porpoises with Pork.—A nautical Joke.—Beware of the *Tigress.*—Fog at Sea.—Naïve Comments on Icebergs.—Tender Hearts among the Blue-jackets.—Illusions.—Aurora.—Whistling to frighten the Bergs.—Splendid Northern Lights.—Heavy Gales.—The Doctor's Clerk.—Two old Whalers.—We leave Night behind us.—Poor Hans's Affliction.—Family returned to Greenland.—The *Tigress* pitching and rolling.—The Fog-blanket.—Cheese for Bait.—An Iceberg turns a Somersault.—A beautiful Display.—A slight Accident.—Meet the Steam-launch.—Official Correspondence with Commander Greer.—Ashore at Littleton Island and Life-boat Cove.—Sounding for the foundered *Polaris.*—Abundance of Food abandoned by the *Polaris* Survivors.

"*Saturday, July* 12, 1873. The Secretary of the Navy, Hon. George M. Robeson, visited the *Tigress* this afternoon, put the ship in commission, and made a short address to the crew. At 5 A.M. got under way for the purpose of correcting the deviation of the compass at the buoys at Sandy Hook. The weather fine and pleasant, with light breeze from the south-west. Several of the men quite nervous this morning, the vessel making considerable water. As I am requested to stand a watch (though no part of my duty), I do so to give others an easier time.

"*July* 13. Started at 1.28 P.M. to return to the Brooklyn Navy Yard. I feel that we are too late getting off to accomplish the object for which we are going. The *Juniata* will most likely succeed, having so much the start of us.

"*July* 14, 5.10 P.M. Started from the Brooklyn Navy Yard. The United States steamship *Brooklyn,* and also the United States steamship *Vermont,* manned their yards and rigging as we left, while cheer upon cheer rent the air all along up the East River; from every ferry-boat and steamer which we passed we received a salute. The waving of handkerchiefs by the lady passengers, and the 'hurras' of the men, showed the deep feeling of the people, and the interest which they took in our mission. It was as much as to say, 'God be with you; rescue those people, and a safe return home to you;' so at least I felt it. The cheering from passing boats and the waving of handkerchiefs continued until darkness closed around us.

"*July* 15. Light breeze from the south-west. We are pressing along under both steam and sail as fast as possible, but our vessel is far from fast. I hope, if slow, she will be on the 'sure' principle, and arrive in time to get those people before the *Juniata*. It is extremely hot this morning; very little wind, scarcely enough to ruffle the quiet waters of Long Island Sound.

"6 P.M. We are now between Block Island and Point Judith, going along about seven knots an hour. Hans and his family were the centre of attraction yesterday, but this morning, on going into the apartments they occupy, the stench drove every one out; so their attractions are much diminished to-day.

"We had 'fire-training' this evening. All were stationed at quarters, my station being at the magazine. I stood at my post boldly and fearlessly, knowing it did not contain any thing more explosive than canned meats and fruits!

"*July* 16. Fair, with just enough wind from the north to kick up a lively sea. The pilot left at 8 A.M., and we are now fairly at sea. I have sent the last good-bye to wife and family for a time (God willing). There are many sick this morning among the men, and it is laughable to see how they try to hide their weakness. They bear up against it manfully, put on a stern and sober face, and say, 'I'm not sick; but I don't care for any breakfast this morning.' Some are inclined to sit on or near the rail, and every once in a while their heads will disappear over the side. What engages their attention just then I am not anxious to observe. Hans's family are all sick too, and every body else is sick of them. We are going along this evening at the rate of five knots, through the stormy tide rips of George Shoals.

"*July* 17. It is a fine pleasant day, the sun shining brightly; a cool but gentle breeze from the N.N.E. All now appear happy and contented. Some amuse themselves with the violin, others telling yarns. One by our yeoman, who, by-the-way, is a man of some account, he being no less than a '*Herald* correspondent,' but who shipped before the mast, finding no other way to get a passage on board—he, speaking of tall men, outdid the experience of the others by declaring that an acquaintance of his was so tall that he could stand on one side of a carriage, reach over the top, and open the door on the opposite side. That was rather a long stretch, even for a sailor's yarn. We see many little fishing-smacks, little fellows which look as if one good sea would annihilate them; but they are like the ducks, always on the top of the rising crests. Indeed, accidents seldom happen to them.

"Our hours for meals are: breakfast at 9 A.M.; dinner at 3 P.M.; tea at 6 P.M.

"*July* 18. Cloudy and cool. We are making very slow progress indeed this morning—only two miles an hour. There is a strong east wind, and, as our course is east, it is consequently dead ahead. There is, too, a dirty little short sea running, which hinders us very much. The men are very busy at work, cleaning paint, scraping, and scrubbing decks. Lat. at noon, 47° 17' N., long. 64° 19' W.

"An amusing incident occurred this afternoon. Many porpoises were playing about the ship—quite a novel sight to those on board who have not been to sea before. Our doctor's clerk, or apothecary, was running around the deck preparing a line and hooks. Finally, coming up to Mr. Chipman (assistant ice-pilot), he asked 'what kind of bait they used to catch these porpoises with?' Mr. Chipman told him by all means to 'bait his hook with pork.' To those who understand that porpoises are captured with the harpoon and lance, it will readily be seen where the joke comes in.

"*July* 19. Beautiful weather, with light south-east breeze; numerous porpoises still playing around the *Tigress*, which, now we have got all sail on, is going through the water at the rate of seven knots an hour. This evening the breeze is strong, and it looks like rain. We have all come away without rubber coats. It was supposed that water-proof suits were among the stores, as a supply had been ordered, but they can not be found. So, whenever it rains, whoever has the watch on deck will have to take a good wetting. It has been quite cold the last three days—cool enough to bring out the overcoats. We work hard every day, trying to get the vessel in presentable condition before arriving at St. Johns.

"*July* 20. It has been a wretched day, so far as weather is concerned. It rained through the entire night, and continued up till noon, when the wind changed from S.E. to S.W., bringing a thick fog. We go along carefully, sounding our way; indeed, have done so since leaving Brooklyn. We are now this evening (8 P.M.) three hundred miles from St. Johns. It is densely thick to-night; so the steam-whistle is kept constantly screaming every five minutes, warning all ocean travelers to beware of the *Tigress*. But we do not mind bad weather, for our wardroom is very comfortable. Some of the officers—we are eleven in number—are enjoying themselves with a violin, while others are reading, the propeller thumping away, and the whistle screaming, so that sleep is scarcely obtainable at present.

"*July* 21. It was densely thick last night, and the fog so very wet that we needed the missing rubber outside garments very much. The wind is fair, and we go along at a good jog, with a long, heavy swell from the southward, which keeps the *Tigress* rolling in a very uncomfortable manner. This morning the sun forced his way through the fog and finally dispersed the mist, for which we are truly thankful. Nothing is more annoying than fog at sea. It is worse than the darkness of night—for the eyes become partially accustomed to that sort of darkness, and can see something; but in foggy weather one can discern absolutely nothing a few feet off.

"*July* 22. This morning at 8 A.M. the westerly breeze brought down another thick fog upon us; it continued very dense until 4.10 P.M., when, just as we had stopped the engine, and brought the ship's head to the wind to sound, behold the fog disappeared, blowing away to the eastward, and as it lifted we discovered the coast—the coast of Newfoundland, just above Cape Race, about forty miles from St. Johns. Nearly at the same moment I perceived an iceberg about four miles off to the eastward. This was the first berg seen by us, and the first ever seen by many on board, and it consequently attracted much attention. Some wanted to know 'if it was as large as the one I had drifted on so far?' not realizing the difference between a *floe*, which is comparatively flat, and a *berg*, which is of an elevated structure. Some thought it 'beautiful,' others that it 'looked cold.' They will see many more of these beautiful, cold bergs before they get back. To-night we shall arrive at St. Johns, making our passage nine days from Brooklyn.

"*July* 23. We arrived at St. Johns at 6.30. The weather quite pleasant for this part of the world, although it is cloudy, and some little rain falling. The pilot boarded us as usual, after we got into the harbor. I have been on shore, but soon returned, it being my watch. As we are about to take our departure from St. Johns to the far north, every body is busy writing home. How much the sailors think of home—their hearts are not all hard! The steamer from Europe touched here this afternoon, staying but to take the mail. I hope my letters have gone by her.

"*July* 27. Since 8 A.M. the weather has been beautiful. We left St. Johns last evening at 6 P.M. At eleven this morning, it being Sunday, we had divine service, all the officers attending.

"We have laid in an abundance of stores, such as fresh meat, potatoes, eggs, Bass's ale, chickens, and three little pigs, two of which, however, died last night. We have also a cat, purchased by one of the officers. We lost two men at St. Johns, and shipped three: one of the men who escaped, the carpenter's mate, went off in double-irons, he having been discovered with his trunk packed, preparing for desertion. We have already started our steam in the wardroom, some complaining of the cold already.

"*July* 28. Some little rain this morning, but otherwise pleasant. There is nothing doing but the usual routine on board ship. Some of the boys who indulged too freely of the good things at St. Johns are feeling the effects now. One discovered last night a very large white rat, which, on investigation, proved to be the innocent cat before mentioned. Another, whose watch did not commence until daylight, got up at midnight, and could scarcely be convinced that it was not morning.

"Last night, about low meridian, there was a beautiful aurora, extending from west to east, and covering the whole heavens to the northward. The display was very brilliant, showing all the colors in nature; vivid flashes resembling lightning followed each other in quick succession, the electric clouds encountering and running together as if endowed with life. While writing the above the wind hauled from the south-west to the north-east again, bringing our old enemy, the fog. As soon as it gets foggy there is a great worry in some quarters about bergs. Overcoats are called for, and some have even taken to mittens.

"*Evening.* It is again cool but clear; the wind is ahead, greatly retarding our progress; for without a fair wind and sails to help we get along very slowly. But, slow as we are, we are constantly increasing the length of our day. At this date there is not more than five hours of darkness out of the twenty-four. This evening our latitude is 52° 14' N.

"*July* 29. Foggy again, but thanks for a fair wind. All sail set, and steam propelling us along as fast as possible. We sound our steam-whistle every five minutes, but for what purpose I know not, unless to frighten the bergs; for in this vicinity there are no ships to run into. The sun set this evening at 8.10 P.M. Now it is nine o'clock, and yet there is good light, as the fog has cleared off, leaving us a fresh breeze from the south-west. We are now going along at from seven to eight knots per hour. Lat. at 8 P.M. 55° 18' N.

"*July* 30. It is a cloudy day, the sun showing himself occasionally through the clouds; pleasant breeze from the westward. With all sail and steam on, our little craft gets along merrily—making, from twelve, noon, yesterday to noon to-day, one hundred and ninety-six miles on our northward course. Last night there was another auroral display, covering nearly the whole heavens. The electric flashes were very brilliant; at first the undulations were of a graceful and tremulous motion, and then

> " 'Anon, as if a sudden trumpet spoke,
> Banners of gold and purple were flung out;
> Fire-crested leaders swept along the lines,
> Which from the gorgeous depths, like meeting seas,
> Rolled to wild battle.'

The scene was one which could scarcely be excelled for spirit and beauty.

"*July* 31. We have a very bad morning; dirty, rainy, blowing weather, with a head wind, our little vessel pitching and rolling in a very uncomfortable manner. Steam is on, full power, and we are battling against wind and sea, but making no headway.

"*Aug.* 1. The gale still continues, but the wind has hauled more to the west, and the rain has ceased, but leaves a very bad sea running; it is also quite cold: thermometer $+47°$. The temperature of the sea-water is the same as the air. No accident has yet occurred worthy of note, except during the late gale a slight one, which might have proved serious. The doctor's clerk, who goes by the inelegant sobriquet of the 'puke-jerker,' was lying on the transom locker asleep, when his medicine-chest from above gave a lurch and came down on him. Had it struck him on the head he would certainly have been killed—by his own drugs. It was a narrow escape.

"*Aug.* 2. The weather still remains cloudy, but the wind is fair, and we are making our way northward at the rate of six or seven miles an hour. Before leaving Brooklyn, I had explained to Commander Greer and Executive Officer White what my views were about the position of the *Polaris* at the time of my separation from her. From an examination of the charts, I felt quite sure that Sergeant Meyers had made a mistake, and that my original impressions were correct. I have told Commander Greer that he would not find the *Polaris* at Northumberland, but at Littleton Island.

"10 P.M. Raining again, but the wind is fair, and the coast of Greenland about forty miles distant; but as yet its lofty, rugged mountains can not be seen, on account of cloudy and hazy weather. Lat. 62° 54' N., long. 53° 50' W. at noon, which at our present rate will bring us this evening abreast of Fiscanaes. Many species of gulls are sporting around the little *Tigress*, also whales—the fin, the hump-back, the bottle-nose, and the huge sulphur-bottom—large numbers of them. It reminds me of my old whaling days; and Mr. Chipman, forgetting for the moment that he was in the United States service, and consequently ought to conserve his dignity to the utmost, could not forbear exclaiming, in whaler phrase, as he caught sight of a spout, 'There she blows!' while, I, to foster his momentary illusion, promptly responded, 'Where away?' But dignity and silence were soon resumed.

"We have now left night behind us, though not quite within the Arctic circle yet. But it is twilight at midnight. Poor Hans and his family were subjected to a severe affliction to-day, being compelled to strip off their vermin-infested clothing and put on clean, than which I suppose no greater discomfort could be imposed upon them. The discarded clothing was put into a strong pickle, to destroy the parasites with which it was all infested. Hans and family will have a grand rejoicing on getting once more to 'Greenland's icy mountains,' their 'native heath,' where they can enjoy their dirt, with none to molest or make them afraid.

"*Aug.* 3. This morning, the weather having partially cleared, we sight the coast of Greenland, just abreast of Sukkertoppen. The lofty mountains were covered with snow, with here and there a glacier, making a very picturesque, though a grand and almost terrible, coast scene. We have passed several bergs this evening;

and the weather is wretched again, fog, rain, blowing, and a head wind at that.

"12, *Midnight*. I have just come from deck. The short sea, and pitching and rolling of the vessel, has made it any thing but pleasant; besides, I have had a thorough soaking in the last four hours. I must say we have had a very unpleasant passage as far as weather has been concerned. It has been rainy, foggy, or head-winds a great part of the time. The excitement on board on passing a berg is quite ludicrous, at least it appears so to me. One would think the Day of Judgment was just at hand, to see some of the pale faces.

"*Aug.* 4. The rain has ceased, the fog has cleared, but it is not settled, clear weather; it is so cloudy one can almost feel the heavy cloud-atmosphere enveloping us—much, I should think, as aeronauts describe their sensations on passing through a cloud-belt. The coast of Greenland, in the distance, is covered as with a heavy pall, the mountain peaks occasionally showing themselves above the dark clouds which enshroud the land.

"*Aug.* 9. Well, we have been to Disco—arrived there on the morning of the 6th; coaled the ship, and off again on the 8th. The *Juniata* had gone on north, I suppose, as far as Upernavik, taking a full supply of coal belonging to the *Polaris*, and leaving fifty or sixty tons for the *Tigress*. We had a very pleasant time at Disco, going on shore and having a dance, and the officers of the *Tigress* are very much gratified with their visit. The doctor's clerk has been fishing, using *cheese* for bait!

"At last we have had a beautiful iceberg exhibition. As I was going on shore in the boat, my attention was attracted by the oscillatory motions of a large berg of irregular formation, only a short distance from me. I watched it with much interest, for, appreciating its peculiar movements, I anticipated the result. At first it swayed backward and forward with a gentle inclination for a few moments; then I perceived that it was losing its balance, when, as if endued with the consciousness that it had lost its original poise, and was seeking to re-establish its centre of gravity, it trembled for an instant, as if uncertain which way to plunge, and then turned over toward me in the most graceful of somersaults. As the immense mass struck the water, many large pieces were shaken from the top; the white foam swirled around the vortex formed by the descending mass; there was a shiver and a struggle, as if some sentient creature were in danger of drowning; the waves sent out by the disturbed waters rocked my little boat with a heavy swell. But in less time than I have taken to describe it, the beautiful berg, with somewhat altered proportions, had arisen from its bath, readjusted its centre of gravity, and presented to my view another façade, on which the summer's sun enkindled a marvelous combination of hues, and away sailed my lovely berg, calm and majestic as a queen in new tiring robes. Those on board of the *Tigress* who were on the lookout had the full benefit of the display.

"*Aug.* 10. This morning we arrived at Upernavik; it was cloudy and raining. Many large bergs around, some of huge proportions, and, I must say, some of them very beautiful, particularly several which appeared like artistic architectural structures. Getting near the harbor, the tall spars of the *Juniata* were discovered; also a Danish bark, the *Thorwaldsen*. In a few minutes we were at anchor, and in a few more had all the news. The *Juniata*, or, rather, Commander Braine, has dispatched the steam-launch across Melville Bay.

"*Aug.* 11. Fine weather; preparing to get under way, and it looks like preparing to winter. We left Upernavik at 5 P.M. The crew of the *Juniata* cheered us as we got off. Away we steamed, through islands and bergs—bergs in thousands. Six hours carried us to Tessuisak. On going into harbor, the engine caught on the centre; therefore could not back the vessel, and she ran on the rocks, but was soon got off again, and, having discharged the pilot, started for the farther north. We had gone but a short distance, when the launch of the *Juniata* was discovered heading toward the south. How far they have been I could not ascertain, but I judge not far. They had no intelligence of the *Polaris*.

"*Aug.* 12. Fine weather. We are steaming gayly along, and are about entering Melville·Bay, the dread of whalemen; bergs are in sight, but no pack-ice.

"*Aug.* 13. Clear and pleasant. Last night we saw the midnight sun. This morning sighted Cape York, getting within about six miles of the cape; found it surrounded by a pack. Seeing no signals, the commander concluded to keep on; so we rounded the west end of the pack, and are now steering north again.

"*Aug.* 14. This morning I received the following communication from Captain Greer:

"'United States Steamship *Tigress*, Smith Sound, Aug. 14, 1873.

"'SIR,—Having this morning passed near to Northumberland Island, affording a good view of the same, I desire you to state to me in writing whether in your opinion the said island was the one which was seen by you when you were separated from the *Polaris*, in October, 1872. If in your opinion it is not the island, you will state to me in detail the reasons why you form that opinion.

"'Respectfully, etc.,   JAS. S. GREER,
"'Commandant, commanding.

"'GEO. E. TYSON, Ice-master, United States Navy,
United States Steamship *Tigress*.'

"To which I made the following reply:

"'United States Steamship *Tigress*, at Sea, Aug. 15, 1873.

"'SIR,—My reasons for thinking Northumberland Island is not the island where the *Polaris* separated from me are as follows, viz.: the island is much larger than the one I saw the *Polaris* go behind; and Hakluyt Island, off the north-west end of Northumberland Island, is larger than the island—or rock—off the island where I last saw the *Polaris*; and the surrounding land in the vicinity of Northumberland Island does not correspond with that I saw at the time of separation.

"'Very respectfully,   GEO. E. TYSON,
"'Ice-master, United States Navy.

'Commander JAMES A. GREER, United States Navy,
commanding United States Steamship *Tigress*.'

"*Aug.* 14. Blowing a strong gale from the south-east. We are now abreast of Northumberland Island, and it has turned out as I anticipated—the *Polaris* is, or was, at Littleton Island. We have steamed along the land within three miles of it, since leaving Cape York, colors flying all the way, so as to attract the attention of the wanderers, if they are anywhere within sight; but we have seen no natives, nor any thing denoting the existence of human beings. We are now on the classic ground of Kane and Hayes.

"This evening the weather is much better; snow-squalls, but very little wind. As we approached Littleton Island, I recognized at once the scene of our separation from the *Polaris*. Though the season of the year is so different, being then in almost continuous darkness; though the weather is now comparatively warm, and the rocks bare of snow, yet the shape of the islands, the forms of the rocks, the contour of the 'everlasting hills' had not changed. I perceived without a doubt that it was from this point that I had been floated off on that marvelous God-built raft, which, with his aid, had borne our company, through fearful perils and suffering, to safety, home, and friends.

As we neared the exact spot, I could not but give thanks to God that he had preserved me to come back and see by the light of the summer's sun the scene of desolation where the *Polaris* had drifted away from us. But where was the ship? Nothing was to be seen of her. By 7 P.M. we were steaming round the north side of Littleton Island, which had last shut out the *Polaris* from my sight. The thought came to me, with significant meaning now, of that devilish proposition which I recorded at the time in my private journal. The engines were now stopped, a boat lowered and manned by all the officers of the vessel, including myself; and we had scarcely got clear of the ship, when cries from human beings were heard on shore. We were about one mile and a half from the beach; and as these voices reached our ears, the excitement of anticipation put double speed into the strokes of the oarsmen. Our boat fairly sprang through the water; but on getting nearer to the shore, I was convinced by the actions of the party that they were not those whom we sought, nor white men of any description. Their actions and antics proclaimed them Esquimaux at sight. A few moments more and we were on the rock-strewn beach; and there the first thing which caught my eye was an old hawser, evidently belonging to the *Polaris*, one end fast to a rock, the other afloat in the water.

"The natives—four or five—gathered around us, and I was soon in conversation with them. From these simple-minded but truthful Esquimaux I learned that the fourteen men of the *Polaris* had gone south some two months ago. They had built two boats with material taken from the ship. After Buddington and his party had left, the *Polaris* broke out of the ice, had drifted off and sunk, to the great grief of the chief to whom it had been left as a legacy. We next sighted a large canvas house. On approaching it, we saw the wreck and ruin of the fore-doomed ship—spars, doors, paneling, sails, rigging, stores, pork, meal, tea, corn, potatoes, books, broken compasses, and instruments of various kinds. Returning on board the *Tigress*, we reported the condition of affairs to Commander Greer, who instantly started for shore, returning again in about two hours. The executive officer, Lieutenant White, then went, and I accompanied him, but gained no additional information. We picked up many relics of the *Polaris*, such as books, tools, and manuscripts, which Captain Greer has now in his possession.

LOSS OF THE POLARIS.

"One of the natives informed me that before Captain Buddington left he had presented the *Polaris* to him. The native was much grieved that his prize had sunk and was lost to him. We spent the whole night sounding for the wreck, but sounded in vain; no trace of her could be found. But there, on shore, was evidence enough. All were satisfied that the *Polaris* no longer remained afloat.

"Captain Greer now decided to return at once, as nothing toward the rescue of the party could be done here; so we bore away to the southward. The latitude of Littleton Island is 78° 24′ N.

"There is one thing certain: these men did not suffer from the want of food or fuel, as discarded provisions were lying scattered all among the rocks, and, of course, the natives had eaten all they wanted in the interval besides.

SCENE IN SOUTHERN GREENLAND.

CHAPTER XXXI.

Homeward-bound.—Fire! Fire!—An honored Custom.—Contrast of the Sailor's Life.—A Set-off to the Midnight Sun.—Heavy Gale.—All want to shoot a Bear.—Executive Officer White the "killing" Man.—A narrow Escape.—Thoughts of Home.—At Upernavik for Repairs.—The Danish and half-breed Girls.—Dress.—Dancing.—A startling Record.—At Goodhavn Harbor.—Captain Tyson visits the *Juniata.*—Continued bad Weather.—Sight Cape Mercy.—The Sea sweeps the Galley.—The Cook disgusted.—Effects of the Gale in the Wardroom.—"At home" in Niountelik Harbor, Cumberland Gulf.

"*Aug.* 15. Homeward-bound. The day is fine; all appear in good spirits, Commander Greer especially. He has been very successful, it being but one month since the *Tigress* went into commission before she was lying alongside of Littleton Island. I do not think there is another passage on record equal to it; but our spirits are a little dampened at not finding the survivors. I have no fears but that they are all alive and well; but we should like to have had the pleasure of picking them up. There is no difficulty at this time of the year in working down to Upernavik, or even getting to the whaling-grounds. I feel sure, therefore, that they are either at some of the settlements on the coast of Greenland, or that they have already been picked up by some of the English or Scotch whalers.

"*Aug.* 16. Cold to-day, with a strong easterly wind. At 1.15 P.M., as I was walking the bridge, guiding the vessel clear of the many bergs in the vicinity of Cape York, the boatswain's mate came to me. 'Sir,' said he, 'there is a great deal of smoke coming through the forward bulk-head.' I instantly mistrusted fire. Going immediately to the cabin, I informed Captain Greer. Captain and officers were soon on the berth-deck; the hatches and store-room were opened, but the fire was not there. The fire-bell was then rung, calling all hands to quarters. The main hold was broken out, and the decks were soon covered with boxes, barrels, etc., but the fire was at last found to be in the coal-bunkers—the coal itself was on fire. We immediately commenced hoisting it out on deck, and after shoveling many tons from the starboard to the port side, got at the fire and flooded it with water.

"In forty-five minutes from the time all hands were called to quarters the watch was piped below again. There had been no unseemly excitement; all were calm and cool as a summer's morning after rain. Here we see the true American spirit of self-control. The incident was one to make me feel proud of my countrymen.

"*Aug.* 17. We have a severe gale on us; thick—sometimes hail, then rain. We are making but little headway against wind and sea.

"It is a time-honored custom in the navy to devote Saturday night to drinking to the health of sweethearts and wives. I generally honored this usage 'more in the breach than in the observance,' but last night I joined my brother-officers. With the wild storm raging over these northern seas, we all assembled in the wardroom, and until midnight consoled ourselves for our enforced absence from the kind hearts we left behind us by drinking to their health and happiness, and wishing ourselves a safe and speedy return. But at twelve it was my watch on deck, and out I tumbled, from the light and warmth and joviality of the wardroom, into the cold and darkness of the storm. A very different next four hours awaited me—rain, hail, wind, icebergs, and thick weather. Such are the sudden contrasts of the sailor's life. Before my watch was out, at 4 A.M., I was thoroughly soaked through, and it was with no gentle summer rain either.

"On the 14th inst. we had bidden adieu to the midnight sun. On that occasion, Captain Greer had ordered the American ensign set, and at precisely twelve, low meridian, gave the beautiful luminary a set-off with three hearty cheers. It was a fitting farewell. How few, indeed, have ever seen a midnight sun!

"*Aug.* 17. The gale increases; the sea is white with foam. I have been thinking all day about the loss of the *Polaris;* it appears to me she could have been saved. The more I think over all the circumstances, the more I believe it to have been a premeditated affair.

"*Aug.* 18. The gale has abated, leaving us inclosed in thick fog and drizzling rain. Many bergs to be seen, as the fog occasionally lifts; then all enshrouded again in the dreary black envelop. This morning, about 7.30, as I was walking the bridge, something white appeared under the bows of the *Tigress*. The vessel was going very slowly at the time. I was not long in making out the white object to be a Polar bear. Now Captain Greer, Lieutenant White, and indeed all the officers of the *Tigress*, were extremely anxious to shoot a Polar bear. As soon as I was certain that it was a bear, and no illusion, I rang the bell to stop, and sent the quartermaster to call the captain and Lieutenant White. I also sent a messenger to the wardroom to inform the other officers. Nobody was up as yet, but soon the captain appeared, half dressed; then Commagere, the *Herald* reporter; next the doctor; and in about five minutes there was at least a dozen rifles cracking away at his bearship. The fog now lifted some, but the bear was as yet untouched, although at least fifty rifle-balls had been fired at him. The captain then ordered all to stop shooting. A boat was lowered for Mr. White; he got off in it, and after a good long chase succeeded in shooting the bear, in evidence of which we have had roast bear for dinner.

"This evening we had quite a narrow escape from collision with a berg. There are hundreds of bergs in sight—in fact, there are so many it is impossible to count them, especially as it has been foggy by spells through the day. But this evening, at 7.30, the fog was so thick, as the sailors say, 'you could cut it with a knife.' It was my watch on deck; from six to eight it was very dense, and the vessel going about five knots. I was keeping, too, as sharp a lookout as possible while walking the bridge. I had also a man on the top-gallant forecastle station especially to watch closely, with orders to call out quickly should he discover any thing. I was pacing up and down, thinking of home—thoughts of home are all the pleasure a sailor has in his lonely watches at sea. Suddenly, through the densest of black fogs, I perceived a bright streak. I thought at first it was only the fog breaking away; but I had little time to reflect. 'Hard a starboard!' I thundered out, in tones so loud and sharp as brought the captain and every body on deck in double-quick time. I had rung the warning cry none too soon, for as the vessel swung off, there, high above our masts—the top of the berg was indeed invisible—but as high as we could see, there loomed this gigantic iceberg. We cleared her but by a few yards; it was a narrow shave at that.

"*Aug.* 19. It is once more clear and pleasant. At 2. P.M. entered the harbor of Tessuisak. Mr. Jansen met us just before the vessel arrived at the anchorage. Mr. Jansen, it will be remembered, is 'governor' here—governor of three frame houses, and about fifty Esquimaux, and perhaps the same number of half-breeds. Mr. Jansen informed Captain Greer, on coming on board, that nothing had been seen or heard of the two boats containing the *Polaris* survivors. Consequently, Captain Greer, will not anchor, but, getting a pilot, will at once proceed to Upernavik by way of the inside passage. Governor Jansen also stated that the *Juniata* had left for Disco. So we shall touch at Upernavik, and then go on to Disco, sending home dispatches by the *Juniata*. From Disco the *Tigress* will go over to the west side, and seek out the English and Scotch whalers, hoping to get some information from them of the two boats and their crews.

"*Aug.* 20. Dark and dismal. Fog, wind, rain, and snow-squalls; but we are safely moored in the little harbor of Upernavik; anchor and three hawsers out to secure the vessel; a long heavy swell rolling in from seaward. The weather is looking very bad; the barometer is very low; that, with the thick fog, has induced Captain Greer to stop here two or three days; needing, also, this opportunity to overhaul and repair the boilers, which are sadly in need of it. It is evident to me that Buddington must have got on board of one of the whale-ships. He has now been about two months on his southward passage.

"Here we are at Upernavik, which is next to the most northern settlement (Tessuisak) in the civilized world. Here, too, still remains good old Governor Rudolph, who, though a European, has spent so many winters in this chill, rough, uncongenial part of the world. There are but few buildings here: some six wooden houses, occupied by Danes either as dwellings or store-houses. The natives and half-breeds occupy stone or turf huts, and these huts are but about six feet high in the interior. There they all bundle in together—men, women, and children; here on the lonely shores of Greenland, with scarce a sound to listen to but the crushing and grinding of the grim and ghastly bergs; for in one look seaward hundreds of bergs can be seen from this place, varied only by the beating of the waves against this truly rock-bound coast. Then, too, all through the winter such short days and long nights, and ice everywhere. Yet these people appear happy and contented; yet they see nothing, and scarcely know any thing, of the outside world. But they have their own pleasures and modes of amusement. They love music and they love dancing, and are made supremely happy by the arrival of a chance ship, like the *Juniata* or *Tigress*. Then out come the violins, and the girls dressed in their best. Many of them, especially the half-breeds, are quite pretty, and fair as the fairest of our own belles at home. Some have the light hair and blue eyes of the Danes. And their dress, too, is very picturesque: a pretty little jacket, made, perhaps, of colored calico seal-skin pants prettily trimmed, and, like the jacket, setting tight to the body; white, red, or blue boots complete the costume, while the hair is tied up with bright-colored ribbons, and a necklace of beads is also worn, if it can be obtained. We have just received an invitation to come on shore this evening to a dance.

"*Aug.* 21. Clear and pleasant. Last evening the officers had their dance in the old carpenter-shop of kind Dr. Rudolph. The girls were all decked out in their very best, and they are really excellent dancers; and as for waltzing, they would compare favorably with, if they do not excel, our own beautiful girls. In one thing they certainly have the advantage—the freedom of their limbs is not embarrassed with hoops or long trains.

"*Evening.* We have again danced the German, all joining—Captain Greer, Executive Officer White, Lieutenant Wilkins, and all the other officers of the *Tigress*—and had a very pleasant time, an agreeable break in the monotony of our voyage, and such a lone voyage, on the very outskirts of the world! Certainly, Greenland is but 'half made up.'

"*Aug.* 22. It is a dark, lowering day, and cold; last evening the frost was quite severe. We kept up our dance till midnight, so to-day aching heads and aching limbs are in order.

"*Aug.* 23. Cloudy and cool; light north-west winds. At 2 P.M. we weigh anchor for Disco, the governor firing a salute of three guns on our departure. Our rigging was manned by the crew, who gave three hearty cheers in return. The girls lined the beach, like a fringe, to witness our departure. Poor things! they seldom see any body except those of their own settlement, and must pass very lonely, but I hope not unhappy, lives.

"I had an opportunity last evening of looking over the mutilated diaries and journals left in the deserted hut off Littleton Island. Not one but has the leaves cut out relating to Captain

Hall's death; but in one of them I read, '*Captain Hall's papers thrown overboard to-day.*' Nor do any of them contain any account of my separation from the *Polaris* last October [1871]. It may be that full journals have been carried home. We shall see. There is one thing which surprises me very much in these records, namely, their selfish and egotistical character.

"*Aug.* 24. We have had wretched weather since leaving Tessuisak and Upernavik. Last night snow and sleet, and it has continued up to the present hour (5 P.M.). Disco is in sight. We can see the lofty snow-clad mountains high above the dark vapor which hangs over and surrounds the island.

"*Aug.* 25. Arrived at Goodhavn Harbor at 2.15 A.M. Here Commander Greer reported the discoveries which had been made to Commander Braine, of the *Juniata;* and the latter subsequently signaled the *Tigress* to send me on board, Commander Braine wishing to consult me as to whether the *Juniata* could be serviceable in pursuing the search on the west coast. I advised that the *Juniata* could probably do more good (if the *Polaris* party had not already been rescued) by remaining at Goodhavn to receive them on their arrival there; and also that the *Juniata* was not fitted to contend with the ice and rough weather to be expected on the west coast; but that the *Tigress* was so fitted, and could cover the whole ground.

"Left again at twelve, noon, for the west coast. They are all safe enough in some whaler long before this, so we shall go over to the west side among the whaling-fleet, and may very likely hear of their rescue. Shortly after leaving Disco, encountered a severe blow, accompanied by rain.

"*Aug.* 26. Good weather this morning; many small patches of ice in sight, but nothing as yet looking like a pack.

"This evening we have the west coast in sight, about twenty-five miles distant; many bergs and many streams of ice; so that we are compelled at times to go very far out of our course to get around them. We are in lat. 67° 40′ N.

"*Aug.* 27. Cloudy, and at times rain; it has been blowing heavy through the night; it is a little more moderate to-day, but still very bad weather. We, however, make a smooth sea of it by lying under the lee of the ice, holding the vessel there by steam-power, and waiting for good weather. Our latitude same as yesterday.

"*Aug.* 28. Cloudy, and at times foggy. We have been steaming through the pack all day; but Captain Greer is naturally very nervous, fearing injury to the vessel, and wishes to get out, so the *Tigress* is headed to the eastward again.

"*Aug.* 29. Raining and foggy, and I must say about as mean and dirty weather as one could desire. This morning we got the vessel out of the pack. It was my watch, and a disagreeable watch it was—raining, thick fog, and surrounded by ice: three very unpleasant companions these. The way things are worked here will make it almost impossible for us to ever find the whalers. Our little *Tigress* was built for ice navigation, but now she is handled as if they were afraid of rubbing the paint off her.

"*Aug.* 30. Another stormy night; snow, wind, high sea. Yesterday evening, just at dark, I succeeded in getting out of the pack, but, once from under the shelter of the friendly ice, met a heavy gale from the south-west; thick fog at times, then again thick snow-squalls; fog again, squalls, pitch darkness, with all the time a heavy sea running, made it a very unpleasant night. This morning it is pleasant again; the sea has gone down; the weather quite clear. Some ice is seen on our starboard beam. We are now steaming in toward Cape Mercy, the northern cape of Cumberland Gulf, not far from where I sighted the abandoned British ship *Resolute* in 1855.

"*Aug.* 31. We are again in bad weather, the pest of the northern regions—fog, and, combined with this, ice, snow, sleet, and a heavy sea; and the vessel is rolling badly, one sea this morning making a clean sweep through the galley, much to the disgust of the cook, who was almost submerged under the various compounds he had prepared for breakfast. The ice is scattering, and I hope to-morrow will see the little *Tigress* safely harbored. There is but little use in keeping under way out here, burning coal, and with no hope whatever of finding the Scotch whalers, as we do not look where we should to find them.

"*Sept.* 1. Yesterday and to-day have been about as uncomfortable as days can be. Last night the gale blew very heavy, with a corresponding sea running. Now one can get along with a good stiff gale of wind, if it don't blow too hard; but wind, sea, snow, darkness, icebergs, and, worse still, hummocks and pieces of floe, with scarcely a possibility of perceiving them, make a very disagreeable and not altogether safe combination. The force of the wind beat the sea into a white foam, which made it the more difficult to distinguish ice from the white-caps. The *Tigress* fouled several pieces, but, fortunately, sustained no damage. It is intensely dark to-night, the wind still blowing fresh, and a heavy sea; but it is far better than last night. Last night one could not see for the blinding snow. To-night, as one of those pleasant varieties which make the spice of life, we are indulged with a smart rain.

"*Sept.* 2. Another day of tossing, pitching, rolling, tumbling; in fact, one gets screwed and twisted into all imaginable shapes in such vile weather. The gale still continues, but with mitigated fury, accompanied with a shade lighter snow. Our wardroom is a complete wreck; broken chairs, crockery, with such heaps of wet clothing, and dirty rubbish, as would have a tendency to shock the nerves of the refined.

"CUMBERLAND GULF, *Sept.* 4. This seems like home—it is my old whaling-ground, and here we are, snug and comfortable, in Niountelik Harbor, so familiar to me. We arrived here this morning. The late gale was very severe, even for this region, and we are all glad to get into harbor.

ENCAMPMENT NEAR IVGITUT.

CHAPTER XXXII.

A Change for the better.—Repairing Damages.—Company in the Gulf.—Looking for Scotch Whalers.—The Natives bring Deer-meat to the Ship.—Arctic Birds flying South.—Captain Hall's old Protégés.—Demoralization of the Natives of the west Coast.—Collecting "Specimens."—Bad Case of "Stone Fever."—"Time and Tide wait for no Man."—Billy's Curiosities.—Captain Tyson meets his late Rescuer, Captain Bartlett.—Mica Speculation.—Short of Coal.—How we lost our Dinner.—A saltatory Dining-table.—Sight a Scotch Whaler.—Arrival at Ivgitut, South Greenland.—Meet the *Fox*, of Arctic Fame.—Kryolite, Coal, Fish, and another Gale.—Friend Schnider, the fat Dane.—Canaries, Pigeons, etc., domesticated here.—The Crew overworked.—A Hurricane.—Antics of the Furniture.—Force of Sea-waves.

"SINCE arriving here we have had most splendid weather, calm, clear, and cold. The crew have been continually employed in getting stone from shore for ballast. The engineers are very busy repairing boilers. The little Scotch brig *Alert*, Captain Walker, and the American brig *Helen F.*, Captain Palmer, are here. Captain Walker sailed from home last April; has been in the ice over two months, trying to force his way into the gulf. Reports that he met a great deal of ice, heavy gales of wind, and had many narrow escapes. Captain Palmer, of the *Helen F.*, had two schooners last fall, but lost one, she breaking out of her winter-quarters in December last, and drifting out into the gulf. He abandoned her, and, with his crew, reached the shore. One man perished, and another was severely frost-bitten, but recovered. One of the Esquimaux, who was hunting on the ice last March, got adrift; and the ice carrying him southward, he has never since been heard from.

"I know not what we are going to do on leaving here—whether we are going directly home, or north again, trying to find the Scotch whalers. I think they could be found (if we went to the right place) in about five days, and that would end our cruise.

"*Sept.* 11. Weather extremely bad; gales of wind, snow, and rain. We are now lying with both anchors down—75 fathoms of chain on one, and 45 on the other. Two vessels have arrived since the 7th inst., the *Clara* and the *Perseverance*, both Scotch vessels. An expected American brig, belonging to New London, has not arrived yet, and fears are entertained concerning her. Some of the Esquimaux have got back from hunting, bringing a supply of deer-meat, for the skins of which the officers are ready to trade largely.

"Still at work repairing boilers. We dare not go anywhere, as we have not coal enough to carry us home, but must depend chiefly on our sails.

"*Sept.* 12. Weather good again. The *Clara* and the *Helen F.* left this morning for the other side of the gulf. We are taking in ballast still, and I fear the vessel will be so deep that there will be little comfort in her. I suppose we shall remain here until the captain is sure that news has been received of the *Polaris* crew at home, and that will not be later than the 10th of October, as the Scotch ships arrive about that time in Scotland; and this season they have left for home earlier than usual.

"*Sept.* 13. Dark, dismal, cold day. Ice made last night, and it promises to be an early winter. The Arctic marine birds are already on their way to a more genial climate. It will not be many days before fast ice and the long Arctic night sets in. Several boats' crews of Esquimaux arrived from hunting last evening, bringing considerable deer-meat. They carry most of it to the Scotch vessels, as there they can get red rum in trade for meat and skins.

"Here is 'Blind George,' 'Bob,' and 'Polly,' who are all historically embalmed in Captain Hall's book which describes his Frobisher Strait expedition. Bob's course is nearly run. I found him lying on his back, in dirt and filth, wasted to a skeleton. Polly—sore-eyed Polly—is as talkative as ever, and looks as young as she did twelve years ago; she looked sixty then, and she looks the same now. Many have died. Within the last three years, as near as I can tell, there appears to have been about three deaths among these natives to every two births; so at this rate the Esquimaux will soon be extinct in this region. The great mortality among the Esquimaux is caused, I think, by contact with the whites. They introduced rum, tobacco, and disease among these poor natives, who have no medicines or medical practice to stop the spread of the evil; so they transmit it all unmitigated from one to another. There are no Christian missionaries here. The country is too poor to support them. I have often wondered how it was that the Christian world has so completely ignored the existence of these tribes on the west coast of Davis Strait.

"As we are doing nothing, the officers spend most of their time in collecting specimens. There are many garnets to be found—very imperfect, however; and quartz of two or three kinds: rose and white are very plentiful. Some of the officers claim to have found stones of some value, but I have not seen any such. Mica is here in abundance, and its collection could, I believe, be made to pay. One of our company got the 'stone fever' bad this morning. Taking chisel and hammer, and a good-sized canvas bag, in which to put the precious specimens, he started off to make his fortune. This evening, just at dark, he returned. The officer of the deck heard his hail, but the shore-going boat was at that time away, some of the officers having gone with her, about half a mile to the north, to some of the natives' huts trading. No other boat was allowed to be lowered; so Billy had to wait until it returned. He sat himself down on a rock, which happened to be on an extensive shoal. He was on the outer edge of it, looking wistfully toward the *Tigress*, and as the inside of the shoal was lower than the off shore, and the tide was setting in, he was, without knowing it, very soon surrounded by water. Suddenly discovering that he was on an improvised island of somewhat limited extent, he yelled most lustily for relief, and we finally got a boat out to him. But poor Billy was the most frightened man I have seen for many a day.

"After supper, Billy, who had by this time recovered his spirits, was showing his beautiful specimens, and among them he handed around what he called a fossil; but the laugh was against him, and his countenance blank with surprise, when an Arctic expert informed him, and that truly, that it was a fine specimen of *canine exuviæ*.

"*Sept.* 14. Quite pleasant to-day, and at times calm. The Newfoundland steamer *Hector* arrived this morning from the head of the gulf. Her master was Captain Bartlett, the good old man who picked me off the ice on the last day of April. I was sorry to learn that he came up here expressly to seine white grampus, for it has proved a failure, and Captain Bartlett will consequently be a loser. I sincerely hope that a good and kind Providence will take care of him to the end of his days.

"At 10.30 A.M. all hands were mustered for inspection; at 11, divine service. So we spend our Sundays.

"*Sept.* 15. We are still engaged taking in ballast. Last night considerable snow fell; it is not snowing this morning, but there is plenty of ice. Thick icicles are hanging from the gunwales of the boats.

"The officers who are scientifically inclined are still making their daily excursions on shore, but, remembering the fate of Billy, are very careful of the returning tides. That suffering individual felt so aggrieved by his enforced stay upon his little island, that he made an official report, in the nature of a complaint against the officer of the deck (I happened to be that unfortunate); but he has recovered his temper since then, and now wishes to go into the mica business with me, he having discovered great quantities on shore.

"*Sept.* 20. Fog, rain, and wind. Nothing of consequence has occurred since the 15th. On the 16th we got under way, Captain Greer being determined to still search for the English whalers. He will try to get one hundred or one hundred and fifty tons of coal at Ivgitut, on the coast of Greenland, which we are now approaching. There is considerable ice and very bad weather; our little *Tigress* still rolling and pitching, notwithstanding the quantity of ballast we have taken in.

"*Sept.* 21. Another day of rolling, pitching, and gyrating. We are not the most comfortably situated of human beings just at this moment. Our dining-table, with all the dinner, has just turned topsy-turvy, smashing many of the dishes, and utterly destroying most of the food.

"This morning, about ten o'clock, we sighted a vessel; we are lying to on the starboard tack under the fore and aft sails heading north. The vessel, which I believe to be a Scotch whaler, is directly to leeward. Report is made to Captain Greer of the fact, but no notice is taken of it or the vessel. I suppose we shall make a show of staying three or four days longer, and then for home.

"*Sept.* 29. Here we are safely harbored at Ivgitut, South Greenland, the famous place for kryolite. We arrived here on the 27th. There is an English bark and a Danish barkentine in. Here, also, is the celebrated little steamer *Fox* which Lady Franklin sent out, under command of Captain M'Clintock, in search of her husband.

"There is abundance of coal here, so that can no longer be an

KRYOLITE MINE.

excuse. We have here also an opportunity to get some fresh fish. Cod are quite plentiful, and also another very large fish called by the people here 'cat-fish;' it is excellent eating, and I should think some of them would weigh a hundred pounds or more.

"*Oct.* 3. A splendid day; the land, or rather the mountains, have their white winter coats on. The day before yesterday we had one of the most severe gales we have experienced since leaving home. Although in harbor, our situation was most dangerous, as the anchorage is so very poor. We parted our shore moorings, and expected every moment to part from our port-anchor, the only one then holding the vessel, except one little kedge. Finally we succeeded, in the very teeth of the gale, in carrying two hawsers ashore, one eight inches and the other six inches in circumference. Making these fast, we felt much more secure. The gale abated toward evening, but settled down into heavy rain, with some little sleet, and snow later in the night.

"The next morning proved fair, and we resumed our coaling. In fact we never stopped work, even during the storm; but the crew were kept at work exposed during the whole of the gale. Never in my life have I seen men worked so hard, and that continuously. The officers are worked hard too, but they bear it better than the men.

"Speaking of Ivgitut, the land of kryolite, we have here quite a company of Danes. There is, first, the manager of the scientific department, and the old fat Schnider, who looks more like two barrels of beer than one; there are also the doctor and several other officials, and about a hundred workmen. Fat Schnider I had the pleasure of meeting at Fiscanaes two years ago, when on the *Polaris*, outward-bound. He is the hugest monster in bulk I ever saw, but a jolly good fellow, grotesque withal, especially when laughing, as he then presents a prodigious cavity quite devoid of teeth. Dentists do not abound in these regions.

"*Saturday, Oct.* 4. We are off. All the coal is on board—one hundred and ninety tons in all. We are now, at 2 P.M., heaving up our anchor, and have been engaged at it the last three hours. The water being forty fathoms deep, and our cable getting foul of the rocks, caused a deal of trouble. So we leave the kind and hospitable Danes of Ivgitut. May they always be happy, for they have been most affable and genial, and the interchange of visits with them has enabled us to pass the time quite agreeably.

"Here, in Greenland, one would not expect to see the canary, the pigeon, and our common duck, rabbits, goats, and hogs domesticated; yet all these birds and mammals not only exist, but thrive here, in lat. 61° 10′ N.

"The sea and fiords in this vicinity, though full of floating ice, never freeze over, even in the coldest days of winter. Indeed, it is seldom below zero here.

"As we are going out, the little steamer *Fox* is towing in the bark *Brilliant*. She is after a load of kryolite. And so farewell to the good people in the mountain homes of Ivgitut.

"*Oct.* 6. Yesterday we had what might be called a moderate gale; in fact, we met it on the evening of the 4th on coming out of harbor. The wind was north-west, with hail, snow, and at times rain. This morning the weather is somewhat better. The sea has gone down, and we have had a change of wind to the south-east. Going along finely now, about seven knots an hour, right in the track of the English whalers on their homeward-bound passage.

"We have many of the crew down sick, on account of their late hard work and exposure. The fact of the matter is, the men are worked to death.

"*Oct.* 7. Last night proved one of the most disagreeable of the whole voyage so far. Toward evening the wind increased to a gale, accompanied by squalls of snow. Shortened sail, and hove the vessel to on the starboard tack. Through the night the gale increased to a hurricane; the sea was very heavy, breaking over our little bark, and keeping her decks continually flooded with water. Not only the deck but the ward and engine rooms were flooded. The fire-room took so much water, that there was danger of the fires being put out; but, fortunately, nothing more serious occurred than the carrying away of the lower bobstay to the bowsprit, and every body getting thoroughly soaked. About 7 A.M. the gale abated, and the wind hauled to the westward some four points; its former course was south-east. Wearing ship; we are now standing off shore, as we are not more than fifteen miles from the Greenland coast. The weather is not pleasant yet; a heavy sea is running, but not so bad as last night. It is impossible as yet, however, to get any thing to eat. Tables think nothing of turning over; the sea gayly indulges in the pastime of extinguishing the galley-fires, keeping the cook and steward in a chronic state of exasperation; and such was the power of the seas which broke over us that even the anvil and iron covers to the coal-bunkers—the latter weighing about seventy pounds and the former two hundred—were found floating around the deck, driven hither and thither by the force of the descending waves. Lat. 63° 15′ N., long. 52° 41′ W.

CHAPTER XXXIII.

The Gale abates.—Consultation as to Course.—Useless Cruising.—Start for Home.—More bad Weather.—Land-birds blown out to Sea.—Reminiscences of the Ice-floe Drift.—A narrow Escape.—A black Fog.—Interviewing a Hawk at the Masthead.—Arrive at St. Johns.—News of the *Polaris* Party.—Return to Brooklyn.—What the *Tigress* accomplished.—Lessons in Arctic Navigation.—Bravery of the Officers.—A stormy but agreeable Cruise.

"WELL, the gale is over. All were badly frightened, but, sailor-like, they will soon forget it. Captain Greer thinks of starting for home as soon as the wind comes from a favorable quarter; he consulted me about it, and my advice was to proceed home. Buddington and his party are safe enough on board some whaler, if not long ago in a British port. They may even be home by this time. At St. Johns we shall be sure to hear of them.

"*Oct.* 8. The weather is looking bad again; another south-east gale, I fear. We have been standing off shore from the Greenland coast all yesterday and all night; the sea from the westward being so heavy, we made but little progress—about four knots an hour. The gale of the 6th and 7th was south-east, but as soon as the wind abated, a long heavy swell came from the southward this morning. The captain ordered the vessel's head to be kept north-east again. He is still unsettled in his purpose what to do. Well, I care but little; but I know it is useless cruising here longer—only a waste of time and money, besides destroying the crew. They have been worked hard; many are now sick. The few that are well are busy to-day securing every thing for the next gale.

"On breaking out to-day we find many of our stores damaged, from the water finding its way down the hatches in the late gale. By observation at noon, lat. 63° 29' N., long. 53° 52' W.

"5 P.M. Have started for home; the long-looked-for time has arrived. Captain Greer, after some little conversation with me, has finally concluded to start for home, stopping at St. Johns, Newfoundland, where I think he will hear of the *Polaris* survivors. It is my opinion they are on, or near, the shores of bonnie Scotland ere this.

"*Oct.* 9. Blowing strong from the north-west. Sailing and steaming eight and nine knots an hour. Last night spoke the Scotch brig *Clara*, from Cumberland Gulf, homeward-bound.

"*Oct.* 10. Cloudy, gloomy-looking day, but the wind is light, and the sea quite smooth, so we take some little comfort; but it is quite cold. Last night snow and hail squalls. Lat. at noon 58° 48'—about six hundred miles from St. Johns.

"*Oct.* 11. It is blowing fresh to-day from the eastward; considerable sea running; otherwise it is quite pleasant. The sun is shining brightly, the weather quite warm, and the air dry. Generally the south-east winds are cold, wet, and very disagreeable; this is an exception, so far. Lat. 56° 39' at 8 A.M.

"Our barometer is falling quite fast; and, as I sit writing in the wardroom, I hear the quartermaster, who has just observed it, say to the officer of the deck, 'So we can look for more wind, and perhaps rain.' Well, we certainly have had very bad weather, and abundance of it, this cruise. All the officers appear heartily tired of the Arctic regions; nor is it strange: this country is not very inviting to spend one's time in, unless there is a definite object to be accomplished, and a prospect of success.

"*Oct.* 12. The morning is beautiful; the sun rose bright and clear, and with that peculiar color which denotes fair weather, and is also a forerunner of a southerly wind.

"The Sunday-morning muster of all the crew in clean blue shirts and the morning service is over. We are now (noon) in lat. 55° 19' N., and about ninety miles to the eastward of the Labrador coast. Many little land-birds visit us daily. Poor things, they get blown off shore, and are glad enough to find a resting-place. Some alight on the deck, but only to die; for they are overexhausted by their long flight and struggle with the winds.

"And here *I* am, traveling over the route—part of it—of my long and terrible drift, but under what different circumstances! Then I was starving with hunger and perishing with cold; now I am on a good staunch steamer instead of a bit of ice, with plenty to eat and drink, and, when tired, a decent place to sleep. The misery of that fearful drift will haunt me so long as memory endures.

"*Oct.* 13. Last evening, just before 'eight bells,' or eight by the clock, a heavy, dense black fog came sweeping down from the S.S.W. Standing on the quarter-deck, I could scarcely see the head of the steamer. It continued, thick as ever, until the midnight watch.

"Mr. Chipman, my assistant ice-pilot, had a very narrow escape from colliding with a large berg. He just saw it in time to clear it. These floating ice-hills are very dangerous to navigation. Many a good ship has gone down and never been heard from; but, could these icebergs speak, they might 'a tale unfold' which would solve the mystery.

"The night and the day have been very unpleasant; it has rained violently, and we have been going very slow—two or three knots, and some of the time not more than one. Even a slight head-sea and head-wind entirely destroys the steaming qualities of the *Tigress*.

"*Oct.* 14. The fog, after enveloping us in worse than the blackness of night for upward of thirty-four hours, has cleared up with a westerly breeze. Several large bergs are in sight. The sea is still long and heavy, causing our little vessel to roll and pitch very badly. We have sail up now, and are making good way. God speed her! I am tired, and want rest.

"6 P.M. The wind is increasing, and it again looks like rain. We certainly get our share of bad weather. Yesterday a little land-bird came on board, and, as he has been treated kindly, continues to stay; he is now so tame one can pick him up with the hand. A large hawk, wanting a rest, lighted on our rigging this afternoon. One of the sailors, thinking to capture it, went aloft for the purpose. On getting near the bird, it turned and looked at the man, as much as to say, 'Who are you, and what do you want?' In consequence of this uncanny and unexpected behavior of his hawkship, the sailor hesitated to touch him, and did not venture to put his hand on him, although his messmates from the deck kept calling out to 'catch him, catch him!' There aloft they eyed each other. The bird looked at the man, and the man looked at the bird, and for some time neither moved. At last the hawk concluded he had enough of that sort of visual interviewing, and desiring no better acquaintance with the strange sort of animal that continued to stare at him in a superstitious, frightened sort of way, took wing and made his escape. Coming down empty-handed, the man got well laughed at for his failure.

"Our latitude at noon to-day is 52° 49' N.—a little over three hundred miles from St. Johns, and twelve miles south of where I was picked up by this same *Tigress* on the last day of April, between five and six months ago.

"*Oct.* 15. We have it. It commenced to rain this morning; but we have not the usual accompanying gale of wind which has hitherto added discomfort to every fall of rain or snow.

"*Oct.* 16. This morning, at 4 A.M., sighted Buena Vista light, wind blowing fresh from the north. With wind and steam-power the *Tigress* is making upward of eight knots an hour. At six sighted Bacelhoe light, thirty-eight miles from St. Johns. So, should the wind not fail, we shall reach there this afternoon about three or four o'clock. I have been quite sanguine all the way on our homeward voyage that on our arrival at St. Johns we should hear from the *Polaris* crew from the way of Scotland. I have been in talking to Captain Greer about it, and have expressed very plainly my opinion that we shall certainly hear from the *Polaris* party at St. Johns. He does not share my confidence in the safety of the *Polaris* party; but he soon will.

"3.15 P.M. Made St. Johns Harbor. The first words which the pilot uttered were, 'The *Polaris* party are safe.' A look of intense relief passed over Commander Greer's countenance, and questions and replies soon made us acquainted with the general outline of the story of the relief of Buddington and his party. It had all happened just as I expected, with the difference that they had got to Scotland even earlier than I had supposed. Our cruise is now ended, and 'homeward-bound' is the word. We shall stop only long enough to make necessary repairs, and then sail for Brooklyn Navy Yard.

"*Sunday, Nov.* 9. Arrived at the Brooklyn Navy Yard this morning; and thus ends our cruise. Successful only in part, because success was rendered impossible by the lateness of the season when the *Tigress* sailed, Captain Buddington's party having been picked up by the *Ravenscraig* three weeks before the *Tigress* was ready for sea. But the voyage was not eventless or useless: the discoveries made at Littleton Island, Life-boat Cove, on shore, at the winter-quarters of the survivors of the *Polaris*, and the information gained from the natives as to the premature abandonment of that vessel, all were needful to elucidate doubtful points in the history of the *Polaris* expedition; as also in determining the topography of the coast, and the correctness of observations made in winter and darkness by the light and certainty of the summer's sun.

"It has also given an opportunity to many worthy officers of the United States Navy to see something of Arctic service, to which some of them may again be called; for though our voyage was brief as to the time occupied, and was favored as to the season of the year, yet it proved exceptionally stormy, and was well calculated to test the courage and endurance of officers as well as the crew; and it is one of the pleasantest duties of my life to here record the perfect harmony and good-fellowship which prevailed, the courage, manliness, and endurance exhibited; and it is a satisfaction to feel that, though Providence had forestalled the chief object of our voyage, there was not an officer on board who was not ready and willing to winter in the Arctic regions had there been the shadow of a reason for so doing. Nor would Captain Greer have turned his vessel's head to the southward had I not positively assured him that, from my experience in those regions, the search beyond a certain date was absolutely useless.

"I shall always remember with pleasure and satisfaction the brief period of my association with the commander and officers of the United States steamship *Tigress** on her searching cruise."

---

* The *Tigress* was subsequently repurchased for the sealing business by Messrs. Harvey & Co., of St. Johns, Newfoundland. On the 2d of April, of the present year (1874), as the steamer was working through the ice in the prosecution of her usual occupation, an explosion occurred, which instantly killed ten of her crew—eleven others being so badly injured that they died the next day. Captain Bartlett fortunately escaped unhurt. The vessel was badly damaged.

CHAPTER XXXIV.

THEORY OF NORTH POLAR CURRENTS.

The Hydrography of Smith Sound.—The Currents forbid the Theory of an "Open Polar Sea."—Movements of the Ice.—A northern Archipelago a reasonable Supposition.—Velocity of Current along the east and west Coasts.—No Current in the Middle.—Experience of the *Polaris*.—Absence of large Bergs in Smith Sound.—Open nearly all Winter.—Radiant Heat preserved by Cloud Strata.—Deflection of the Current at Cape York.—Robeson Channel described.—Land seen from the Mast-head both east and west.—Coast-line beyond Cape Union.—Two Headlands to the east-north-east of Repulse Harbor.—Absence of Snow on Coast of North Greenland above Humboldt Glacier.—Elevated Plateaus in the Interior.—The Land around Polaris Bay.—Clam-shells at an Elevation of two thousand Feet.—Variegated but odorless Flora.—Animal Life.—Insects.—Skeletons of Musk-cattle.

IN describing the hydrography of Smith Sound I free myself entirely of the geographers, chart-makers, and romancers, and relate only what I observed with my own eyes.

Through the fall and winter of 1871-72 I noticed, while on the ice in the mouth of Newman Bay, when on my boat-expedition, that the current through Robeson Channel was about one mile an hour; this, however, was very much accelerated by the freshets of the spring and summer. I saw ice drifting at the rate of four miles an hour, driven along by the heavy north-east gales supplementing the natural velocity of the current.

I do not think there is enough current in Robeson Channel to warrant the theory of an open Polar Sea. Neither such a sea, nor a portion of such a sea, could empty itself through such a narrow channel as Robeson. Any large sea to the northward coursing through such a contracted passage would cause such a powerful current as to make it unsafe for navigation by any but the most powerful steamers, yet the *Polaris* overcame it without difficulty.

I carefully watched the movements of the ice during the winter, and in fact all the time from my first entering Smith Sound, and I became fully convinced that there was no Polar Sea through or beyond Robeson Channel. I believe the space northward of that to be occupied by an archipelago, for, as I said before, the current, uninfluenced by winds or freshets, is but about one knot an hour; and at that rate the current continues through Robeson Channel, Polaris Bay, and Kennedy Channel, the current setting uniformly from the north. This state of the current I observed on coming southward in August, 1872. But on entering Smith Sound, I found little or no current in the middle of the sound; but on the *west shore*, on my passage up, I observed considerable current, which convinced me that constantly setting from the north, after entering Smith Sound, it diverged toward the shores, to both the east and west sides. This view was confirmed on the *Polaris* getting beset in the ice in the vicinity of Cape Frazier, about twenty-five miles to the east-north-east. Our drift averaged from the first from one to four miles daily, and nearly due south—a very little easterly.

On approaching the east coast in the vicinity of Rensselaer Harbor—Kane's winter-quarters—which we drifted by at about five miles from shore, our progress was more rapid, but at that time the wind was very strong from the north, which greatly accelerated our drift, the fall gales having set in. But even then I detected a stronger current near the shore than in the middle of the sound.

Another point which attracted my attention, and interested me greatly, was the total absence of large icebergs in Smith Sound, and from thence as far north as $82°\ 16'$ the bergs were very few and very small, while in Baffin and Melville bays, and all along the shores of North Greenland, huge bergs abound, especially between Disco and Kane's winter-quarters. But north of that the ice formations should be called hummocks, rather than bergs.

Smith Sound is open nearly all winter, seldom closing until February or March. I know it to be so by the heavy north and north-east gales which prevail all winter. They blow with indescribable fury; and I have seen Robeson Channel, Polaris Bay, and as far as the eye could reach, north and south, with the ice cleaned out from shore to shore during the months of November, December, January, and February. In March we had a long spell of quiet weather, and very cold; then the ice formed rapidly. In the absence of the sun the weather was chiefly cloudy. In fact, I never saw so much cloudy weather, nor so many gales of wind in any previous winter.

When the sun was absent the temperature was usually more

moderate than when it shone brightly. It was not so low by many degrees as I have found it twenty degrees south of Polaris Bay—the thermometer often above zero, seldom 15° below. On the return of the sun, in March and April, it was very cold, the lowest we experienced reaching to 58° below zero, with very little wind during those months. The fact appears to be that a dense strata of cloud over the land intercepts and retains the radiant heat of the earth; which, when the clouds are removed, readily escapes into the upper atmosphere and is lost, the direct rays of the sun not compensating in warmth for this cooling of the earth by radiation.

After my separation from the *Polaris*, I found that, in the vicinity of Cape York, the current suddenly deflects to the southwest, and continues south-west down Baffin and Melville bays, Davis Strait, and so along the coast of Labrador. I have other reasons for thinking that Robeson Channel does not lead to a Polar sea. When the *Polaris*, under the command of Captain Hall, penetrated these icy solitudes to lat. 82° 16' N.—but four hundred and sixty-four miles from the geographical pole—I was much of the time at the mast-head, looking out for whatever was to be seen, and I certainly saw plenty of water to the north. Robeson Channel was at that time much obstructed, even blocked, with ice, but beyond it there was free water; and had we got through the channel, we should have been clear of ice for a long distance. But I am sure that on the west side, above what is called "Grinnell Land," far north of Hayes's Cape Union, there is land; for I could trace the coast leading north as far as the eye, with the assistance of a spy-glass, could see.

This coast-line north of Cape Union is to the westward of the cape, but runs N. or N.N.E.; and I think that between the Cape Union of Hayes's and the land seen beyond to the westward there is another strait or sound. On the east side there was a dark, heavy bank, denoting water in that direction; and I am inclined to think that the land trends to the eastward at about 82° 20' or 25'; but the appearance may indicate only a deep circuit or bay, as I am quite sure that from the mast-head when we were farthest north I saw two headlands to the north and east. However, as we never attained the same latitude again, probably those who saw only the masses of wide-spreading water before them will still cling to the theory of an open

Polar sea; but I have the evidence of my eyes, long trained to observation in the Arctic regions, to the contrary.

I have seen a public statement by one of our number (Mr. Chester), that on Captain Hall's last sledge-expedition, Mr. C. being one of the party, he went north of Newman Bay, up to Repulse Harbor; but Captain Hall, on his return, told me that they went no farther than Cape Brevoort, in lat. 82° 1' N., long. 61° 20' W.; and from there his latest dispatch—in it misprinted 82° 3'—was dated. Joe, the Esquimau, also accompanied Captain Hall, and when he, with Mr. Meyers and myself, were off hunting and taking observations, Joe pointed out the place where Hall's party had stopped the fall before. Mr. C. must, therefore, be mistaken in thinking he went to Repulse Harbor, or that the land terminated just above there. It is not so. The land certainly trends to the eastward a little north-east of Repulse Harbor; but *it is again to be seen trending north*.

The absence of snow along the entire coast of North Greenland is worthy of remark; while as far north as Humboldt Glacier one can see the snow and glacier clad mountains, standing but little back of the sea-coast, some, indeed, approaching closely to the coast; but they are chiefly in the interior—the icebergs finding their way to the sea by the many glacier-fed fiords with which the coast of Greenland abounds. These fiords have probably been formed by ancient, and some by extinct, glaciers.

On getting north of Humboldt Glacier, I was astonished at the entire absence of snow on both sides of the strait. The land looked dried up; even the ravines, as I found on landing at Polaris Bay, had exhausted themselves, showing plainly that the summer had been very warm. The coast is high and rugged, composed principally of limestone and slate-rock; toward the interior is an elevated plateau, level and firm under foot, so that traveling is quite easy. But these elevated plains, or plateaus, are cut through in many places by ravines or water-courses. The soil is a light clay, which absorbs the moisture very fast.

The whole land around Polaris Bay, and above and beyond, both the plains and the highest mountains, have at some remote period been an ocean-bed; for the entire land is covered with marine fossils. I found three fossilized sea-snails, one on the top of a mountain, near our winter-quarters; and at the height of two thousand feet the clam-shells were so thickly scattered that one could not put his foot down without crushing them. Even some of the ponds are salt, though far away from the sea.

The flora found at Thank God Harbor, Newman Bay, and vicinity I can not classify, not having preserved any; but there were many species, highly variegated, and of most beautiful colors, but odorless. They grow in small patches or clumps, and in many spots these groups look like little fairy gardens, and are in pleasing contrast to the generally rugged scenery of the country. In taking up a moss-like substance from a fresh-water pool, and bringing a microscope to bear on it, the most beautiful vegetation was descried. Not only were the forms most elegant and graceful, but the colors were as brilliant as tropical flowers.

In the farthest north to which the *Polaris* reached birds of various kinds abound. There are brent-geese, eider-ducks, ivory and burgomaster gulls, ptarmigan, plover, mollemokes, and several others which I had not an opportunity to examine. In early summer these birds are seen flying to the north-east. So I am convinced there is a continuation of the land in that direction, as they go there to breed; and in the fall they go southward again.

Of animals I saw the white hare and the little lemming, the heavy musk-cattle, and the Polar bear, with his usual follower, the white fox. Seals were found as far north as we went. The fauna of the Arctic regions plainly shows that the vicinity of the Pole can not be destitute of land, nor without some sort of vegetation which will sustain animal life.

The insects I discovered were large blue-flies, and a smaller species very similar to our common house-fly. There were also butterflies, some quite brilliant, and of course caterpillars. Bees, mosquitoes, and spiders were also seen. On the plain abreast of the ship there were many skeletons of musk-oxen, killed by the Esquimaux who formerly inhabited these regions.

CHAPTER XXXV.

How to reach the North Pole.—Smith Sound the true Gate-way.—This course offers the Alternative of Land Travel.—Plenty of Game in Summer.—April and May the Months for Sledging.—Proper Model of Vessel's Hull.—Twenty-five Men enough.—A Tender necessary.—A Dépôt at Port Foulke with a detail of Men.—Ice at Rensselaer Harbor.—Avoid Pack-ice in Smith Sound.—Go direct for west Coast.—Form *Caches* at intervals of fifty Miles.—Deposit Reserve Boats. —Style of Traveling-sledge.—Native preferred.—Selecting Dogs.—Keep them well fed.—Keep Stores on Deck.—Winter as far north as the Ship can get.—How to get out of a Trap.—Provision a Floe, and trust to the Current.—Take your Boats along.—Replenish at *Caches.*—Two Months from a high Latitude sufficient.—It will yet be done.

It is not to be supposed that the search for the pole terminates with the loss of the *Polaris* and the death of her commander; and as I have given considerable thought to the subject, I will here make some suggestions in regard to the furnishing and equipment of any future expedition.

There is no doubt in my mind that Smith Sound is the true "gate-way to the pole," if for no other reason, because there is land to operate on. To the Spitzbergen route is the fatal objection of ice-beset seas, without the choice of land to resort to in case of need; but the Arctic explorer by Smith Sound can at least avoid the experience of Parry, in his journey on the ice in the eastern Arctic seas north of Spitzbergen, in which he was carried south by the ice faster than he traveled north with his sledges.

Another advantage of the Smith Sound route is the abundance of game to be found during a considerable portion of the year. A party traveling on the land, if good hunters, could shoot enough to keep both themselves and dogs from starving, and in summer could secure enough to save some for future use; for in many localities the musk-cattle are quite numerous, with hares and ptarmigans in plenty. They are found as far north as our winter-quarters, and probably farther, certainly at Newman Bay, until late in the fall, and as soon as the sun re-appears in the spring they are discovered again.

On the land north of Washington Land, "Hall Land," both of which are but extensions of Greenland, though the coast is rugged and mountainous, a short distance inland it becomes quite level—is, in fact, so far as I traversed it, an elevated plateau, over which a party with sledges, starting in early spring, about the 1st of April, would have two full months to explore before the melting snows caused them any serious inconvenience. There is this advantage, also, in selecting this period of the year—it is continually light; no darkness to contend with, which, in this region, is a serious obstacle to travel, either by land or water.

The vessel—a steamer, of course—which is expected to prosecute with any hope of success the search for the pole must be built as strong as wood and iron, properly combined, can make her: sharp bows, and stem sloping, so that on striking ice she will run out on it. If the stem is straight or perpendicular, the vessel brings up with a heavy thud, which is very damaging to her. The hull should be so modeled as to allow the vessel to rise or lift up in case of severe pressure by ice; neither should it fall in above the water-line, or be wall-sided.

Twenty-five picked men would be enough to man her, and she should leave New York, at the latest, by the middle of June, so as to give time to stop at the Greenland ports for dogs, and such other things as are needed and can not be obtained here.

Any exploring vessel ought to be accompanied a certain distance by a tender, with coal, extra provisions, and articles for which there may not be room in the steamer. A schooner capable of carrying about three hundred tons of coal, and a small frame house, packed in sections, would answer. There are many articles which would be useful which such a tender might convey, but which would needlessly encumber a steamer. This tender should go as far as Port Foulke—Hayes's winter-quarters—and there leave her coal, house, and stores, the latitude of that harbor being 78° 17' 41" N., and the longitude 72° 30' 57" W., and twenty miles south of Rensselaer Harbor, Kane's winter-quarters. The steamer should be at least capable of steaming six to eight knots an hour, and should make under sail the same, and be bark-rigged. That, I think, is the best rig for such service.

At Port Foulke the frame house should be erected, and the coal landed, with provisions enough to last four men three years, and also to last the whole party going north in the steamer six months. There should also be deposited here two boats. Four men should be detached from the company to remain here to take charge of the property, to take observations, and to assist any returning party. These should be reliable, well-tried men, or they would weary of waiting, and be tempted to leave. Supposing this dépôt to be established, and the four men to be taken from the steamer's company of twenty-five, would still leave twenty-one to man the steamer, which, with care and good discipline, would be sufficient. All these arrangements should be completed so that the steamer could leave Port Foulke and go on her northward course by the 15th of August.

As above Port Foulke it is almost impossible to penetrate the ice on the east side, we must now head to the westward, to Cape Isabella or to Cape Fraser. The ice lies inshore, like a wall, around the vicinity of Rensselaer Harbor, Kane's winter-quarters, extending from the harbor across toward Cape Isabella. So any vessel hoping to make any progress must at this point go as direct as possible across to the west coast, along the shores of which they will find water, though they may not be able to see it until they get nearly over. On no account must they take the pack-ice in Smith Sound at any considerable distance from the shore; for, once getting beset among the large, heavy floes which are always found in the sound, it would be almost impossible to extricate the vessel during the same season, and the consequence would be a long winter drift to the southward, such as delayed and exasperated Captain M'Clintock, in the *Fox*, in 1857-'58. Go direct for the west coast above Port Foulke, and if you can not get through along the shore, be quite sure you can not get through at all; for the fall winds, the currents, and the pack-ice settle that. But, unless the season is very unfavorable, you *can* work up on the west side. In going north, then, through this westerly channel—from one to four miles wide—at every fifty miles' advance there should be *caches* formed, in which should be deposited a certain amount of bread and pemmican. Either bury it, if the soil permits, or put it in the crevices of the rocks, crevices which can always be found on these coasts, so deep and narrow that the bears can not get at it. On the surface, it is scarcely possible to protect food from the bears. Of course some recognizable mark should indicate the spot.

There should also be two boats left at some more northern

point on the coast, for relief, in case of disaster to the ship. Boats and provisions deposited along the route in this manner would, in case of necessity, make a retreat comparatively safe and easy; and would, in any event, if not needed, have a beneficial moral effect on the crew, giving them courage and hope, and inspiriting them to do their utmost, knowing they would have these reserves to fall back upon in case of being compelled to abandon the ship at a high northern latitude. Especially would this keep up the spirits of the men, feeling assured that, however hard their journey back, they would find also succor, stores, and shelter at Port Foulke. Then, with good skin clothing and sleeping-bags, which may be made from sheep-skins in the States, no one need fear to make their way north.

For traveling-sledges I prefer those made in the simple fashion of the natives; they are not so liable to break in traveling over the rough ice as the more complicated inventions of civilization. In regard to dogs, two rules should be observed: first, to *get the best*; and, secondly, to *feed them well* on good wholesome food when you have them. Better have ten picked dogs, healthy and of the best breed, than forty young, weak, sickly mongrels, who consume food and take up room, and do little or no service. It is a great mistake, too, to feed these creatures, as many do, on putrid meat, and keep them half starved at that. They are like human beings to this extent, that ill usage incapacitates them for work and breeds disease. They will eat almost any thing but iron or rock, if they are hungry; but their owners pay the penalty in final loss and disaster. The rabies which decimated Kane's dogs had no doubt some such origin; they had suffered from scanty and improper food.

Having, like a good general, made arrangements for an orderly and safe retreat if Fate compels, it is next in order to get your ship as far north as possible, and not be afraid of wintering at the extremest latitude which can be attained by steam. But during the whole progress of the vessel, after entering the regions of ice, provisions, ammunition, extra clothing, matches, medicines, and, indeed, every thing absolutely necessary to sustain life and health, should be kept on deck, with the boats in a position to be lowered at a moment's notice; for in navigating these Arctic seas a vessel is liable to be nipped so suddenly that there is no opportunity to save any thing which is below. Only that which is on deck, and in condition to be thrown overboard, can be relied on. Hence every thing which may be supposed to be needed in such an emergency should be carefully boxed, and ready to be hove overboard at a moment's notice.

Now, supposing our vessel bound for the North Pole, and her commander fortunately hits on a favorable season, and is thus enabled to penetrate far beyond even the latitude made by the *Polaris*, he will, of course, in the fall put his ship into winter-quarters, without thinking of turning south to look for a harbor. He may thus possibly find himself in a trap, and must pass one or two seasons watching for an opportunity to get out—availing himself, of course, of chances to make land or ice explorations as the season permits. On the worst supposition, if, after waiting as long as his commissariat warrants, and he feels that he can not risk the health and safety of those who depend upon his judgment any longer, but that the time has come for him to extricate them from a perilous position, what shall he do if his vessel is still beset and can not be saved?

I will tell you what I would do under such circumstances. I would get every thing together that was necessary for a long sojourn upon the ice, including boats, and, picking out a large, strong floe, I would not hesitate to trust myself to that comparatively frail and ofttimes treacherous support; for, frail as it might be, as long as there was room to stand upon it I know that it would carry me southward, for all the currents set from the north. Then, knowing that along the coast I have provisions *cached* at intervals of about fifty miles, and at Port Foulke comfortable quarters, could I reach there, almost any decent company of men might make the journey without cause for despair.

These plans might, of course, all fail through unforeseen accident; but at least all the pre-arrangements would give a good basis for hope, and the journey I believe to be not only possible, but that in all probability it could be performed in safety. And I know of no reason why a ship need be abandoned; but I have simply planned what could be done at the worst. The chances appear to me reasonable that a ship might go far higher than we did with safety and return, and that by ship or sledges the pole might be reached with proper management in two seasons as well as in two centuries. A good ship, a united company, and a calm, courageous leader will yet do this thing.

CHAPTER XXXVI.

THE FATE OF THE POLARIS.

The *Polaris* Survivors.—Ship driven to the North-east.—Her Position on the Night of October 15.—Darkness and Confusion.—Anchors and Boats gone.—The Leak gains.—Steam up.—Roll-call on Board.—Lookout for the Floe Party.—Storm abated.—Inspection of Stores.—The *Polaris* fast to grounded Hummocks.—"Let her fill!"—Life-boat Cove.—The *Polaris* left a Legacy to an Esquimau Chief.—She founders in his Sight.

THE south-west gale of October 15, which culminated in the breaking up of the ice which surrounded the *Polaris*, and in leaving the party under Captain Tyson adrift on the floe, was almost equally disastrous to the vessel. All of a sudden, snap went the bow-hawser, "like a pack-thread," slip went the anchors, and away went the *Polaris* none knew whither; for the darkness and the blinding, drifting snow prevented all observation.

The floe remained grounded during the night between the heavy bergs, which by their pressure had disrupted the ice surrounding the *Polaris;* but the vessel was driven by the force of the wind in a north-easterly direction; and on the morning of October 16 they found their position to be "a little north of Littleton Island, in Smith Sound—having been exactly abreast of Sutherland Island during a portion of the night."

It will be remembered, as Captain Tyson has described in his journal, that all had been excitement and confusion on board; the darkness and the disappearance of the floe party greatly adding to the very natural anxiety for their own safety; for, in following the fortunes of the vessel (Oct. 15), it was related that large quantities of provisions, stores, and clothing had been thrown overboard; and at first it was not known on the *Polaris* if sufficient of these things had been retained for their own use; for much of that which was put on the ice was known to have been lost. Boxes and bags had gone overboard, without discrimination as to whom they belonged. The clothing of those who were on board was much of it thrown out, as all expected that they might have to leave the ship either that night or the next morning. The wind was blowing with a velocity of forty miles an hour, and at half-past seven the pressure of the heavy floe on the starboard side of the ship had keeled her over to port, so that it was no longer easy to keep footing on the deck. The boats, too, had been lowered—the only two remaining whale-boats—and the little flat-bottomed square-ended scow. Hence two fears oppressed the *Polaris* company: one that the water would reach the fires; and the other that the party on the floe, with all the boats, would be lost. The anchors, too, were gone, and vessel and crew were at the mercy of the gale. All that could be done was to try and reduce the water in the hold. The bilge-pump was kept at work, with the alley-way pump, but the water still gained, and every probability was against saving the vessel unless the steam-deck pumps could be started. The pipes were frozen, and the pumps chocked with ice; but after a time, the fires being renewed, sufficient hot water was procured from the boiler to thaw out the deck-pumps. Bucketful after bucketful was poured down, and finally they were started.

Every one felt that life depended on those pumps, and all in turn worked with an energy commensurate with the gravity of the occasion. At last Mr. Schuman reported "steam up," and with this hopeful assurance all watched and waited through the night, thankful that they were still afloat. As the dim daylight of October 16 dawned upon the *Polaris*, the reduced company of fourteen souls found themselves to the north of Littleton Island, in Smith Sound, having drifted through the night, or rather been forced by the violence of the wind, beyond the head of Baffin Bay.

Counting heads, it was found that there remained on board the following officers and men:

S. O. Buddington, captain; H. C. Chester, chief mate; William Morton, second mate; Emil Schuman, chief engineer; A. A. Odell, assistant engineer; W. F. Campbell, fireman; J. W. Booth, fireman; N. J. Coffin, carpenter; H. Sieman, H. Hobby, N. Hays, Joseph B. Mauch, seamen; Emil Bessel, chief of the Scientific Corps; and R. W. D. Bryan, astronomer and chaplain.

The cook and steward were both gone, as well as the Esquimaux, the assistant navigator, the meteorologist, and six seamen; also some of the dogs.

Mr. Chester[*] went up to the mast-head, and looked around in

---

[*] See letter of Mr. Bryan's in Appendix.

all directions, but reported that he could see nothing of the lost party or the boats. The general opinion was, however, that they had probably saved the boats, as it was known that they had been dragged back to what seemed a safe distance from the ship. It was now calm, the gale having abated.

The next thing was to see how much provision was left on board, and if there was enough fuel to get through the winter. The inspection was re-assuring, as of food plenty remained; and as it had been determined by Captain Buddington to abandon the vessel and make winter-quarters on shore, there could be no scarcity of fuel while a plank or a timber of the *Polaris* remained afloat. There was also a large quantity of coal still in the bunkers.

Toward noon a breeze sprung up from the north, and a lead opened inshore to the eastward. The vessel, which was no longer under perfect control, was drifting, with the set of the current out of the strait. Fortunately the lead toward shore opened out wide enough to admit the vessel, and then, putting on full steam, and also setting sail to assist her forward, the *Polaris* was run as near shore as the ice permitted; and the ground-tackle, and even the ice-hooks, being lost, she was made fast by lines to some heavy grounded hummocks in nine feet of water at high tide, but aground at low. At six P.M. orders were given to stop the steam-pumps and let her fill.

Life-boat Cove—Kane's Life-boat Cove! It seemed to give new spirits to the party that Providence had guided them to the shelter which had protected, under equally perilous, though different circumstances, the beloved Arctic explorer whose fame is dear to every American. This cove is in lat. 78° 23' 30" N., and long. 73° 21' W.

On October 17 Captain Buddington surveyed the ship, and found that her stem was entirely broken off below the six-foot mark, with other serious injuries—so serious that he considered it was only wonderful that she had kept afloat so long; and concluding that she could not be repaired, preparations commenced for permanently abandoning her. There was no difficulty in doing this, as the *Polaris* lay so near shore.

The few succeeding days were passed in bringing from the *Polaris* all the food, fuel, and the most necessary articles with which to build a shelter for the party, and to sustain them through the winter. This done, the vessel was left to her fate.

LAST OF THE POLARIS. 401

In the early summer, for favors received, Captain Buddington conveyed all the title to the *Polaris* he was able to give on a native Esquimau chief; but shortly after she drifted out of the cove, and foundered in the sight of her last *quasi* owner, as described in the cruise of the *Tigress*. And so ends the story of the good ship *Polaris*, on which so many unfulfilled hopes had been centred.

A SUMMER ENCAMPMENT.

CHAPTER XXXVII.

THE FORTUNES OF THE POLARIS SURVIVORS.

Life on Shore.—A House built.—Visitors.—Womanly Assistance.—Scientific Observations.—Amusements.—Old Myoney.—Hunting.—Boat built.—Starting for Home.—A Summer-trip.—Sight a Vessel.—Rescue by Captain Allen, of the *Ravenscraig*.—Romance of the *Polaris* Expedition.—Safe Arrival of all the Survivors at New York.—Consul Molloy.

As soon as practicable a house was erected on the shore, composed principally of portions of the *Polaris*, the spars of which had all been hauled ashore, the bulk-heads of the state-rooms removed, and the sails of the good ship assisted in forming a roof for the new residence. The house was quite commodious, being twenty-two feet in length by fourteen in width, and was perfectly water-tight. It was very comfortable, but there was no attempt to keep up any class distinctions. It was simply one open space inclosed by four walls and a roof, without any subdivisions for privacy or state. Like a ship in Arctic winter-quarters, it was banked around with snow to keep out the cutting winds, and was soon made warm and cheerful by the introduction of a stove from the ship. A galley was established, and implements for cooking secured. Bunks, or sleeping-berths, were built around the sides of the house, fourteen in number, to accommodate the whole company. A table, lamps, and whatever else the *Polaris* afforded which could increase the comfort of the party were brought ashore and placed in position; and all being thus happily arranged, there was nothing more that could be done until spring came; for that they must wait before they could hope to leave their present quarters.

The party were not long without visitors. Five or six days after the *Polaris* was abandoned, a party of native Esquimaux, in five sledges, came to the encampment, and greatly assisted the crew in getting things out of the vessel, in cutting fresh-water ice, and hauling it to the house on their sledges, and in various other ways showing their friendliness to the party.

Of course their assistance was rewarded with presents of knives, needles, pieces of wood and iron, and such little things as they most valued; and after a short time they returned to the settlement of Etah, from whence they came. But ere long others appeared upon the scene, and finally two or three families built their huts in the vicinity, and prepared to spend the winter as friends and neighbors of the whites. The women of these families soon made themselves very useful by making and repairing clothing, and performing other feminine courtesies for the men; and as the season advanced, and game became more plenty, the native hunters brought to the house fresh meat, which is always so greatly prized during an Arctic sojourn. The walrus hunters, in particular, having good success, often brought to the house a feast of walrus liver, and by their good-natured friendliness greatly aided the party, not only to maintain a condition of physical health, but encouraged them through the Arctic night by the feeling that, though cut off from the civilized world, they were surrounded by friends, and not enemies. Among the Esquimaux visitors was old Myoney, who had also visited Dr. Kane on board the *Advance*. But he died during the winter.

Once established, and the routine of their winter life commenced, Dr. Bessel and Mr. Bryan resumed their scientific observations with such limited appliances as remained to them; while the rest of the party amused themselves as best they might in reading, writing, making up journals, and playing at chess, draughts, and cards. Then there was house-work to be done, ice-blocks to be cut for water, fires to be made, lamps to be trimmed, food to be cooked; and when, in February, the coal gave out, wood for firing was to be cut from the *Polaris*, and brought on shore. And then, as spring approached and light replaced the prevailing darkness, favorable days would come, when some of the party would tramp off in search of game, or set traps for foxes or nets for seal. But, happily, there was no lack of food, and no suffering from cold. They were well clothed, well fed, and well sheltered; and though ice-bound, there was really nothing serious in their position—a striking contrast to the fate of their late companions adrift on the ice-floe.

But one thing they lacked—namely, boats. But they had seamen to direct, a carpenter to execute, and plenty of materials wherewith to build and rig one or more. As the season advanced, and the topic of returning home began to be discussed as an event to prepare for, it was decided that at least two boats must be built to convey the party and necessary stores for consumption on the way.

The sun had re-appeared on February 15, and from that time forward the prospects and preparations for traveling were an almost daily topic of conversation. As the light increased hunting-parties went out, and a large number of foxes and some hares were killed, but not much other game was secured. There were a great many deer seen, but all, with one exception, escaped the aim of the huntsmen.

As April came in, Mr. Chester, with the aid of the carpenter and others, commenced to build two boats. Each was twenty-feet five in length, square fore and aft, and five feet beam, capable of carrying seven men with provisions for about two months, in which time it was calculated the party could reach a latitude where assistance might reasonably be expected. The material used was, of course, taken from the *Polaris*, and chiefly from the ceiling of the alley-ways and after-cabin, as most of the light material had been already consumed for fuel. Notwithstanding the disadvantage under which the building progressed, two very serviceable boats were produced, and also a third, a smaller one, which was presented to the natives, who had been most friendly in bringing fresh meat to the party.

It was the end of May before the condition of the ice was such as to promise success to boat travel. Previous to the final arrangements for leaving, Captain Buddington says, "A deposit of certain valuables in boxes was made on the north shore of Lifeboat Cove, and these were protected with rocks," and there necessarily left to the mercy of the natives and the elements.

On the 3d of June, a little after 1 A.M., the boats were laden with food, and other necessary articles for the journey; the party equally divided into seven for each boat, and launching out into the opened waters of Smith Sound, the survivors of the still floating *Polaris* bade farewell to their winter home, and turned their faces southward, with high hopes and confident expectation of a timely rescue.

With the exception of slight indications of scurvy in a few of the men, all had retained their health through the Arctic winter and the early spring. It was now summer, and continuous day.

Neither cold nor darkness benumbed the frame or obscured the vision. The aquatic birds, seals, and other game could now be had in abundance wherever they put ashore; the eggs of the eider and other ducks gave a pleasant variation to their diet; and, but for the occasional interruption of their course by the pack-ice, nothing occurred to discourage or dishearten them.

On the way they touched at the deserted native settlement of Etah-y-tancy, and at Hakluyt Island, and subsequently landed on the west shore of Northumberland Island. There the pack-ice prevented their leaving until the 10th inst., and then entering a lead toward Cape Parry, they were subsequently drifted back by the pack-ice to the place from whence they had started. On the 12th a better prospect offered, and they started again, and, crossing the southern part of Murchison Sound, rounded Cape Parry, and pulled up on Blackwood Point, near Fitz-Clarence Rock. Proceeding again the next day, they reached and landed on Dalrymple Island; from thence reached Wolstenholme Island, Conical Rock, and Cape York.

Thus far the course had been comparatively easy; but they must now face the ice of the glacier-fed Melville Bay; and here considerable more exertion was required, the leads sometimes closing so that they had to haul their boats on the ice and over it to get another lead, and so on. But their troubles were not to be of long duration. On the twentieth day after leaving Life-boat Cove, and soon after entering on the ice-beset waters of Melville Bay, their eyes were gladdened by the sight of a steamer in the distance. They were at the time twenty-five miles southeast of Cape York.

True, they perceived that the vessel was beset and could not come to them, and she was some ten miles away. But being beset, she was sure to remain until they could get to her, and the relief appeared all the more timely, since one of the boats had been injured in its contact with the ice, and only about one week's provisions remained. The party had apparently overeaten their rations, or had not rightly estimated them.

Two men were sent forward toward the steamer, but had traversed only a portion of the distance when they were met by a body of eighteen men from the ship, which proved to be the *Ravenscraig*, of Dundee, Captain Allen; lying in lat. 75° 38′ N., long. 65° 35′ W., Cape York being to the north-west, at about twenty-five miles distance.

The party on the ice had been sighted by the lookout on the vessel at about 1 A.M. (it being light all the time then); they were at that time about fourteen miles off, and were supposed to be Esquimaux. By nine o'clock it was observed that the party were moving toward the vessel, but very slowly, not having made more than two miles since first seen; and it was now discovered that they were not natives, but white men. This naturally increased the interest on board. It was perceived that they had two boats, and their colors on a pole. Volunteers were now ready to go to their relief, and eighteen picked men were chosen for the purpose, Captain Allen also hoisting his ensign as an encouragement to the wanderers.

Captain Buddington and his party were intensely gratified to see that they had been noticed, and all watched with the greatest anxiety the progress of the two men who had gone forward toward the vessel. But when the rescuers were seen returning with them every heart was relieved, and weariness gave place to the joy of anticipated security.

The boats had been considerably injured by contact with the rough, hummocky ice, and one of them was slightly stove, but had been repaired. The fatigue of dragging boats over such ice may be partly imagined when we find that it took the combined party of thirty-two from 6 P.M. until midnight to get to the vessel—a rate of two miles an hour. The difficulty had been greatly increased by a deep, slushy snow, which was spread over the entire surface, and which was not only heavy and disagreeable to wade through, but was not without its real dangers, as more than one found by suddenly sinking into some treacherous hole which was concealed by it. One of the men had great difficulty in extricating himself from one of these hidden pitfalls; indeed, without assistance the accident might have proved fatal.

Captain Allen received the weary men with open-hearted hospitality, such as a Scotch whaler knows how to render to any shipmate hoisting a flag of distress. Through him the party learned, with unmingled astonishment, of the safety of the ice-floe party — most, if not all, of whom they had supposed numbered with the dead.

The *Polaris* expedition had indeed proved exceptionally prolific of startling and exciting incidents. From the time when Captain Bartlett picked up the exhausted waifs of the ice-floe, until the last scene in this thrilling drama was enacted, public expectancy had been kept continually on the *qui vive* by the progress of events connected with the story of these Arctic explorers: the sailing and return of the *Frolic* with the nineteen waifs; the developments of the examination at Washington; the lamentable and melancholy death of Captain Hall; the dispatch of the *Juniata;* the purchase and fitting out of the *Tigress;* the finding of Buddington's camp; the story of the foundered *Polaris,* and the mutilated log-book; and the rescue of Buddington and his men by a Scottish whaler, were events which succeeded each other with such rapidity that it almost seemed as if an accomplished stage-manager was working the machinery toward some rapid and astonishing transformation scene, and, possibly, unexpected *denouement.*

After what had happened, the country would scarcely have been surprised had the buried commander arisen from his frozen grave and haunted some of the fugitives on their flight through the Arctic zone, across the Atlantic waves, to finally confront them in the very place and in the very presence where his great hopes had been so nobly helped and cherished.

On September 19, 1873, the New York papers published a telegram from London, stating that the day before "the Dundee whaling-steamer *Arctic* had arrived at Dundee, having on board Captain Buddington and the remainder of the *Polaris* crew." Later information disclosed the fact that it was not the *Arctic* which had picked them up, but the whaling-ship *Ravenscraig,* of the same port, on the 23d of July; and that she had transferred eleven of the party to the *Arctic,* and three of the *Polaris* survivors to the whaler *Intrepid.* The names of those who had already arrived in the *Arctic* are as follows:

Sidney O. Buddington, sailing and ice master; Hubbard C. Chester, chief mate; William Morton, second mate; Emil Schuman, chief engineer; A. A. Odell, assistant engineer; Dr. Emil Bessel, chief of the Scientific Corps; Nathan J. Coffin, carpenter; Herman Sieman, Henry Hobby, Noah Hays, seamen; W. F. Campbell, fireman.

These were all reported well, and that the parties who were on the *Intrepid* were

Mr. R. W. D. Bryan, astronomer and chaplain; Joseph B. Mauch, seaman; John W. Booth, fireman.

This information was received at Washington through the

United States Consul at St. Johns, Newfoundland, Mr. Molloy, to whom the first communication from Dundee was made—Dr. Bessel, mindful of his European friends, announcing his safety by telegram to Dr. Peterman, of Gotha, Germany. United States Consul Molloy, as soon as he received the information, without waiting to telegraph to Washington for orders, took the responsibility of hiring a swift steamer, the *Cabot*—and went in pursuit of the *Juniata*, which had sailed from St. Johns but a few hours before the telegram from Dundee arrived, intending to still prosecute the search.

Mr. Molloy knew that it was the intention of Commander Braine to proceed, in the first instance, toward Cumberland Sound, and from thence to take a northerly direction; he had, therefore, no difficulty in following his track. He had wisely provided himself with rockets and signal-lights to use, if necessary. At 11 P.M. of the same day the *Cabot* overtook the *Juniata*, attracting the attention of those on board by throwing up rockets and displaying all the light possible.

Commander Braine naturally thought that the vessel was the *Tigress*, which was thus endeavoring to communicate with him, and concluded that the latter vessel had picked up the party, or at least had information of them. The night was dark and cloudy; but as soon as the signals were observed the boat of the *Juniata* was lowered, and, in charge of Lieutenant De Long, she came, like a streak of light marking the waves, toward the *Cabot*, which had been only seven hours in the pursuit.

As soon as the information was conveyed to Commander Braine that the *Polaris* survivors were all safe, he ordered the head of the *Juniata* to be turned toward St. Johns, and both vessels were soon again in the port they had so recently left.

Mr. William Reed, Vice United States Consul at Dundee, calling the attention of the State Department, by telegram, to the fact that the *Polaris* survivors were entirely destitute of means, the Secretary of the Navy promptly responded by requesting that they be supplied by the consulate, and sent home as passengers by the first steamer. On Tuesday, September 23, Captain Buddington and his ten companions sailed from London for New York in the steamer *City of Antwerp*, and arrived safely at New York October 4, 1873.

It was some weeks later before Mr. Bryan and two others, who

NAVAL BOARD OF INQUIRY. 409

had been transferred to the *Intrepid*, and from that to the *Erick*, reached their homes, Mr. Bryan having asked and received a fortnight's furlough.

As in the case of the ice-floe waifs, a naval Board of Inquiry was convened in Washington, to which all the survivors were summoned, the results of which will be found in the Appendix.

This examination failed to elicit any additional facts of interest beyond the information summarized in the preceding pages. The testimony was so conflicting on certain points, that it was obvious there was falsehood somewhere; and as no good could result from its publication, the Secretary of the Navy has very prudently limited its circulation to a few official copies.

THE LATEST STYLE.

CHAPTER XXXVIII.

SCIENTIFIC NOTES.

The Pacific Tidal Wave.—Meteorological and Magnetic Records.—Glaciers' Fauna.
—Entomology.—Flora.

THE scientific observations made at winter-quarters during 1871–'72 included records of the barometer, the force and velocity of the wind, the temperature of the air, and also the tension of the aqueous vapor in the atmosphere (by the aid of Professor August's psychrometer), and attention was paid to frequent measurements of atmospheric precipitation and tidal registration. Through the latter, Dr. Bessel states, it was discovered that the Pacific tidal wave reaches Robeson Channel, and extends south as far as Cape Hatherton, the proof being that the tides rose earlier at Newman Bay than south of it, a convincing proof of the free circulation of the Polar waters from west to east above the latitude of 82° N., if true. Captain Tyson does not coincide in this view, nor as to the reported fact of the earlier rise of the tides at Newman Bay.

The astronomical observations made by Mr. Bryan were very numerous, and were first directed to obtaining absolute certainty as to the meridian of Thank God Harbor. On this meridian all subsequent surveys were based, except such as were obtained by triangulation; and these were also repeated with the utmost carefulness and exactitude. In connection with the astronomical observations, the pendulum observations were constantly recorded; and during the months of February, March, April, and May hourly observations were made on the declination of the needle and its variation, with other magnetic phenomena, such as the dip, the horizontal intensity, and the moment of inertia.

Every favorable opportunity was improved to examine into the movements of glaciers; and as Greenland, and the whole coast north and east of it, abounds in these rivers of ice, many interesting observations were made, especially on the limits of *néve*.

. The fauna observed has been incidentally mentioned in Captain Tyson's journal and elsewhere; and the principal novelties of animal life, it will be seen, consisted in the rediscovery of the musk-cattle, which were supposed to be extinct in that country; the other mammals and birds but little known were the lemmings, a small rodent, and the Arctic raven and falcon; Sabine's gull, and Tueston's sanderlings.

Fish is comparatively scarce; but one specimen was secured of the salmon species. Marine invertebrates of new and interesting forms were also obtained.

In entomology Dr. Bessel added largely to his collections; and among others which have been incidentally mentioned in the preceding pages were several varieties of ichneumon.

Of botanical specimens a full and curious collection was made; fourteen varieties of *phanerogamic* plants (those having visible flowers, stamens, and pistils) were classified, and many species of mosses, lichens, and fungi—the effort being, of course, to secure specimens of those not before obtained, or of which but few specimens have been previously secured. Some of the fungi under the microscope reveal a beauty of form and brilliancy of coloring perfectly astonishing, suggestive rather of tropical suns than Arctic cold. Much of Dr. Bessel's collection was lost, and the more bulky articles were necessarily abandoned on taking to the boats; but his records being preserved, the scientific world will yet receive a full detailed description of his work and discoveries.

CONCLUSION.

The *Polaris* expedition, combined with the pre-Arctic experiences of Captain Tyson, prove conclusively several points:

1st. That white men can safely winter at as high a latitude as 81° 38′ N.;

2d. That a well-built and properly officered steam-vessel can sail to 82° 16′ N. with no insuperable impediment to farther progress;

3d. That no degree of cold yet encountered by Arctic travelers impedes explorations, if the party is well fed and clothed;

4th. From the experience of Captain Tyson on his whaling voyages, with his observations of others engaged in the same pursuit, it is evident that there exists in the United States a large body of men sufficiently acclimated to Arctic cold to make excellent Polar explorers, without submitting themselves to greater hardships than are common incidents of whaling voyages;

5th. That it is of the utmost importance in exploring expeditions to have homogenity, not only of feeling but of race, in the members of the party;

6th. That there should be but one head or commander; and that all others, including the Scientific Corps, if any are present, should be subordinate to the commander in chief;

7th. That definite and severe penalties should be attached, by competent authority, to the crime of willfully impeding or obstructing the purposes of an exploring expedition by any member of it;

8th. Considering that the public interest centres in the discovery of the pole and the settlement of that elusive mystery, we believe that not only geographical but all affiliated sciences would be in the end promoted by making the next expedition *purely geographical*, and leaving it unembarrassed by a possibly conflicting corps of officers under the name of "scientific." An expedition organized for the sole purpose of reaching an objective point would be far more likely to accomplish it if unimpeded by discursive projects, however valuable *per se*. Let a good ice navigator find the pole first, and then all the scientific men who are so disposed can avail themselves of the opened pathway to make what investigations they please.

We have seen clearly where the weak points of the *Polaris* expedition marred its efficiency. The members sailed out of the Brooklyn Navy Yard already disintegrated by lack of mutual respect, and divided by the line of native and foreign interests; the commander also knowing that part of his company were not to be relied on in an emergency; and from the fact of having so large a number of foreigners on board, he had not that moral support to lean upon which could alone nerve him to the exercise of his ultimate authority in quelling insubordination. No commander is justified in proceeding on an expedition with such clearly defined elements of weakness patent to his observation.

Another weak point, and a great incumbrance in case of disaster, was the presence of women and children; these should find no place in an exploring expedition; they are drag-anchors, at the best, and the source of discord and demoralization in almost every instance.

So far as the vessel and her equipments were concerned, the expedition reflects the highest credit upon the liberality of the

Government and the judicious arrangements of the Secretary of the Navy. The fact that not one of the members were afflicted with scurvy, or any serious illness, proves the excellence of the provisions furnished and the discrimination with which they were selected; and it is the plainest and most obvious deduction, that if this party had worked harmoniously together, and the commanding officer had remained to winter another season, by detaching exploring parties, and establishing *caches* of provisions at distances of forty or fifty miles apart, that the intervening distance—only four hundred and sixty miles from the geographical pole, a distance no greater than from New York to Halifax—could have been reached, and the great Polar mystery solved.

In the possibilities which this expedition has revealed, there is more encouragement to future explorers than can be derived from that of any preceding it; and the lonely grave of the brave commander, instead of proving a warning beacon, will be but the beckoning signal to his successor, pointing the way whither he himself would have gone had not a treacherous fate struck him down where he lies.

ARCTIC CHRONOLOGY.

From 1496 to 1857 there were one hundred and thirty-four voyages, land journeys, and trans-glacial expeditions made to the Polar regions. Of these, sixty-three went to the north-west, twenty-nine *via* Behring Strait, and the balance to the north-east or due north.

Within the same period there were published two hundred and fifty-seven volumes on Arctic research; to which may now be added a host of American, English, and other foreign writers, with a long list of scientific and popular works germane to Polar matters.

To those of our readers who desire an intimate acquaintance with the scientific results of exploring-parties, and what was accomplished in this respect by the Scientific Corps of the *Polaris*, we refer them to the publications of the Smithsonian Institution, which, either in their "Reports" or "Contributions," conserves all information of value relating to the meteorology, astronomical observations, experiments on magnetic influences within the Arctic circle, with descriptions of the fauna and flora of those regions, giving the details of work which would not comport with the popular character of this book.

We also think that it would be wearisome and useless to burden our chronology with every name and date connected with Arctic research. We have, therefore, carefully selected such as will give the inquirer a general continuous outline and digest of facts; sufficiently full, however, to enable any would-be student of minutiæ to fill up the history by personal research with very little trouble.

Gunniborn, a Norseman, from Iceland, visits Greenland............................	872
Eric the Red, son of a Norwegian jarl, settled in Greenland near present site of Julianashaab...	983
Eric visits Iceland, and returns with a number of emigrants, with the purpose of settling a colony in Greenland, at Brattahlid......................................	985
Lief, son of Eric, visits Norway, receives Christianity, returns, and proclaims it in Greenland toward the close of the ninth century, about......................	998
Thjodhilda, wife of Eric, builds a church..	1002
Bishop Eric, a Christian prelate, visits Greenland...................................	1120
Bishop Arnold founds an Episcopal see at Garder, Greenland, and builds a cathedral..	1126
Colony attacked by the Skraellings, or Esquimaux, who burn and pillage the settlement, killing many inhabitants...	1349
Voyage of the two Venetian brothers Zeni, who reported land in the north-west.	1380
Bishop's See abandoned in Greenland, and Eric's descendants utterly wiped out and exterminated soon after by the Esquimaux..	1409
Voyages of John and Sebastian Cabot, passing the Arctic circle to the north-west...	1496–'98
The brothers Cortereal made three voyages to the north-west..................	1500–'03
Polar expedition, under Sir Hugh Willoughby, discovered part of Nova Zembla; the whole party subsequently frozen to death on the coast of Lapland..........	1503
J. Cartier, a French navigator, made several voyages of discovery to the north and west...	1534–'42
Sir Humphrey Gilbert, on return from north-west voyage of exploration, foundered at sea...	1578
Captain John Davis explored the east and west coasts of Davis Strait.......	1585–'88
William Barentz made three Arctic voyages to the north-east.................	1594–'96
Captain Weymouth sailed from England, under a contract to find a north-west passage to China, or forfeit all pay for his voyage.....................................	1602
Henry Hudson made four voyages of discovery, sailing due north, north-east and north-west; found new and valuable lands, extending from New York to north of Hudson Bay. On last voyage, deserted by mutinous crew, set adrift in a small boat with six sick sailors, and voluntarily accompanied by his carpenter, John King, he and his seven companions perished at sea.	1607–'10
Jan Mejan, a Dutch navigator, discovers Arctic island of that name............	1611
Sir Thomas Button, the first to sail across Hudson Bay from east to west. His name obliterated from modern maps..	1612

William Baffin and Fotherby, in 1614, and Baffin and Bylot, in 1616, sailed

through Baffin Bay to mouth of Lancaster Sound. Their reported discoveries treated as myths.. 1614–'16
Captain Luke Fox discovered Fox Channel, and penetrated other waters to the north and west; could have accomplished much more had he not been trammeled by official "orders." Captain James sailed on a similar expedition the same day in May .... 1631, '32
A Dutch navigator discovers and names, after himself, Gillies Land.............. 1707
Vitus Behring, a Russian naval officer, a Dane by birth, discovers Behring Strait and Behring Island... 1741
Middleton discovers Wager Bay... 1742
The British Parliament offered £20,000 for the discovery of a north-west passage to the Pacific *via* Hudson Bay.. 1743
A private expedition to the north-west, under Captain Charles Swayne, sails in the *Argo* from Philadelphia.. 1754
Hearne made three land-journeys north of American continent, discovered the Coppermine River, which he traced to its source.................................. 1772
A private expedition, under Captain Wilder, in the brig *Diligence*, sails from Virginia in search of the north-west passage.. 1772
Captain Phipps (Lord Mulgrave) makes a voyage of Polar discovery to the north-east.. 1773
British Parliament offers £20,000 for the discovery of any through passage to the North Pacific, and £5000 to any party getting within one degree west of the magnetic pole... 1776
Captain Cook, the circumnavigator, attempts the passage by Behring Strait, without success... 1776
Mackenzie found and traced the river of that name................................. 1789
William Scoresby, a Greenland whaler, makes a remarkable voyage due north. Reports open water "beyond the ice-barrier" 1806
Scoresby stimulates Arctic research by numerous publications .... 1806–'18
Expeditions under Captain Ross and Lieutenant Parry 1818
Captain Buchan and Lieutenant (afterward Sir J.) Franklin sail with a thoroughly equipped expedition, the first scientific party sent out by the British Government... 1818
Captain Parry and Lieutenant Lyon, in the *Hecla* and *Griper*; Parry sailed north-west through Barrow Strait and beyond, and claimed the Government reward of £5000... 1820
Remarkable land-journeys of Lieutenant Franklin and Dr. Richardson, also Midshipmen Hood and Back (afterward Sir George), from York Factory to Cape Turnagain... 1819–'22
Baron Von Wrangel makes his famous sledge-journey, and reported open water in high northern latitude, known as "Wrangel Sea"........................... 1820–'23
Clavering, with Colonel Sabine, go to Spitzbergen and Greenland in 1823
Captain Beechey, in the *Blossom*, goes through Behring Strait, and follows the coast easterly to Barrow Point... 1823, '24
Captain Parry made improvement in compasses to be used in Arctic navigation 1824
Franklin skirts the north coast of America as far west as Return Reef........... 1826
Franklin winters with Dr. Richardson at Great Bear Lake; made interesting experiments on terrestrial magnetism, etc... 1826, '27
Captain Beechey makes a second attempt to meet Franklin from the Pacific side, and fails.. 1827

Parry makes another voyage to the north-east; travels in sledges north of Spitzbergen, and drifts faster south than he travels north. 1827

A private expedition, under patronage of Sir Felix Booth, sails under Captain John Ross in the *Victory*, in which steam was first used in Arctic exploration... 1829

Sir John Ross finds and fixes magnetic pole; English "union-jack" planted upon it by his nephew, James C. Ross... 1831

Captain Ross abandons his vessel, after wintering three years in the Arctic regions, builds boats, drags them overland to the coast; put to sea; picked up by a whaler in .... July, 1833

Lieutenant Back and Dr. King go overland from Fort Resolution in search of Ross... 1833

Lieutenant Back discovers and traces the Great Fish, or Back, River...... 1833–'35

The Hudson Bay Company send out Messrs. Dease and Simpson, who make valuable discoveries on a land and boat journey............................. 1837, '38

Sir John Franklin, in the *Erebus*, and Captain Crozier, in the *Terror*, sail in the spring of... 1845

The *Erebus* and *Terror* last seen by Captain Dannet, master of the whaler *Prince of Wales*, in Baffin Bay, near Lancaster Sound .... July 26, 1845

Hudson Bay Company send Dr. John Rae to ascertain if Boothia is an island or peninsula.. 1846

The *Plover*, Commander Thomas Moore, and the *Herald*, Captain Kellet, with Mr. Robert Sheddon in his pleasure-yacht, the *Nancy Dawson*, sail to Behring Strait, and make boat-journeys eastward, searching for Sir J. Franklin.. 1848–'50

A searching expedition for Sir J. Franklin sails, under Sir James C. Ross, in 1848

The *North Star* is sent out with supplies..................................... 1849

The British Government offers a reward of £20,000 to any party, of any nation, relieving Sir J. Franklin's expedition.. 1849

The British Government sends out eight vessels, with several tenders, to continue the search.. 1850

Mr. Henry Grinnell, of New York, furnishes and equips the *Advance* and *Rescue*, to aid in the search for Franklin. The United States Government orders Lieutenants De Haven and Griffith to command the vessels................. 1850

Captain M'Clure, in the *Investigator*, and Captain Collinson, in the *Enterprise*, go on the search through Behring Strait. The north-west passage solved by M'Clure (by observation)... October 31, 1850

The *Lady Franklin* is sent out by wife of Sir J. Franklin, under Captain Penny 1850

Lady Franklin organizes another expedition to sail in the yacht *Prince Albert*, of 89 tons. Mr. William P. Snow, of New York, goes to Aberdeen, and sails in her as amateur explorer, with Captain Forsyth............................ 1850

Captain Ommany, of the *Assistance*, discovers the first traces of Sir J. Franklin's party at Cape Riley.. August 23, 1850

At Beechey Island Lieutenant Sherrard Osborne first found *débris* of Franklin's first winter encampment, and three graves of sailors belonging to *Erebus* and *Terror*.. August 25, 1850

Lieutenant De Haven, in *Advance*, arrived at Beechey Island........ August 27, 1850

Ten of the searching vessels, drawn as by a common instinct, without appointed rendezvous, met at Beechey Island... August 29, 1850

Leigh Smith, in English yacht, reaches lat. 81° 13,' sailing north-east.......... 1851

ARCTIC CHRONOLOGY.

Captain Wilkes, United States Navy, memorialized Congress for appropriation of $50,000 to fit out a sledge-expedition to aid in the search...... 1851
Captain Kennedy, with young French volunteer, Bellot, sails in *Prince Albert*. 1851
Captain William Penny discovers sea to the north of Wellington Channel; names Grinnell Land Albert Land, thinking it unknown........ 1851
Sir Edward Belcher sails with a fleet of five vessels, to continue the search...... 1852
Captain Inglefield, with René Bellot, sails in the *Phœnix*.......... 1852
Dr. E. K. Kane, late surgeon of the *Advance*, is sent out in that vessel, fitted up at expense of Mr. H. Grinnell, of New York, and Mr. George Peabody, of London; the latter paid £10,000.......... 1853
Captain M'Clure, in the *Investigator*, from Behring Strait, meets Lieutenant Pim near Dealy Island, the latter having entered the Arctic regions through Baffin Bay ...... April 19, 1853
British searching ship *Resolute*, of Sir Edward Belcher's fleet, abandoned........ 1853
Captain Collinson, in the *Enterprise*, completes the passage (solving northwest) *in his ship* twenty days after M'Clure; turns to the south-east, makes many discoveries, and brings home relics of Sir John Franklin's party...... 1850–'54
Sir John Franklin's name stricken from the Navy List...... March 13, 1854
Sir Edward Belcher orders five good ships to be abandoned.......... 1854
The *Resolute*, one of Sir Edward Belcher's fleet, starts on a drift of a thousand miles, from near Dealy Island to Cape Mercy.......... 1854, '55
Sir Edward Belcher and officers court-martialed in England; all honorably acquitted, except Sir Edward, "whose sword was returned to him in significant silence "...... 1854, '55
Captain M'Clure knighted, and Captain Collinson receives medal of honor...... 1854
Dr. Kane, in brig *Advance*, explores east coast of Smith Sound; discovers and names Humboldt Glacier; surveys eight hundred miles of coast of Greenland and Washington Land, which he finds and names; abandons *Advance*; comes down with crew to Upernavik in small boats, which he reaches August 6; announces discovery by Morton of the "Polar Sea"........ 1853–'55
Lieutenant Hartstene, in United States ship, searches for Dr. Kane, passing above Rensselaer Harbor; returns to Upernavik, and takes on board Kane and his company.......... 1855
Messrs. Anderson and Stuart find relics of the Franklin expedition at Montreal Island.......... 1855
Lieutenants Meacham and M'Clintock make separate Arctic land-journeys of nearly fifteen hundred miles each.......... 1854, '55
Captain Tyson, then boat-steerer on board bark *George Henry*, of New London, Captain James Buddington, first sighted the *Resolute* near Cape Mercy, and visited it with three companions, bringing back relics to his vessel.. August, 1855
The *Resolute* taken possession of by Captain Buddington, and brought to the United States.......... 1855
Resolute refitted, and presented to Queen Victoria by Lieutenant Hartstene, representing the United States, on December 16, "in the name of the American people".......... 1856
Lady Franklin sends out the steam-yacht *Fox*, Captain M'Clintock, to make a final search for Sir John.......... 1857
Private expedition of James Lamont, F.G.S., to the north-east.......... 1858
The *Fox* is beset in the Melville Bay ice-pack, August and September, 1857, and drifts southward until.......... April, 1858

Lieutenant Hobson, of the *Fox*, found the record of the death of Sir John Franklin in a cairn at Victory Point. Date of death, June 11, 1847 1858
Captain M'Clintock finds two skeletons in a boat, members of Franklin's party; collects numerous relics, and returns.. 1859
Dr. I. I. Hayes sailed in the steamer *United States*, from Boston; made extended land-journey on west coast of Smith Sound and north of it; made extensive discoveries of new lands and connecting waters; planted the American flag on the most northern latitude attained on foot up to that date.. 1860, '61
Captain Parker Snow sailed for Bellot Strait and King William Land............ 1861
A Government Swedish expedition, under Professor Torell, thoroughly fitted out, and having on board a large number of scientists, naturalists, and students, sailed for the seas north of Spitzbergen............................... May 9, 1861
Blomstand finds the sea free of ice to the north of Spitzbergen...... August 10, 1861
CHARLES FRANCIS HALL, an amateur explorer, sailed from New London, Connecticut, in the brig *George Henry*, to continue the search for survivors of Franklin's party; lost his vessel, the *Rescue;* explored Frobisher Strait; found the "strait" to be a bay; brought back many relics of the old navigator .... 1860–'62
Charles Francis Hall makes a second voyage to Hudson Bay, north shore, in the bark *Monticello*, with only two Esquimaux companions; increases his native company; adds five white sailors; explores to the north and west, and gains much information respecting the Franklin party........................... 1864
Edward Whymper, a member of the London "Alpine Club," went to Greenland, and made interesting explorations.. 1867
Baron Schilling projected an expedition by the Behring Strait route, expecting to skirt the Siberian coast, and find the Polar Sea, or a Polar continent, which he believes exists... 1867
Captain Long, of bark *Nile*, in lat. 70° 40′ N., 178° 15′ W., explored over 3° of an extensive land, and examined an extinct volcano 2480 feet high............ 1867
Captain Raynor, of ship *Reindeer*, explored the same land for over 5° of longitude; thought it extended for more than 8° of longitude, and north for 120 miles. The south-west cape of this land he reports 25 miles from the Asiatic coast.. 1867
Captain Lewis, of the *Corinthian*, landed on this coast in August; found flowers and birds, and indications of coal... 1867
Lord Dufferin, in schooner-yacht *Foam*, made an Arctic voyage to north of Spitzbergen, etc... 1867
A Swedish expedition, under Professor Nordenskiold, made interesting discoveries in natural science to the north-east... 1868
A Russian merchant—Sidgeoff—sent out a scientific exploring expedition in a screw-steamer... June, 1868
A private expedition, sent out by M. A. Rosenthal, a merchant of Bremen, with scientific corps, with eminent astronomer, Dr. J. S. Doest, of Jülich.... 1868
Captain Blowen, of bark *Nautilus*, explored north of Spitzbergen to 72° N.; observed land extending west as far as he could see............................... 1868
The Gotha expedition, forwarded by Mr. Rosenthal, with large screw-steamer *Albert*, a walrus-hunter, with a crew of fifty-five men, provisioned for fifteen months, with scientific corps, under Dr. Emil Bessel (late of *Polaris*), returned after an absence of only four months.. 1869
Charles Francis Hall returns from his Franklin search expedition, after five

years' residence with the Esquimaux, with 150 relics of the Franklin party.. September 1, 1869
Seven Arctic expeditions organized and forwarded from different parts of Europe to Arctic regions 1869
Dr. A. Peterman organized a party which sailed in the *Germania*, Captain Hegeman; and sailing-vessel *Hansa*, Captain Koldewy—thirty-one officers and men, and six scientists. Left Bremerhaven for North Pole, via east coast of Greenland, provisioned for two years.. 1869
Captain Palliser, private English gentleman, goes to the north, between Spitzbergen and Nova Zembla... 1869
J. Lamont, F.G.S., author of "Arctic Zoology" and M.P. for Buteshire, fits out his own steam-yacht, at a cost of nearly $50,000; sailed for the north from Caledonian Canal.. 1869
Mr. Robert Brown, an English naturalist, explored extensively within Arctic circle in Greenland. Printed report on Arctic fauna............................ 1869
Steamship *Panther*, from Boston, with Bradford, the artist and photographer, Dr. Hayes, and others, penetrate the Melville ice-pack, in pursuit of artistic icebergs... 1869
Dr. Hayes, in the steamer *Panther*, in boats, and on foot, makes interesting archæological discoveries relating to the early Norse settlements in Greenland, summer of 1869
A French expedition went from north of Europe to Arctic regions, to observe and collect facts relating to the aurora borealis, with the following eminent savants: MM. Lottin, Bravais, Lillebook, and Silgestrom, with M. Bevalet as artist. 1868, '69
The sailing vessel *Hansa*, of the German expedition, was lost on the east coast of Greenland, in lat. 70° 50', with a valuable collection of fauna and flora and scientific records, on the 23d of October, 1869. Her captain and crew, 14 in number, had collected provisions and fuel, with three boats, on the ice, on which also was a small house. They saw the *Hansa* sink; then the ice drifted with them to the south. On January 2, 1870, the ice-floe was broken up in a storm, and greatly reduced in size. The party divided, each taking a boat on January 11, to be ready for emergencies. They took to the sea and worked southward in their boats, after drifting 193 days, and crossing over 9° of latitude. Early in June they rounded Cape Farewell, and reached the Danish mission station, at Friedrichsthal, and from thence obtained passage home to Bremen .... 1869, '70
An excellently planned French Arctic exploring expedition, under the savant Gustave Lambert, was prevented from sailing by the outbreak of war with Germany; its projector, Lambert, killed in battle 1870
United States Congress makes appropriation for outfit of the United States North Polar expedition, under Captain C. F. Hall, the authorizing act being signed by President Grant .... July 12, 1870
Polaris leaves Washington June 10; arrives at the Brooklyn Navy Yard, to be fitted for the voyage... June 14, 1871
United States steamship *Polaris* sails from Brooklyn June 29; touches at New London, Connecticut; St. Johns, Newfoundland; and several Greenland ports; and meets the United States steamship *Congress*, with supplies, at Goodhavn.. August 10, 1871
A French gentleman, Octave Pavy, from San Francisco, sailed to go to the Polar Sea, via the *Kuro Siwo*.. 1871

The *Polaris* reaches the highest northern latitude ever attained by any vessel.. August 30, 1871
Goes into winter-quarters at Thank God Harbor, September 3, in highest winter-quarters made by Arctic explorers 1871
Captain Hall starts on a sledge-journey to the north; is absent two weeks; returns in good health to the *Polaris*... October 24, 1871
Is immediately taken ill, partially recovers, relapses, and dies November 8, and is buried on the 11th... 1871
Captain Altman came near King Carl Land; saw no ice at 79° N................ 1871
A Norwegian whaleman, Elling Carlsen, circumnavigated Nova Zembla; anchored in Ice Haven, 74° 40', on the south-east shore of most eastern island; found the house erected by William Barentz, the Dutch explorer, two hundred and eighty-seven years before.. 1871
A Russian Government expedition started from Archangel, and another from the Yenisei River... 1871
James Lamont, of England, made three Arctic voyages........................ 1869–'71
West Indian fruits and drift-wood found north of Nova Zembla by whalers. 1871, '72
Swedish Government expedition, under Professor Nordenskiold, sailed in the *Polheen*, an iron steamship, with a steam consort, and a brig, in the summer of.. 1872
Lieutenant Payer, of the Austrian army, and Lieutenant Weyprecht, of the Austrian navy, hired a Norwegian sailing-vessel, and sailed in June for King Carl Land; in 78° found the sea open, and made other valuable discoveries... 1872
An Italian Government steamship accompanied the above expedition to the north cape of Nova Zembla... 1872
Captain Nils Jansen, a Norwegian whaler, in a vessel of 26 tons, sailed east of Spitzbergen to bay of King Carl Land; from top of high mountain saw the open water east and north-east; no ice; to the N.N.W. land was visible (the Gillis Land of the old geographers); saw birds, seals, large reindeer, and quantities of drift-wood. Anchored in 79° 8' N............................. 1872
Arctic exploring ship *Polaris* breaks, during a violent storm, from the floe to which she is anchored, and is driven in a north-easterly direction by the wind, leaving nineteen persons adrift on the ice........................ October 15, 1872
Captain Buddington beaches the *Polaris* (October 16) at Life-boat Cove, near Littleton Island; abandons the ship, and winters on the main-land....... 1872–'73
Captain Tyson with a party of eighteen souls drift away on an ice-floe; lose sight of the *Polaris*, and continue to drift a S.S.W. course from October 15, 1872, until the 30th of April, 1873, without serious sickness or loss of life.. 1872, '73
Captain Tyson and party picked up by sealer *Tigress*, Captain Bartlett.. April 30, 1873
The United States steamship *Frolic* sent to bring the party to Washington; took them on board at St. Johns, Newfoundland; sailed thence May 28; arrived at Washington... June 5, 1873
Official examination of the officers, crew, and Esquimaux rescued by *Tigress* before a Naval Board of Inquiry, held on board the United States steamship *Tallapoosa* at Washington, concluded.. June 16, 1873
Imperial Geographical Society of Russia sent sledge exploring party, under the experienced Siberian traveler, M. Tschekanowski, with a two-years' outfit, to survey the coast of the Polar Ocean in Arctic Siberia..................... 1873

ARCTIC CHRONOLOGY.

Captain Buddington and party picked up on June 23, by the Scotch whaler *Ravenscraig*, Captain Allen, twenty-five miles south-west of Cape York....... 1873

Captain Allen transfers Captain Buddington, Emil Bessel, and nine others of the party to the whaler *Arctic*, of Dundee, which arrived with them at that port.. September 18, 1873

Mr. Bryan, the chaplain and astronomer of the *Polaris* expedition, with two others of the rescued party, transferred to the Scotch whaler *Intrepid*, and from that to the *Erick* 1873

News received from the Swedish expedition which sailed in 1872; the spectrum analysis applied by this party to the aurora borealis; wintered in lat. 79° 53', proceeding north in July.. 1873

Professor Nordenskiold's expedition beset in the ice at Mosel Bay; relieved by Leigh Smith's party in the summer of... 1873

The United States steamship *Juniata*, Commander Braine, fitted out by Secretary of the Navy, and ordered to sail as a tender and store-ship to the coast of Greenland, with supplies for the *Tigress*.............................. June 12, 1873

The sealer *Tigress*, having been purchased by the United States, and fitted up to go in search of the *Polaris* and the party remaining in her, sailed from the Brooklyn Navy Yard, under Commander Greer, accompanied by Captain (*pro tem.* lieutenant United States Navy) Tyson as ice-pilot................. July 14, 1873

The *Tigress* and *Juniata* meet at Upernavik August 10; the *Tigress* sails north on the 11th; meets the steam-launch *Little Juniata* in Melville Bay.......... 1873

Commander Greer, of the *Tigress*, finds Captain Buddington's deserted winter-camp, August 14, near Littleton Island, at Life-boat Cove. Finds *Polaris* sunk one mile and a half from shore... 1873

The searching steamship *Tigress* returned, touching at Tossac, Upernavik, and Goodhavn for news. August 25, sailed to the west coast to intercept whalers; put into Cumberland Gulf.. 1873

The whaler *Arctic*, having on board Captain Markham, B.R.N., and Dr. Emil Bessel, visited Fury Beach; found wreck of the British ship *Fury*, lost by Captain Parry in 1824; also canned provisions in good preservation, and two English muskets bearing date of 1850, probably left by Captain Penny in 1851..August, 1873

Commander Braine, of the *Juniata*, sends exploring party to north-west side of Disco Island; coal found in abundance................................ September, 1873

Leigh Smith's English expedition in large screw-steamer *Diana*, at Trenerenberg Bay, July 4. Returned to Scotland in September. *Discovered that North Cape is an island*.. 1873

Captain Buddington, Emil Bessel, and party arrive at New York October 4; proceed under orders in the United States steamship *Tallapoosa* to Washington; examination of the party by Naval Board............. October–December, 1873

The *Juniata* returns to New York, arriving............................. October 24, 1873

Steamship *Tigress* left Cumberland Gulf September 16; four days later experienced a heavy gale, which continued three days; made Cape Desolation September 24; driven to sea in another gale; next day anchored in a small Fiord; repaired engine; took native pilot; made Ivgitut Fiord on the 27th; refitted, and repaired boilers and engine; sailed October 4; struck by heavy gale on the 5th, which lasted till the 8th; after a short abatement, another gale. Returned to St. Johns, Newfoundland........................ October 16, 1873

Tigress arrived at Brooklyn Navy Yard.. November 9, 1873

APPENDIX.

Extract from Letter of Captain EDWIN W. WHITE, *of Groton, Connecticut, to a Friend in Brooklyn, New York, September 30, 1873.*

"AFTER having ten years' experience in the Arctic regions with Captain Tyson, I will say that I have always found him the best man to consult with that I have ever met.

"I have also made several sledge-journeys with him, and have always found his power of endurance ahead of any one I ever traveled with.

"Yours truly,   EDWIN W. WHITE."

To the Honorable Secretary of the United States Navy, GEORGE M. ROBESON.

Sixth Snow-house Encampment, Cape Brevoort, north side entrance to Newman Bay, lat. 82° 3′ N., long. 61° 20′ W., October 20, 1871.

Myself and party, consisting of Mr. Chester, first mate, my Esquimau, Joe, and Greenland Esquimau, Hans, left the ship in winter-quarters, Thank God Harbor, lat. 81° 38′ N., long. 61° 44′ W., at meridian of October 10, on a journey by two sledges, drawn by fourteen dogs, to discover, if possible, a feasible route inland for my sledge-journey next spring to reach the North Pole, purposing to adopt such a route, if found better than a route over the old floes and hummocks of the strait, which I have denominated Robeson Strait, after the honorable Secretary of the United States Navy.

We arrived on the evening of October 17, having discovered a lake and a river on our way; the latter, our route, a most serpentine one, which led us on to this bay fifteen minutes distant from here, southward and eastward. From the top of an iceberg, near the mouth of said river, we could see that this bay—which I have named after Rev. Dr. Newman—extended to the highland eastward and southward of that position about fifteen miles, making the extent of Newman Bay, from its headland or cape, full thirty miles.

The south cape is a high, bold, and noble headland. I have named it Sumner Headland, after Hon. Charles Sumner, the orator and United States Senator; and the north cape, Brevoort Cape, after J. Carson Brevoort, a strong friend to Arctic discoveries.

On arriving here we found the mouth of Newman Bay open water, having numerous seals in it, bobbing up their heads; this open water making close both to Sumner Headland and Cape Brevoort, and the ice of Robeson Strait on the move, thus debarring all possible chance of extending our journey on the ice up the strait.

The mountainous land (none other about here) will not admit of our journeying farther north; and as the time of our expected absence was understood to be for two weeks, we commence our return to-morrow morning. To-day we are storm-bound to this our sixth encampment.

From Cape Brevoort we can see land extending on the west side of the strait to the north 22° W., and distant about seventy miles, thus making land we discover as far as lat. 83° 5′ N.

There is appearance of land farther north, and extending more easterly than what I have just noted, but a peculiar, dark nimbus cloud that constantly hangs over what seems may be land prevents my making a full determination.

On August 30 the *Polaris* made her greatest northing, lat. 82° 29′ N.; but after several attempts to get her farther north she became beset, when we were drifted down to about lat. 81° 30′. When an opening occurred, we steamed out of the pack and made harbor September 3, where the *Polaris* is. [Corner of the manuscript here burned off.]

Up to the time I and my party left the ship all have been well, and continue with high hopes of accomplishing our great mission.

We find this a much warmer country than we expected. From Cape Alexander the mountains on either side of the Kennedy Channel and Robeson Strait we found entirely bare of snow and ice, with the exception of a glacier that we saw, covering about lat. 80° 30′ east side the strait, and extending east-north-east direction as far as can be seen from the mountains by Polaris Bay.

We have found that the country abounds with life—seals, game, geese, ducks, musk-cattle, rabbits, wolves, foxes, bears, partridges, lemmings, etc. Our sealers have shot two seals in the open water while at this encampment. Our long Arctic night commenced October 13, having seen only the upper limb of the sun above the glacier at meridian, October 12. This dispatch to the Secretary of the Navy I finished this moment, 8.23 P.M., having written it in ink in our snow-hut, the thermometer outside —7°. Yesterday all day the thermometer —20° to 23°; that is, —20° to —23° Fahrenheit.

[Copy of dispatch placed in pillar, Brevoort Cape, October 21, 1871.]

To the Hon. CHARLES P. DALY, *President of the American Geographical Society.*

Washington, April 9, 1871.

DEAR SIR,—Continued occupation since my return has prevented me from giving you, as you requested, an account in detail of what I have observed in respect to the geography of the Arctic regions.

With my first voyage you are sufficiently familiar, and I have nothing to add to what is contained in the volume published by the Harpers. During the last five years that I have spent in the Arctic regions I availed myself of every opportunity afforded me for accurate observation, and I give you the results.

APPENDIX. 425

You will remember that Wager Bay is an old discovery of Middleton's in 1742, when he was in search of a north-west passage. The general outline or rough sketch then made remains unimproved to the present day. I explored this inlet for sixty miles, up and down, to its junction with Rowe Welcome, and made a series of observations from astronomically determined positions.

Repulse Bay, though visited by Middleton, and afterward by Parry and by Rae, still remains but imperfectly defined. I have from my own observations the data for a more accurate delineation of the outline of this bay. I discovered and surveyed a new inlet, in lat. 67° N., long. 84° 30′ W., a few miles north of Norman Creek, of which it may be said to be a counterpart, running from Lyon Inlet to the eastward. I may be excused for expressing to you the gratification I felt in making this discovery, remembering that Parry, in 1821, when exploring and surveying the opening to which he gave the name of Lyon Inlet, determined, as he says in his narrative, to leave no opening or arm unvisited; and yet, with all his care, and the aid of his officers and four boats' crews, he overlooked the new inlet I found, from the fact that a high island shut out from his view the entrance to it.

I discovered a bay on the west side of Fox Channel, lat. 69°, long. 81° 30′ W., which makes west-south-west for fifteen miles. This Parry also missed, which is not remarkable when we consider that his was a marine survey, along the west side of Fox Channel to Igloolik, an island near the eastern end of Fury and Hecla Strait.

I discovered an important lake, twenty-five miles in length, in lat. 68° 45′ N., long. 82° W. I call it important, as it abounds in salmon of large size —some being six feet in length. It contains also many other species of fish, some of which, I think, have been hitherto unknown. Also another lake, lat. 69° 35′ N., running parallel with Fury and Hecla Strait, about fifty miles in length. It has two outlets. I followed up Crozier River, the mouth of which Parry discovered, and found its source to be the lake described. At the west end of the lake is another outlet, forming a river, which I followed down to the Gulf of Boothia, where the river discharges itself into a fine bay—another discovery.

It fell to my lot, also, to ascertain the north-western part of Melville Peninsula, at and below the western outlets of Fury and Hecla Strait, which may be said to complete the discovery of the American continent.

I discovered a long island, lying to the north-west and westward of the western outlet of Fury and Hecla Strait, and also the coast of the main-land on the north side of the above-mentioned outlet of the strait; and I found that the "Jesse Isle" laid down and so named on Dr. Rae's chart, at the north of Parry Bay, lat. 69° 30′, long. 85° 10′, is not there.

Although Parry had his vessels, the *Fury* and the *Hecla*, near to Amherst Island, in 1822, and sent out from there exploring and surveying parties, directing them to search, if possible, for the western outlet of Fury and Hecla Strait, they were unable to find it. In the following spring, 1823, while his vessels were in harbor at Igloolik Island, lat. 69° 21′ N., long. 82° W., Lyon, Parry's associate, undertook to reach the western outlet of the strait by means of sledges and dogs; but, after journeying for nineteen days, he failed to accomplish it. In 1847 Dr. Rae left his head-quarters at Fort Hope, at the head of Repulse Bay, with the intention of reaching the outlet of Fury and Hecla Strait; but before he could get there his provisions gave out, and he was compelled to turn back. I had some reason, therefore, to feel gratified when I found myself traversing the very region that such intrepid explorers as Parry, Lyon, and Rae had attempted to reach in vain.

The next important contribution to geography was my discovering an important island north of Ormond Island, at the east end of Fury and Hecla Strait. What Parry has put down upon his chart as the main-land north of Ormond Island is an island, but somewhat less in size than Ormond Island.

I think that if Parry had known of the existence of the channel which is on the north side of the new island that I refer to, he would have succeeded in getting his vessels much farther to the westward in the strait than he did. By passing through this new channel, and by keeping close to the land on the north side of the strait, Fury and Hecla Strait, like the passage leading into Wager Bay, and like the Hudson Strait, in the navigable season, may be penetrated by keeping on the north side, while the opposite or south side is encumbered by heavy ice.

From intelligent Esquimaux whom I met at Igloolik, I obtained information about, and sketches of, the west coast of Fox's Farthest, lat. 66° 50' N., up to what Parry calls "Murray Maxwell Inlet," which is near the east end of Fury and Hecla Strait. Murray Maxwell Inlet is in reality a sound, or strait, that sweeps around to the eastward, forming a large island. If you take your pencil, and continue the so-called Murray Maxwell Inlet to the eastward, and to the blank in Parry's chart, you will have the delineation of the island that is there. To the eastward of the Calthorpe Isles and Cape König you will find the broken lines of the land that Parry discovered. He could not determine whether it consisted of islands, or formed a part of the main-land. From the Esquimaux who had been there I learned that it consisted of two islands. The nearest approach I made to them was on my visit to Fern Island, which you will find upon Parry's chart, attached to the narrative of his second voyage.

At Igloolik I met Esquimaux who were natives of Cumberland Sound, sometimes called Cumberland Inlet, which you know is on the west side of Davis Strait, above Frobisher Bay. These natives made their way to Igloolik by first making a portage from Cumberland Sound to a large lake, called upon the charts Kennedy Lake, and which, by-the-way, I may remark, no white man has ever yet seen, and then launching their oomiaks (women's boats) upon the lake, which they traversed westward, entering a large river, and drifting down it with a swift current to Fox's Farthest, where the river enters the sea. From there they turned north, and coasted along up to the Calthorpe Isles, and from there crossed over to Igloolik.

From Esquimaux at Igloolik I also obtained important information of a new bay, that will not only be of interest to geographers, but must, I think, eventually be of great value to our commerce. The entrance to this bay has only been seen, and is indicated upon the Arctic charts as Admiralty Inlet. Nothing has been known, however, by civilized man of this bay or of its character. The entrance is from Barrow Strait, lat. 73° 43′ N., long. 83° W., and the bay extends very nearly in a southern direction to about 71° N. lat. The west side [east?] has a coast-line on a gradual curve from Barrow Strait to near its limit — the concave on the east, while the west side has many bays, or fiords, with some good harbors in them. The bay is free from ice every summer, and none of the ice from Barrow Strait ever finds its way into it. This bay abounds in whales (*Balæna mysticetus*, or smooth-back, the most important to civilized man), in narwhals (the sea-unicorn), and in seals. So abundant are the whales that the natives sometimes kill, in their rude way, as many as five large ones in a few days.

The information which I derived from the Esquimaux has convinced me that this new bay will prove as valuable to whalers as Cumberland Sound. From 1840 to the present time the product of whalebone and oil from Cumberland Sound, by English and American whalers, has amounted to $15,000,000; and as the area of the whale-fishery is gradually diminishing, the fact of the existence of this bay I regard as of great value, as opening up a new ground for the prosecution of this important industry.

I also obtained valuable information from the Esquimaux at Igloolik respecting Pond Bay, the western prolongation of which upon our present Arctic charts is miscalled Eclipse Sound. If the testimony of the Esquimaux can be relied on—and I place the fullest confidence in it—Pond Bay terminates in long. 81° W. (approximately); and the representation upon the Arctic charts of a strait from Pond Bay to Prince Regent Inlet, on the northern part of the Gulf of Boothia, is erroneous.

It has been the supposition of geographers that Davis Strait and Baffin Bay are connected with Fox Channel by straits. This is not the fact. All the intelligent Esquimaux that I have met in my two voyages assert that the land bounded on the north by Barrow Strait, upon the east by Baffin Bay, and Davis Strait, on the south by Hudson Strait, and on the west by Fox Channel and Prince Regent Inlet, is one land, or *one great* island. They know of a much smaller island that has Pond Bay on its south side, Navy Board Inlet, or more properly strait, on its west, Lancaster Sound on its north, and Baffin Sea on its east side.

My other contributions to geography are, that Dr. Rae's Colville Bay, in lat. 68° N., long. 88° 20′, is not a bay, but very low land; that his Grinnell Lake and Simpson Lake, which he delineates as one continuous lake, are in fact three distinct lakes; and, lastly, that his Shepherd Bay extends northerly about twelve miles beyond the limit he has assigned to it.

This, my dear sir, embraces all I have to communicate. You will remember that I went out with very limited resources, and was more circumscribed for the want of means than almost any Arctic explorer. Should I again go out, as I trust to do, I hope to extend the area of geographical discovery, and accomplish something that may redound to the credit of our common country. Very respectfully yours, C. F. HALL.

Act Authorizing North Polar Expedition.

Under a general appropriation act "for the year ending the thirtieth of June, eighteen hundred and seventy-one," we find the Congressional authority for the outfit of the "UNITED STATES NORTH POLAR EXPEDITION."

"SEC. 9. *And be it further enacted,* That the President of the United States be authorized to organize and send out one or more expeditions toward the North Pole, and to appoint such person or persons as he may deem most fitted to the command thereof; to detail any officer of the public service to take part in the same, and to use any public vessel that may be suitable for the purpose; the scientific operations of the expeditions to be prescribed in accordance with the advice of the National Academy of Sciences; and that the sum of fifty thousand dollars, or such part thereof as may be necessary, be hereby appropriated out of any moneys in the treasury not otherwise appropriated, to be expended under the direction of the President."

Instructions to Captain C. F. HALL, Commander of the United States North Polar Expedition.

Navy Department, June 9, 1871.

SIR,—Having been appointed by the President of the United States commander of the expedition toward the North Pole, and the steamship *Polaris* having been fitted, equipped, provisioned, and assigned for the purpose; you are placed in command of the said vessel, her officers, and crew, for the purposes of the said expedition. Having taken command, you will proceed in the vessel, at the earliest possible date, from the Navy Yard in this city to New York. From New York you will proceed to the first favorable port you are able to make on the west coast of Greenland, stopping, if you deem it desirable, at St. Johns, Newfoundland. From the first port made by you on the west coast of Greenland, if farther south than Holsteinborg, you will proceed to that port, and thence to Goodhavn (or Lively), in the island of Disco. At some one of the ports above referred to you will probably meet a transport, sent by the Department, with additional coal and stores, from which you will supply yourself to the fullest carrying capacity of the *Polaris.* Should you fall in with the transport before making either of the ports aforesaid, or should you obtain information of her being at, or having landed her stores at, any port south of the island of Disco, you will at once proceed to put yourself in communication with the commander of the transport, and supply yourself with the additional stores and coal, taking such measures as may be most expedient and convenient for that purpose. Should you not hear of the transport before reaching Holsteinborg, you will remain at that port, waiting for her and your supplies as long as the object of your expedition will permit you to delay for that purpose. After waiting as long as is safe, under all the circumstances as they may present themselves, you will, if you do not hear of the transport, proceed to Disco, as above provided. At Disco, if you hear nothing of the transport, you will, after waiting as long as you deem it safe, supply yourself, as far as you may be able, with such supplies and articles as you may need, and proceed on your expedition without further delay. From Disco you will proceed to Upernavik. At these two last-named places you will procure dogs and other Arctic outfits. If you think it of advantage for the purpose of obtaining dogs, etc., to stop at Tossac, you will do so. From Upernavik, or Tossac, as the case may be, you will proceed across Melville Bay to Cape Dudley Digges, and thence you will make all possible progress, with vessels, boats, and sledges, toward the North Pole, using your own judgment as to the route or routes to be pursued and the locality for each winter's quarters. Having been provisioned and equipped for two and a half years, you will pursue your explorations for that period; but, should the object of the expedition require it, you will continue your explorations to such a further length of time as your supplies may be safely extended. Should, however, the main object of the expedition, viz., attaining the position of the North Pole, be accomplished at an earlier period, you will return to the United States with all convenient dispatch.

There being attached to the expedition a Scientific Department, its operations are prescribed in accordance with the advice of the National Academy of Sciences, as required by the law. Agreeably to this advice, the charge and direction of the scientific operations will be intrusted, under your command, to Dr. Emil Bessel; and you will render Dr. Bessel and his assistants all such facilities and aids as may be in your power, to carry into effect the said further advice, as given in the instructions herewith furnished in a communication from the President of the National Academy of Sciences. It is, however, important that objects of natural history, ethnology, etc., etc., which may be collected by any person attached to the expedition, shall be delivered to the chief of the Scientific Department, to be cared for by him, under your direction, and considered the property of the Government; and every person be strictly prohibited from keeping any such object. You will direct every qualified person in the expedition to keep a private journal of the progress of the expedition, and enter on it events, observations, and remarks, of any nature whatsoever. These journals shall be considered confidential, and read by no person other than the writer. Of these journals no copy shall be made. Upon the return of the expedition you will demand of each of the writers his journal, which it is hereby ordered he shall deliver to you. Each writer is to be assured that when the records of the expedition are published he shall receive a copy; the private journals to be returned to the writer, or not, at the option of the Government; but each writer, in the published records, shall receive credit for such part or parts of his journal as may be used in said records. You will use every opportunity to determine the position of all capes, headlands, islands, etc., the lines of coasts, take soundings, observe tides and currents, and make all such surveys as may advance our knowledge of the geography of the Arctic regions.

You will give special written directions to the sailing and ice master of the expedition, Mr. S. O. Buddington, and to the chief of the Scientific Department, Dr. E. Bessel, that in case of your death or disability—a contingency we sincerely trust may not arise—they shall consult as to the propriety and manner of carrying into further effect the foregoing instructions, which I here urge must, if possible, be done. The results of their consultations, and the reasons therefor, must be put in writing, and kept as part of the records of the expedition. In any event, however, Mr. Buddington shall, in case of your death or disability, continue as the sailing and ice master, and control and direct the movements of the vessel; and Dr. Bessel shall, in such case, continue as chief of the Scientific Department, directing all sledge-journeys and scientific operations. In the possible contingency of their non-agreement as to the course to be pursued, then Mr. Buddington shall assume sole charge and command, and return with the expedition to the United States with all possible dispatch.

You will transmit to this Department, as often as opportunity offers, reports of your progress and results of your search, detailing the route of your proposed advance. At the most prominent points of your progress you will erect conspicuous skeleton stone monuments, depositing near each, in accordance with the confidential marks agreed upon, a condensed record of your progress, with a description of the route upon which you propose to advance, making *caches* of provisions, etc., if you deem fit.

In the event of the necessity for finally abandoning your vessel, you will at once endeavor to reach localities frequented by whaling or other ships, making every exertion to send to the United States information of your position and situation, and, as soon as possible, to return with your party, preserving, as far as may be, the records of, and all possible objects and specimens collected in, the expedition.

All persons attached to the expedition are under your command, and shall, under every circumstance and condition, be subject to the rules, regulations, and laws governing the discipline of the navy, to be modified, but not increased, by you as the circumstances may in your judgment require.

To keep the Government as well informed as possible of your progress, you will, after passing Cape Dudley Digges, throw overboard daily, as open water or drifting ice may permit, a bottle, or small copper cylinder, closely sealed, containing a paper, stating date, position, and such other facts as you may deem interesting. For this purpose you will have prepared papers containing a request, printed in several languages, that the finder transmit it by the most direct route to the Secretary of the Navy, Washington, United States of America.

Upon the return of the expedition to the United States, you will transmit your own and all other records to the Department. You will direct Dr. Bessel to transmit all the scientific records and collections to the Smithsonian Institution, Washington.

The history of the expedition will be prepared by yourself, from all the journals and records of the expedition, under the supervision of the Department. All the records of the scientific results of the expedition will be prepared, supervised, and edited by Dr. Bessel, under the direction and authority of the President of the National Academy of Sciences.

Wishing for you and your brave comrades health, happiness, and success in your daring enterprise, and commending you and them to the protecting care of the God who rules the universe,

I am, very respectfully, yours,   GEO. M. ROBESON,
                                                                                 Secretary of the Navy.

CHAS. F. HALL, Commanding Expedition toward the North Pole.

Letter of Professor JOSEPH HENRY (*President of the National Academy of Sciences*), *with Instructions for the Scientific Operations of the Expedition.*

                                                               Washington, D.C., June 9, 1871.

SIR,—In accordance with the law of Congress authorizing the expedition for explorations within the Arctic circle, the scientific operations are to be prescribed by the National Academy; and in behalf of this society I respectfully submit the following remarks and suggestions:

The appropriation for this expedition was granted by Congress principally on account of the representations of Captain Hall and his friends as to the possibility of improving our knowledge of the geography of the regions beyond the eightieth degree of north latitude, and more especially of reaching the pole. Probably on this account and that of the experience which Captain Hall had acquired by seven years' residence in the Arctic regions, he was appointed by the President as commander of the expedition.

In order that Captain Hall might have full opportunity to arrange his plans, and that no impediments should be put in the way of their execution, it was proper that he should have the organization of the expedition and the selection of his assistants. These privileges having been granted him, Captain Hall early appointed, as the sailing-master of the expedition, his friend and former fellow-voyager in the Arctic zone, Captain Buddington, who has spent twenty-five years amidst Polar ice; and for the subordinate positions, persons selected especially for their experience of life in the same regions.

It is evident from the foregoing statement that the expedition, except in its relations to geographical discovery, is not of a scientific character, and to connect with it a full corps of scientific observers, whose duty it should be to make minute investigations relative to the physics of the globe, and to afford them such facilities with regard to time and position as would be necessary to the full success of the object of their organization, would materially interfere with the views entertained by Captain Hall, and the purpose for which the appropriation was evidently intended by Congress.

Although the special objects and peculiar organization of this expedition are not primarily of a scientific character, yet many phenomena may be observed and specimens of natural history be incidentally collected, particularly during the long winter periods in which the vessel must necessarily remain stationary; and therefore, in order that the opportunity of obtaining such results might not be lost, a committee of the National Academy of Sciences was appointed to prepare a series of instructions on the different branches of physics and natural history, and to render assistance in procuring the scientific outfit.

Great difficulty was met with in obtaining men of the proper scientific acquirements to embark in an enterprise which must necessarily be attended with much privation, and in which, in a measure, science must be subordinate. This difficulty was, however, happily obviated by the offer of an accomplished physicist and naturalist, Dr. E. Bessel, of Heidelberg, to take charge of the scientific operations, with such assistance as could be afforded him by two or three intelligent young men that might be trained for the service. Dr. Bessel was the scientific director of the German expedition to Spitzbergen and Nova Zembla, in 1869, during which he made, for the first time, a most interesting series of observations on the depths and currents of the adjacent seas. From his character, acquirements, and enthusiasm in the cause of science, he is admirably well qualified for the arduous and laborious office for which he is a volunteer. The most important of the assistants was one to be intrusted, under Dr. Bessel, with the astronomical and magnetic observations, and such a one has been found in the person of Mr. Bryan, a graduate at Lafayette College, at Easton, Pennsylvania, who, under the direction of Professor Hilgard, has received from Mr. Schott and Mr. Keith, of the Coast Survey, practical instructions in the use of the instruments.

The Academy would, therefore, earnestly recommend, as an essential condition of the success of the objects in which it is interested, that Dr. Bessel be appointed as sole director of the scientific operations of the expedition, and that Captain Hall be instructed to afford him such facilities and assistance as may be necessary for the special objects under his charge, and which are not incompatible with the prominent idea of the original enterprise.

As to the route to be pursued with the greatest probability of reaching the pole, either to the east or west of Greenland, the Academy forbears to make any suggestions, Captain Hall having definitely concluded that the route through Baffin Bay, the one with which he is most familiar, is that to be adopted. One point, however, should be specially urged upon Captain Hall, namely, the determination with the utmost scientific precision possible of all his geographical positions, and especially of the ultimate northern limit which he attains. The evidence of the genuineness of every determination of this kind should be made apparent beyond all question.

On the return of the expedition, the collections which may be made in natural history, etc., will, in accordance with a law of Congress, be deposited in the National Museum, under the care of the Smithsonian Institution; and we would suggest that the scientific records be discussed and prepared for publication by Dr. Bessel, with such assistance as he may require, under the direction of the National Academy. The importance of refusing to allow journals to be kept exclusively for private use, or collections to be made other than those belonging to the expedition, is too obvious to need special suggestion.

In fitting out the expedition, the Smithsonian Institution has afforded all the facilities in its power in procuring the necessary apparatus, and in furnishing the outfit for making collections in the various departments of natural history. The Coast Survey, under the direction of Professor Peirce, has contributed astronomical and magnetical instruments. The Hydrographic Office, under Captain Wyman, has furnished a transit instrument, sextants, chronometers, charts, books, etc. The Signal Corps, under General Myer, has supplied anemometers, thermometers, aneroid and mercurial barometers, besides detailing a sergeant to assist in the meteorological observations. The members of the committee of the Academy, especially Professors Baird and Hilgard, have, in discussing with Dr. Bessel the several points of scientific investigation, and in assisting to train his observers, rendered important service.

The liberal manner in which the Navy Department, under your direction, has provided a vessel and especially fitted it out for the purpose, with a bountiful supply of provisions, fuel, and all other requisites for the success of the expedition, as well as the health and comfort of its members, will, we doubt not, meet the approbation of Congress, and be highly appreciated by all persons interested in Arctic explorations.

From the foregoing statement it must be evident that the provisions for exploration and scientific research in this case are as ample as those which have ever been made for any other Arctic expedition; and should the results not be commensurate with the anticipations in regard to them, the fact can not be attributed to a want of interest in the enterprise or to inadequacy of the means which have been afforded.

We have, however, full confidence, not only in the ability of Captain Hall and his naval associates, to make important additions to the knowledge of the geography of the Polar region, but also in his interest in science and his determination to do all in his power to assist and facilitate the scientific operations.

Appended to this letter is the series of instructions prepared by the committee of the Academy, viz.: the instructions on astronomy, by Professor Newcomb; on magnetism, tides, etc., by Professor J. E. Hilgard; on meteorology, by Professor Henry; on natural history, by Professor S. F. Baird; on geology, by Professor Meek; and on glaciers, by Professor Agassiz.

I have the honor to be, very respectfully, your obedient servant,

JOSEPH HENRY,
President of the National Academy of Sciences.

Hon. GEO. M. ROBESON, Secretary of the Navy.

INSTRUCTIONS.

GENERAL DIRECTIONS IN REGARD TO THE MODE OF KEEPING RECORDS.

Record of Observations.—It is of the first importance that in all instrumental observations the fullest record be made, and that the original notes be preserved carefully.

In all cases the actual instrumental readings must be recorded, and if any corrections are to be applied, the reason for these corrections must also be recorded. For instance, it is not sufficient to state the index error of a sextant; the manner of ascertaining it and the readings taken for the purpose must be recorded.

The log-book should contain a continuous narrative of all that is done by the expedition, and of all incidents which occur on shipboard, and a similar journal should be kept by each sledge-party. The actual observations for determining time, latitude, the sun's bearing, and all notes having reference to mapping the shore, soundings, temperature, etc., should be entered in the log-book or journal in the regular order of occurrence. When scientific observations are more fully recorded in the note-books of the scientific observer than can be conveniently transcribed into the log-book, the fact of the observation and reference to the note-book should be entered.

The evidence of the genuineness of the observations brought back should be of the most irrefragable character. No erasures whatever with rubber or knife should be made. When an entry requires correction, the figures or words should be merely crossed by a line, and the correct figures written above.

ASTRONOMY.

Astronomical Observations.—One of the chronometers, the most valuable, if there is any difference, should be selected as the standard by which all observations are to be made, as far as practicable. The other chronometers should all be compared with this every day at the time of winding, and the comparisons entered in the astronomical note-book.

When practicable, the altitude or zenith distance of the sun should be taken four times a day—morning and evening for time; noon and midnight for latitude. The chronometer or watch times of the latitude observations, as well as of the time observations, should always be recorded. Each observation should always be repeated at least three times in all, to detect any mistake.

When the moon is visible, three measures of her altitude should be taken about the time of her passage over each cardinal point of true bearing, and the chronometer time of each altitude should be recorded.

As the Greenwich time deduced from the chronometers will be quite unreliable after the first six months, it will be necessary to have recourse to lunar distances. These should be measured from the sun, in preference to a star, whenever it is practicable to do so.

If a sextant is used in observation, a measure of the semi-diameter of the sun or moon should be taken every day or two for index error.

The observations are by no means to be pretermitted when lying in port, because they will help to correct the position of the port.

The observations should, if convenient, be taken so near the standard chronometer that the observer can signal the moment of observation to an assistant at the chronometer, who is to note the time. If this is not found convenient, and a comparing watch is used, the watch-time and the comparison of the watch with the chronometer should both be carefully recorded.

The observations made by the main party should be all written down in full in a continuous series of note-books, from which they may be copied in the log. Particular care should be exercised in always recording the *place*, *date*, and *limb* of sun or moon observed, and any other particulars necessary to the complete understanding of the observation.

Observations at Winter-quarters.—The astronomical transit instrument will be set up in a suitable observatory. A meridian mark should be established as soon as practicable, and the instrument kept with constant care in the vertical plane passing through the mark, in order that all observations may be brought to bear, on determining the deviation of that plane from the meridian of the places. The transits of circumpolar stars, on both sides of the pole, and those of stars near the equator, should be frequently observed.

Moon culminations, including the transits of both first and second limbs, should be observed for the determination of longitude independently of the rates of the chronometers. Twelve transits of each limb is a desirable number to obtain—more, if practicable. If any occultations of bright stars by the moon are visible, they should be likewise observed.

The observations for latitude will be made with the sextant and artificial horizon, upon stars both north and south of the zenith.

All the chronometers of the expedition should be compared daily, as nearly as practicable about the same time.

Whenever a party leaves the permanent station for an exploration, and immediately upon its return, its chronometer should be compared with the standard chronometer of the station.

Observations during Sledge or Boat Journeys.— The instruments to be taken are the small Casella theodolite, or a pocket-sextant and artificial horizon, one or more chronometers, and a prismatic compass, for taking magnetic bearings of the sun. In very high latitudes the time of the sun's meridian altitude is not readily determined; it will be advisable, therefore, to take altitudes when the sun is near the meridian, as indicated by the compass, with regard to the variations of the compass, as derived from an isogonic chart. The time when the observation is taken will, of course, be noted by the chronometer. Altitudes should be taken in this way, both to the south and north of the zenith; they will enable the traveler to obtain his latitude at once very nearly, without the more laborious computation of the time.

The observations for time should be taken as nearly as may be when the sun is at right angles to the meridian, to the east and west, the compass being again used to ascertain the proper direction. This method of proceeding will call for observations of altitude at or near the four cardinal points, or nearly six hours apart in time.

When the party changes its place in the interval between their observations, it is necessary to have some estimate of the distance and direction traveled. The ultimate mapping of the route will mainly depend upon the astronomical observations, but no pains should be spared to make a record every hour of the estimated distance traveled—by log, if afloat—of the direction of the route, by compass, and of bearings of distant objects, such as peaks, or marked headlands, by which the route may be plotted.

In case of a few days' halt being made when a very high latitude has been reached, or at any time during the summer's explorations, a special object of care should be to ascertain the actual rate of the chronometers with the party. To this end, a well-defined, fixed object, in any direction, should be selected as a mark, the theodolite pointed on it, and the transit of the sun over its vertical observed on every day during the sojourn at the place. If the party be only provided with a sextant, then the same angular distances of the sun from a fixed object should be observed on successive days, the angles being chosen so as to be between 30° and 45°. For instance, set the sextant successively to 40°, to 40° 20', 40° 40', etc., and note the time when the sun's limb comes in contact with the object. The same distances will be found after twenty-four hours, with a correction for change in the sun's declination. The sun's altitude should be observed before and after these observations, and its magnetic bearing should be noted, as well as that of the mark. The altitude of the mark should also be observed, if practicable, either with the sextant or clinometer, but this is not essential.

MAGNETISM.

On the voyage and sledge-journey, at all times when traveling, the *declination* or variation of the compass should be obtained by observing the magnetic bearing of the sun, at least once every day on which the sun is visible. On shipboard or in boats the azimuth compass is to be used; on land the small theodolite will be found preferable.

When afloat, no valuable observations of the magnetic *dip* and *intensity* are practicable. On the sledge-journey the dip-circle may be carried, and when halts are made longer than necessary to determine the place by astronomical observations, the *dip* and relative *intensity*, according to Lloyd's method, should be ascertained.

At winter-quarters, in addition to the above-mentioned observations, those of *absolute horizontal intensity* should be made with the theodolite magnetometer, including the determination of moment of inertia. Also with the same instrument the absolute declination should be determined.

The least that the observer should be satisfied with is the complete determination of the three magnetic elements,—namely, declination, dip, and horizontal intensity. At one period, say within one week, three determinations of each should be made.

It is advisable that the same observations be repeated on three successive days of each month during the stay at one place; and that on three days of each month, as the 1st, 11th, and 21st, or any other days, the variation of the declination-magnet be read every half-hour during the twenty-four hours; also that the magnetometer, or at least a theodolite with compass, remain mounted at all times, that the variation of the needle may be observed as often as practicable, and especially when unusual displays of aurora borealis take place.

In all cases the *time*, which forms an essential part of the record, should be carefully noted.

Not long before starting on a sledge-journey from a winter station, and soon after returning, the observations with the loaded dipping-needles for relative intensity should be repeated, in order to have a trustworthy comparison for the observations which have been made on the journey.

FORCE OF GRAVITY.

As the long winter affords ample leisure, pendulum experiments may be made to determine the force of gravity, in comparison with that at Washington, where observations have been made with the Hayes pendulum lent to the expedition. The record of the Washington observations, a copy of which is furnished, will serve as a guide in making the observations. Special care should be taken while they are in progress to determine the rate of the chronometer with great precision, by observations of numerous stars with the astronomical transit instrument, the pointing of which on a fixed mark should be frequently verified.

OCEAN PHYSICS.

Depths.—Soundings should be taken frequently, when in moderate depths, at least sufficiently often to give some indication of the general depth of the strait or sound in which the vessel is afloat at the time. If an open sea be reached, it should be considered of the greatest importance to get some measure of its depth, and since no bulky sounding apparatus can be carried across the ice barrier, the boat party should be provided with one thousand fathoms of small twine, marked in lengths of ten fathoms. Stones, taken on board when the boat is launched, may serve as weights.

Bottom should be brought up whenever practicable, and specimens preserved. Circumstances of time and opportunity must determine whether a *dredge* can be used, or merely a *specimen-cup*.

Temperature of the sea should be observed with the "Miller protected bulb thermometer," made by Casella, near the surface, about two fathoms below the surface, and near the bottom. When time permits, observations at an intermediate depth should be taken. These observations have a particular bearing on the general circulation of the ocean, and are of great importance.

Tides.—Observations of high and low water, as to time and height, should be made continuously at winter-quarters. The method adopted by Dr. Hayes is recommended. It consists of a graduated staff anchored to the bottom, directly under the "ice-hole," by a mushroom-anchor, or heavy stone and a chain, which is kept stretched by a counter-weight attached to a rope that passes over a pulley rigged overhead. The readings are taken by the height of the water in the "ice-hole." In the course of a few days' careful observations, the periods of high and low water will become sufficiently well known to predict the turns approximating from day to day, and subsequently, observations taken every five minutes for half an hour, about the anticipated turn, will suffice, provided they be continued until the turn of tide has become well marked.

Tidal observations taken at other points, when a halt is made for some time, even if continued not longer than a week, will be of special value, as affording an indication as to the direction in which the tide-wave is progressing, and inferentially, as to the proximity of an open sea. If, as the expedition proceeds, the tide is found to be later, the indication is that the open sea is far distant, if indeed the channel be not closed. But if the tide occurs earlier, as the ship advances, the probability is strongly in favor of the near approach to an open, deep sea, communicating directly with the Atlantic Ocean.

In making such a comparison, attention must be paid to the semi-monthly inequality in the time of high water, which may be approximately taken from the observations at winter-quarters. Observations made at the same age of the moon, in different places, may be directly compared.

When the water is open, the tide may be observed by means of a graduated pole stuck into the bottom; or, if that can not be conveniently done, by means of a marked line, anchored to the bottom, and floated by a light buoy, the observation being taken by hauling up the line taut over the anchor.

Currents.—It is extremely desirable to obtain some idea of the currents in the open Polar sea, if such is found. No special observations can be indicated, however, except those of the drift of icebergs, if any should be seen.

Density.—The *density* of the sea-water should be frequently observed with delicate hydrometers, giving direct indications to the fourth decimal. Whenever practicable water should be brought up from different depths, and its density tested. The specimens should be preserved in carefully sealed bottles, with a view to the subsequent determination of their mineral contents.

METEOROLOGY.

The expedition is well supplied with meteorological instruments, all the standards, with the exception of the mercurial barometers, manufactured by Casella, and compared with the standards of the Kew Observatory under the direction of Professor Balfour Stewart. Dr. Bessel is so familiar with the use of instruments, and so well acquainted with the principles of meteorology, that minute instructions are unnecessary. We shall, therefore, merely call attention, by way of remembrance, to the several points worthy of special notice.

Temperature.—The registers of the temperature, as well as of the barometer, direction of the wind, and moisture of the atmosphere, should, in all cases in which it is possible, be made hourly, and when that can not be done, they should be made at intervals of two, three, four, or six hours. The temperature of the water of the ocean, as well as of the air, should be taken during the sailing of the vessel.

The minimum temperature of the ice, while in winter-quarters, should be noted from time to time, perhaps at different depths, also that of the water beneath.

The temperature of the black-bulb thermometer *in vacuo* exposed to the sun, and also that of the black-bulb free to the air, should be frequently observed while the sun is on the meridian, and at given altitudes in the forenoon and afternoon, and these observations compared with those of the ordinary thermometer in the shade.

Experiments should also be made with the thermometer in the focus of the silvered mirror, the face of which is directed to the sky. For this purpose the ordinary black-bulb thermometer may be used as well as the naked-bulb thermometer. The thermometer thus placed will generally indicate a lower temperature than one freely exposed to radiation from the ground and terrestrial objects, and in case of isolated clouds will probably serve to indicate those which are colder and perhaps higher.

Comparison may also be made between the temperature at different distances above the earth, by suspending thermometers on a spar at different heights.

The temperature of deep soundings should be taken with the thermometer, with a guard to obviate the pressure of the water. As the tendency, on account of the revolution of the earth, is constantly to deflect all currents to the right hand of the observer looking down stream, the variations in temperature in connection with this fact may serve to assist in indicating the existence, source, and direction of currents.

The depth of frost should be ascertained, and also, if possible, the point of invariable temperature. For this purpose, augers and drills with long stems for boring deeply should be provided.

Pressure of Air.—A series of comparative observations should be made of the indications of the mercurial and aneroid barometers. The latter will be affected by the variation of gravity as well as of temperature, while the former will require a correction due only to heat and capillarity.

As it is known that the normal height of the barometer varies in different latitudes, accurate observations in the Arctic regions with this instrument are very desirable, especially in connection with observations on the moisture of the atmosphere, since, to the small quantity of this in northern latitudes the low barometer, which is observed there, has been attributed. I think, however, it will be found that the true cause is in the rotation of the earth on its axis, which, if sufficiently rapid, would project all the air from the pole.

In the latitude of about 60°, there is a belt around the earth in which the barometer stands unusually high, and in which violent fluctuations occur. This will probably be exhibited in the projection of the curve representing the normal height of the barometrical column in different latitudes.

Moisture.—The two instruments for determining the moisture in the air are the wet and dry bulb thermometer and the dew-point instrument, as improved by Regnault. But to determine the exact quantity in the atmosphere in the Arctic regions will require the use of an aspirator, by which a given quantity of air can be passed through an absorbing substance, such as chloride of calcium, and the increase of weight accurately ascertained. It may, however, be readily shown that the amount is very small in still air.

A wind from a more southern latitude will increase the moisture, and may give rise to fogs. Sometimes, from openings in the ice, vapor may be exhaled from water of a higher temperature than the air, and be immediately precipitated into fog.

The inconvenience which is felt from the moisture which exhales with the breath in the hold of the vessel may, perhaps, be obviated by adopting the ingenious expedient of one of the Arctic voyagers, namely, by making a number of holes through the deck and inverting over them a large metallic vessel like a pot. The exterior of this vessel being exposed to the low temperature of the air without, would condense the moisture from within on its interior surface, and thus serve, on the principle of the diffusion of vapor, to desiccate the air below.

The variation of moisture in the atmosphere performs a very important part in all meteorological changes. Its effects, however, are probably less marked in the Arctic regions than in more southern latitudes. The first effect of the introduction into the atmosphere of moisture is to expand the air and to diminish its weight; but after an equilibrium has taken place, it exists, as it were, as an independent atmosphere, and thus increases the pressure. These opposite effects render the phenomena exceedingly complex.

Winds.—As to these, the following observations are to be regularly and carefully registered, namely: The average velocity, as indicated by Robinson's anemometer; the hour at which any remarkable change takes place in their direction; the course of their veering; the existence at the same time of currents in different directions, as indicated by the clouds; the time of beginning and ending of hot or cold winds, and the direction from which they come. Observations on the force and direction of the wind are very important. The form of the wind-vane should be that of which the feather part consists of two planes, forming between them an angle of about 10°. The sensibility of this instrument, provided its weight be not too much increased, is in proportion to the surface of the feather planes. Great care must be taken to enter the direction of the wind from the true meridian, whenever this can be obtained, and in all cases to indicate whether the entries refer to the true or magnetic north. Much uncertainty has arisen on account of the neglect of this precaution.

In accordance with the results obtained by Professor Coffin, in his work on the resultant direction of the wind, there are, in the northern hemisphere, three systems roughly corresponding with the different zones—namely, the tropical, in which the resultant motion is toward the west; the temperate, toward the east; and the Arctic, in which it is again toward the west.

In the discussion of all the observations, the variation of the temperature and the moisture will appear, in their connection with the direction of the wind. Hence the importance of simultaneous observations on these elements, and also on the atmospheric pressure.

Precipitation.—The expedition will be furnished with a number of rain-gauges, the contents of which should be measured after each shower. By inverting and pressing them downward into the snow, and subsequently ascertaining, by melting in the same vessel, the amount of water produced, they will serve to give the precipitation of water in the form of snow. The depth of snow can be measured by an ordinary measuring-rod. Much difficulty, however, is sometimes experienced in obtaining the depth of snow on account of its drifting, and it is sometimes not easy to distinguish whether snow is actually falling or merely being driven by the wind.

The character of the snow should be noted, whether it is in small rounded masses, or in regular crystals; also the conditions under which these different forms are produced.

The form and weight of hailstones should be noted, whether consisting of alternate strata, the number of which is important, of flocculent snow, or solid ice, or agglutinations of angular crystals, whether of a spherical form, or that of an oblate spheroid.

The color of the snow should be observed in order to detect any organisms which it may contain, and also any sediment which may remain after evaporation, whether of earthy or vegetable matter.

Clouds.—The character of the clouds should be described, and the direction of motion of the lower and the higher ones registered, at the times prescribed for the other observations. Since the expedition is well supplied with photographic apparatus, frequent views of the clouds and of the general aspect of the sky should be taken.

Aurora.—Every phase of the aurora borealis will of course be recorded; also the exact time of first appearance of the meteor, when it assumes the form of an arch or a corona, and when any important change in its general aspect takes place. The magnetic bearing of the crown of the arch, and its altitude at a given time, should be taken; also, if it moves to the south of the observer, the time when it passes the zenith should be noted. The time and position of a corona are very important.

Two distinct arches have sometimes been seen co-existing—one in the east and the other in the west. In such an exhibition, the position of the crown of each arch should be determined. Drawings of the aurora, with colored crayons, are very desirable. In lower latitudes a dark segment is usually observed beneath the arch, the occurrence of which, and the degree of darkness, should be registered. It also sometimes happens that a sudden precipitation of moisture in the form of a haziness is observed to cover the face of the sky during the shooting of the beams of the aurora. Any appearance of this kind is worthy of attention.

Wave motions are sometimes observed, and it would be interesting to note whether these are from east to west or in the contrary direction, and whether they have any relation to the direction of the wind at the time. The colors of the beams and the order of their changes may be important in forming a theory of the cause of the phenomena. Any similarity of appearance to the phenomena exhibited in Geissler's tubes should be noted, especially whether there is any thing like stratification.

The aurora should be frequently examined by the spectroscope, and the bright lines which may be seen carefully compared with one of Kirchoff's maps of the solar spectrum.

To settle the question as to the fluorescence of the aurora and its consequent connection with the electric discharge, a cone of light reflected from the silver-plated mirror should be thrown on a piece of white paper, on which characters have been traced with a brush dipped in sulphate of quinine. By thus condensing the light on the paper, any fluorescence which the ray may contain will be indicated by the appearance of the previously invisible characters in a green color.

Careful observations should be made to ascertain whether the aurora ever appears over an expanse of thick ice, or only over land or open water, ice being a non-conductor of electricity.

The question whether the aurora is ever accompanied with a noise has often been agitated, but not yet apparently definitely settled. Attention should be given to this point, and perhaps the result may be rendered more definite by the use of two ear-trumpets, one applied to each ear.

According to Hansteen, the aurora consists of luminous beams, parallel to the dipping-needle, which at the time of the formation of the corona are shooting up on all sides of the observer, and also the lower portions of these beams are generally invisible. It is, therefore, interesting to observe whether the auroral beams are ever interposed between the observer and a distant mountain or cloud, especially when looking either to the east or west.

The effect of the aurora on the magnetism of the earth will be observed by abnormal motion of the magnetic instruments for observing the declination, inclination, and intensity. This effect, however, may be more strikingly exhibited by means of a galvanometer, inserted near one end of a long insulated wire extended in a straight line, the two extremities of which are connected with plates of metal plunged in the water, it may be through holes in the ice, or immediately connected with the ground.

To ascertain whether the effect on the needle is due to an electrical current in the earth, or to an inductive action from without, perhaps the following variation of the preceding arrangement would serve to give some indication. Instead of terminating the wire in a plate of metal plunged in the water, let each end be terminated in a large metallic insulated surface, such, for example, as a large wooden disk, rounded at the edges and covered with tin-foil. If the action be purely inductive, the needle of the galvanometer inserted, say, near one end of the wire, would probably indicate a momentary current in one direction, and another in the opposite, at the moment of the cessation of the action. For the purpose of carrying out this investigation, the Smithsonian Institution has furnished the expedition with two reels of covered wire, each a mile in length, one of which is to be stretched in the direction, perhaps, of the magnetic meridian, and the other at right angles to it. It would be well, however, to observe the effect with the wires in various directions, or united in one continuous length.

Electricity.—From the small amount of moisture in the atmosphere, and the consequent insulating capacity of the latter, all disturbances of the electrical equilibrium will be seen in the frequent production of light and sparks on the friction and agitation of all partially non-conducting substances. Any unusual occurrences of this kind, such as electrical discharges from pointed rods, from the end of spars, or from the fingers of the observer, should be recorded.

A regular series of observations should be made on the character and intensity of the electricity of the atmosphere by means of an electrometer, furnished with a polished, insulated, metallic ball, several inches in diameter, and two piles of deluc to indicate the character of the electricity, whether + or —, and also supplied with a scale to measure, by the divergency of a needle, the degree of intensity. This instrument can be used either to indicate the electricity of the air by induction or by conduction. In the first case it is only necessary to elevate it above a normal plane by means of a flight of steps, say eight or ten feet, to touch the ball at this elevation and again to restore it to its first position, when it will be found charged with electricity of the same character as that of the air. Or the ball may be brought in contact with the lower end of an insulated metallic wire, to the upper end of which is attached a lighted piece of twisted paper which had been dried after previous saturation in a solution of nitrate of lead.

Thunder-storms are rare in the Arctic regions, although they sometimes occur; and in this case it is important to observe the point in the horizon in which the storm-cloud arises; also the direction of the wind during the passage of the storm over the place of the observer; and also the character of the lightning—whether zigzag, ramified, or direct; also its direction—whether from cloud to cloud, or from a cloud to the earth.

Optical Phenomena.—Mirage should always be noted, as it serves to indicate the position of strata of greater or less density, which may be produced by open water, as in the case of lateral mirage, or by a current of wind or warmer air along the surface.

The polarization of the light of the sky can be observed by means of a polariscope, consisting of a plate of tourmaline with a slice of Iceland-spar, or a crystal of nitre cut at right angles to its optical axis, on the side farthest from the eye. With this simple instrument the fact of polarization is readily detected, as well as the plane in which it is exhibited.

Halos, parhelia, coronæ, luminous arches, and glories should all be noted, both as to time of appearance and any peculiarity of condition of the atmosphere. Some of these phenomena have been seen on the surface of the ice by the reflection of the sun's beams, from a surface on which crystals had been formed by the freezing of a fog simultaneously with a similar appearance in the sky, the former being a continuation, as it were, and not a reflection of the latter.

In the latitude of Washington, immediately after the sun has sunk below the western horizon, there frequently appear faint parallel bands of colors just above the eastern horizon, which may very possibly be due to the dispersion of the light by the convex form of the atmosphere, and also, at some times, slightly colored beams crossing the heavens like meridians, and converging to a point in the eastern horizon. Any appearance of this kind should be carefully noted and described.

Meteors.—Shooting-stars and meteors of all kinds should be observed with the spectroscope. The direction and length of their motion should be traced on star maps, and special attention given at the stated periods in August and November. A remarkable disturbance of the aurora has been seen during the passage of a meteor through its beams. Any phenomena of this kind should be minutely described.

Ozone.—The expedition is furnished with a quantity of ozone test-paper, observations with which can only be rendered comparable by projecting against the sensitized paper a given quantity of atmospheric air. For this purpose an aspirator should be used, which may be made by fastening together two small casks, one of which is filled with water, with their axes parallel, by means of a piece of plank nailed across the heads, through the middle of which is passed an iron axis, on which the two casks may be made to revolve, and the full cask may readily be placed above the empty, so that its contents may gradually descend into the latter. During the running of the water from the upper cask, an equal quantity of air is drawn through a small adjutage into a closed vessel and made to impinge upon the test-paper. The vessel containing the test-paper should be united with the aspirator by means of an India-rubber tube.

Miscellaneous.—The conduction of sound during still weather, through the air over the ice, through the ice itself, and through the water, may be studied.

Evaporation of snow, ice, and water may be measured by a balance, of which the pan is of a given dimension.

Experiments on the resistance of water to freezing in a confined space at a low temperature may be made with small bomb-shells closed with screw-plugs of iron. The fact of the liquidity of the water at a very low temperature may be determined by the percussion of a small iron bullet, or by simply inverting the shell, when the ball, if the liquid remains unfrozen, will be found at the lowest point. It might be better, however, to employ vessels of wrought iron especially prepared for the purpose, since the porosity of cast-iron is such that the water will be forced through the pores, *e. g.*, the lower end of a gun-barrel, which, from the smallness of its diameter, will sustain an immense pressure, and through which the percussion of the inclosed bullet may be more readily heard. Water, in a thin metallic vessel, exposed on all sides to the cold, sometimes gives rise to hollow crystals of a remarkable shape and size, projecting above the level surface of the water, and exhibits phenomena worthy of study.

Experiments may be made on regulation, the plasticity of ice, the consolidation of snow into ice, the expansion of ice, its conducting power for heat, and the various forms of its crystallization. The effect of intense cold should be studied on potassium, sodium, and other substances, especially in relation to their oxidation.

The melting point of mercury should be observed, particularly as a means of correcting the graduation of thermometers at low temperatures. The resistance to freezing of minute drops of mercury, as has been stated, should be tested. Facts long observed, when studied under new conditions, scarcely ever fail to yield new and interesting results.

NATURAL HISTORY.

Objects of natural history of all kinds should be collected, and in as large numbers as possible. For this purpose all on board the vessel, both officers and sailors, should be required to collect, upon every favorable opportunity, and to deliver the specimens obtained to those appointed to have charge of them.

Zoology.—The terrestrial mammals of Greenland are pretty well known, but it is still desirable that a series, as complete as possible, of the skins should be preserved, great care being taken to always indicate, upon the label to be attached, the sex and probable age, as well as the locality and date of capture. The skeleton, and, when it is not possible to get this complete, any detached bones, particularly the skull and attached cervical vertebræ, are very desirable. Interesting soft parts, especially the brain, and also embryos, are very important. If it should be considered necessary to record measurements, they should be taken from specimens recently killed.

Of walruses and seals there should be collected as many skeletons as possible, of old and young individuals; also skins, especially of the seals. Notes should be made regarding the habits in general, food, period of copulation, duration of gestation, and time of migration, it being desirable to find out whether their migrations are periodical.

Of the *Cetacea*, when these are too large to be taken on board the vessel, the skull and cervical vertebræ, the bones of the extremities and penis, and whatever else may be deemed worthy of preservation, should be secured. All the animals should be examined for ecto- and ento-parasites, and the means by which they become affixed to the animals noted.

Collect carefully the species of *Myodes* (lemmings), *Arctomys*, and *Arvicola*, so as to determine the variations with locality and season. The relationship of two kinds of foxes, the blue and white, should be studied to determine their specific or other relationship. Any brown bears should be carefully collected, both skin and skeleton, to determine whether identical or not with the Old World *Ursus Arctus*.

Reference has already been made to the seals and cetaceans; of these the *Phoca cristata* (the white whale), *Beluga*, and the *Monodon* are particularly desired.

What has been said in regard to the mammals will apply equally well to the *birds*—skins and skeletons being equally desirable. It is especially important that the *fresh colors* of the bill, cere, gums, eyes, and feet, or caruncles, or bare skin, if there be any, should be noted, as the colors of these parts all change after the preparation of a specimen.

Of birds, the smaller land species are of the greatest interest, and complete series of them should be gathered. The northern range of the insectivorous species should be especially inquired into. The Arctic falcons should be collected in all their varieties, to ascertain whether there are two forms, a brown and a white, distinct through life, or whether one changes with age into the other.

Inquiry should be directed to the occurrence of *Bernicla leucopsis*, *Anser cinereus*, or other large gray geese, and the *Camptolæmus Labradora;* and a large number of specimens, of the latter especially, should be obtained. Indeed the geese and ducks generally should form subjects of special examination. Among the *Laridæ* the most important species is the *Larus rossii* or *Rhodostethia rosea*, scarcely known in collections. A large number of skins and eggs will be a valuable acquisition. *Larus eburneus* is also worthy of being collected. The *Alcidæ* should be carefully examined for any new forms, and inquiries directed in regard to the *Alca impennis*.

Of all birds' eggs an ample store should be gathered; and the skeletons of the *Arctic raptores* and the *Natatores* generally.

It will be a matter of much importance to ascertain what is the extreme northern range of the continental species of birds, and whether, in the highest latitudes, the European forms known to occur in Greenland cross Baffin Bay.

Eggs and nests of birds, in as large numbers as possible, should be procured, great care being taken, however, in all cases to identify them by the parents which may be shot, and some portion, if not all of them, preserved, if not recognized by the collector. All the eggs of one set should be marked with the same number, that they may not be separated; the parent bird, if collected, likewise receiving the same number. It should also be stated, if known, how long the eggs have been sat upon, as incubation influences very much their color; the situation of the nest, also, is very important. Notes on the manner of nesting, localities selected, and other peculiarities of breeding, should be carefully kept; whether they are polygamous, whether there are struggles between the males, and the manner in which the old birds feed their young; and whether these remain helpless in the nest for a given time, or whether they accompany the parents from birth. A journal of the arrival and departure of the migratory species should also be kept, to find out whether those which leave latest return earliest, and *vicè versa*.

Of fishes that are obtained, the best specimens should be photographed, the fresh colors noted, and then they should be preserved in alcohol or carbolic acid.

Among the fishes the *Salmonidæ*, *Cottidæ*, *Gadidæ*, and *Clupeidæ* will be of most interest, and good series should be secured.

The terrestrial inferior animals should be all collected, each class in its appropriate way.

Try to get larvæ of insects, and observe their life, whether they are well adapted to their surroundings; for in proportion to the insects are the number of insectivorous animals; and for that reason the struggle for life would be more energetic, and, therefore, only those insects which are best adapted to the conditions will survive.

Inferior marine animals are usually collected by two methods, viz., with a pelagic net and by a dredge. Both these methods should be employed whenever practicable. Especial attention should be paid to the larvæ, of which sketches should be made. The results of the dredging should be noted in blanks printed for this purpose, the specimens to be preserved as their constitutions require. Muller's liquor, glycerine, solution of alcohol and sugar, etc.

It would be of peculiar interest to study the several deep regions, admitted by Forbes and others, to ascertain if in the Arctic regions the intensity of color increases with the depth, as has been stated to be the case with red and violet, which, if true, would be just the contrary to what is observed in the temperate and tropical regions.

Of shells two sets should be preserved, one dry and the other *with the animal*, in alcohol; the dry shell is necessary from the fact that the alcohol, by the acetic acid produced, is apt to destroy the color.

It is particularly important to get as full a series as possible of the members of the smaller families, with a view to the preparation of monographs.

There should be paid as much attention as possible to the fauna of freshwater lakes to ascertain whether they contain marine forms, as has been found to be the case with some of those in North America, Scandinavia, Italy, and other countries. From this important conclusions regarding the rising of the coast may be arrived at.

Botany.—Plants are to be collected in two ways. Of each species some specimens should be put in alcohol to serve for studying the anatomy; the others to be dried between sheets of blotting-paper. The locality of each specimen should be noted, also its situation, the character of the soil and height above the sea, the season, and whether there is *heliotropismus*, etc., etc. In the general notes there should be remarks on the horizontal and vertical distribution.

GEOLOGY.

The most important point in the collection of geological specimens, whether they consist of rocks, minerals, or fossils, is, that on breaking or digging them from the matrix or bed, each individual specimen should be carefully wrapped separately in pliable but strong paper, with a label designating the exact locality from which it was obtained. If two or more beds of rock (sandstone, limestone, clay, marl, or other material) occur at the locality from which specimens are taken, the label should also have a number on it corresponding to the particular bed in which it was found, as designated in a section made on the spot in a note-book. This should be done in order that the specimens from each bed may be separated from those found in others, whether the beds are separable by differences of composition or by differences in the groups of fossils found in each; and it is, moreover, often important that this care should be observed, even when one or more of the beds are of inconsiderable thickness, if such beds are characterized by peculiar fossils. For in such cases it often happens that what may be a mere seam at one place may represent an important formation at another.

Specimens taken directly from rocks in place are, of course, usually more instructive than those found loose; but it often happens that much better specimens of fossils can be found already weathered out, and lying detached about an outcrop of hard rock than can be broken from it. These can generally be referred to their place in the section noted at the locality, by adhering portions of the matrix, or from finding more or less perfect examples of the same species in the beds in place; but it is usually the better plan to note on the labels of such specimens that they were found loose, especially if there are any evidences that they may have been transported from some other locality by drift agencies.

All exposures of rocks, and especially those of limestone, should be carefully examined for fossils, for it often happens that hard limestones and other rocks that show no traces of organic remains on the natural surfaces (covered, as they often are, with lichens and mosses), will be found to contain fossils when broken into. In cases where fossils are found to exist in a hard rock, if time and other circumstances permit, it is desirable that it should be vigorously broken with a heavy hammer provided for that purpose, and as many specimens of the fossils as possible (or as the means of transportation will permit) should be collected.

Fossils from rocks of all ages will, of course, be interesting and instructive, but it is particularly desirable that organic remains found in the later tertiary and quaternary formations of these high northern latitudes, if any such exist there, should be collected. These, whether of animals or plants, would throw much light on the question respecting the climatic conditions of the Polar regions at or just preceding the advent of man.

Specimens illustrating the lithological character of all the rocks observed in each district explored should also be collected, as well as of the organic remains found in fossiliferous beds; also, of all kinds of minerals. Those of rocks and amorphous minerals should be trimmed to as nearly the same size and form as can conveniently be done—say three by four inches wide and long and one and a quarter inches in thickness. Crystalline minerals ought, of course, to be broken from the matrix, rather with the view of preserving the crystals, as far as possible, than with regard to the size or form of the hand specimens; and the same remark applies equally to fossils.

On an overland journey the circumstances may not *always* be such as to allow the necessary time to wrap carefully and label specimens on the spot where they were collected; but in such cases numbers, or some other marks, should be scratched with the point of a knife, or other hard-pointed instrument, on each, by means of which the specimens collected at different times and places during the march can be correctly separated, labeled, and wrapped when the party stops for rest.

All specimens should be packed tightly in boxes as soon as enough have been collected to fill a box, and a label should be attached to each box indicating the particular district of country in which the collections were obtained. For this purpose, empty provision boxes or packages can generally be used.

In examining sections or exposures of rocks along a shore or elsewhere, it is a good plan to make a rough sketch in a note-book, thus: "Section 1:

5	Clay.	8 feet.
4	Shale.	7 feet.
3	Clay.	12 feet.
2	Sandstone.	12 feet.
1	Limestone.	10 feet.

APPENDIX.

Then, on the same or following pages, more particular descriptions of the nature and composition of the several beds should be written, referring to each by its number. Sections of this kind should be numbered 1, 2, 3, and so on, in the order in which they were observed, and the specimens from each bed ought also to be numbered on its label, so as to correspond. That is, specimens from the lowest bed of the first section should be, for instance, marked thus: "Section No. 1, bed No. 1," and so on. The name of the locality, however, should also, as already suggested, be written on the labels as a provision against the possible loss of note-books.

It generally happens that an outcrop will show only a part of the beds of which it is composed, thus:

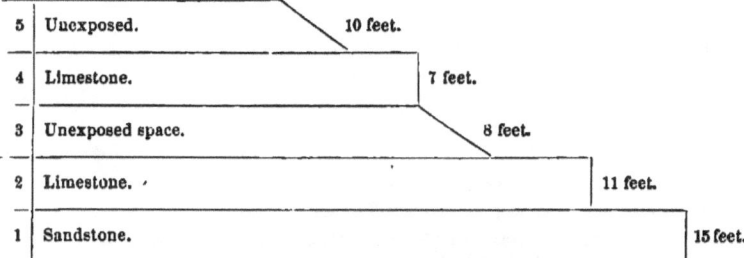

In such a case the facts should be noted exactly as seen, without any attempt to guess at the nature of the material that may fill the unexposed spaces; but generally, by comparing different sections of this kind taken in the same region, the entire structure of a district may be made out.

The dip and strike of strata should also be carefully observed and noted, as well as the occurrence of dikes or other outbursts of igneous rocks, and the effects of the latter on the contiguous strata.

All evidences of the elevation or sinking of coasts should likewise be carefully observed and noted.

Especial attention should be given to glacial phenomena of every kind, such as the formation, size, movements, etc., of existing glaciers, their abrading and other effects upon the subjacent rocks, their formation of moraines, etc.; also, the formation, extent, and movements of icebergs, and their power of transporting masses of rock, etc.

At Cape Frazier, between lat. 80° N. and long. 70° W., Dr. Hayes found some upper silurian fossils in a hard gray limestone. This rock doubtless has a rather wide extension in the country referred to, as other explorers have brought silurian fossils from several localities farther southward and westward in this distant northern region. Should the party visit the locality from which Dr. Hayes collected his specimens, it is desirable that as complete a collection as possible should be obtained, as most of those found by Dr. Hayes were lost.

For making geological observations and collecting geological specimens, very few instruments are required. For determining the elevations of mountains and the general altitude of the country, a barometer is sufficiently accurate. For local elevations of less extent, a pocket-level (Locke's) should be provided. Tape-lines are also useful for measuring vertical outcrops and other purposes; and a good pocket-compass is indispensable. The latter should have a clinometer attached.

A good supply of well-tempered cast-steel hammers should also be provided. They should be of various sizes and forms, and ought to be made with large enough eyes to receive stout handles, of which a good number, made of well-seasoned hickory, should be prepared. Chisels of different sizes should also be prepared of well-tempered steel.

A pouch of leather or stout canvas, with a strap to pass over the shoulder, will be found useful to carry specimens for short distances.

GLACIERS.

The progress of our knowledge of glaciers has disclosed two sides of the subject entirely disconnected with one another, and requiring different means of investigation. The study of the structure of glaciers as they exist now, and the phenomena connected with their formation, maintenance, and movement, constitute now an extensive chapter in the physics of the globe. On the other hand, it has been ascertained that glaciers had a much wider range during an earlier, but nevertheless comparatively recent geological period, and have produced during that period phenomena which, for a long time, were ascribed to other agencies. In any investigation of glaciers nowadays, the student should keep in mind distinctly these two sides of the subject. He ought also to remember at the outset, what is now no longer a mooted point, that, at different times during the glacial period, the accumulations of ice covering larger or smaller areas of the earth's surface have had an ever-varying extension, and that whatever facts are observed, their value will be increased in proportion as the chronological element is kept in view.

From the physical point of view, the Arctic expedition under the command of Captain Hall may render science great service should Dr. Bessel have an opportunity of comparing the present accumulations of ice in the Arctic regions with what is known of the glaciers of the Alps and other mountainous regions. In the Alps, the glaciers are fed from troughs in the higher regions, in which snow accumulates during the whole year, but more largely during winter, and by a succession of changes is gradually transformed into harder and harder ice, moving down to lower regions where glaciers never could have been formed. The snow-like accumulations of the upper regions are the materials out of which the compact, transparent, brittle ice of the lower glaciers is made. Whatever snow falls upon the glaciers in their lower range during winter melts away during summer, and the glacier is chiefly fed from above and wastes away below. The water arising from the melting of the snow at the surface contributes only indirectly to the internal economy of the glacier. It would be superfluous here to rehearse what is known of the internal structure of glaciers and of their movement; it may be found in any treatise on glaciers. Nor would it be of any avail to discuss the value of conflicting views concerning their motion. Suffice it to say that an Arctic explorer may add greatly to our knowledge by stating distinctly to what extent the winter snow, falling upon the surface of the great glacial fields of the Arctic, melts away during summer, and leaves bare an old icy surface covered with fragments of rock, sand, dust, etc. Such an inquiry will teach us in what way the great masses of ice which pour into the Arctic Ocean are formed, and how the supply that empties annually into the Atlantic is replenished. If the winter snows do not melt entirely in the lower part of the Arctic glaciers during summer, these glaciers must exhibit a much more regular stratification than the Alpine glaciers, and the successive falls of snow must in them be indicated more distinctly by layers of sand and dust than in those of the Alps by the dirt bands. Observations concerning the amount of waste of the glaciers by evaporation or melting, or what I have called *ablation* of the surface during a given time in different parts of the year, would also be of great interest as bearing upon the hygrometric condition of the atmosphere. A pole sunk sufficiently deep into the ice to withstand the effects of the wind could be used as a meter. But it ought to be sunk so deep that it will serve for a period of many months, and rise high enough not to be buried by a snow-storm. It should also be ascertained, if possible, whether water oozes from below the glacier, or, in other words, whether the glacier is frozen to the ground or separated from it by a sheet of water. If practicable, a line of poles should be set out with reference to a rocky peak or any bare surface of rock, in order to determine the motion of the ice. It is a matter of deep interest with reference to questions connected with the former greater extension of glaciers, to know in what manner flat sheets of ice move on even ground, exhibiting no marked slope. It may be possible to ascertain, after a certain time, by the change of position of poles sunk in the ice, whether the motion follows the inequalities of the surface or is determined by the lay of the land and the exposure of the ice to the atmospheric agents, heat, moisture, wind, etc. It would be of great interest to ascertain whether there is any motion during the winter season, or whether motion takes place only during the period when water may trickle through the ice. The polished surfaces in the immediate vicinity of glacier-ice exhibit such legible signs of the direction in which the ice moves, that wherever ledges of rocks are exposed, the scratches and furrows upon their surface may serve as a sure register of its progress; but before taking this as evidence, it should, if possible, be ascertained that such surfaces actually belong to the area over which the adjoining ice moves during its expansion, leaving them bare in its retreat.

The geological agency of glaciers will no doubt receive additional evidence from a careful examination of this point in the Arctic regions. A moving sheet of ice, stretching over a rocky surface, leaves such unmistakable marks of its passage that rocky surfaces which have once been *glaciated*, if I may thus express the peculiar action of ice upon rocks, namely, the planing, polishing, scratching, grooving, and furrowing of their surfaces, can never be mistaken for any thing else, and may everywhere be recognized by a practiced eye. These marks, in connection with transported loose materials, drift, and boulders, are unmistakable evidence of the great extension which glaciers once had. But here it is important to discriminate between two sets of facts, which have generally been confounded. In the proximity of existing glaciers, these marks and these materials have a direct relation to the present sheet of ice near by. It is plain, for instance, that the polished surfaces about the Grimsel, and the loose materials lying between the glacier of the Aar and the Hospice, are the work of the glacier of the Aar when it extended beyond its present limits, and step by step its greater extension may be traced down to Meyringen, and, in connection with other glaciers from other valleys of the Bernese Oberland, it may be tracked as far as Thun or Berne, when the relation to the Alps becomes complicated with features indicating that the whole valley of Switzerland, between the Alps and the Jura, was once occupied by ice. On the other hand, there are evident signs of the former presence of local glaciers in the Jura, as, for instance, on the Dent de Vaulion, which mark a later era in the history of glaciation in Switzerland. Now the traces of the former existence of extensive sheets of ice over the continent of North America are everywhere most plainly seen, but no one has yet undertaken to determine in what relation these glaciated surfaces of past ages stand to the ice-fields of the present day in the Arctics. The scientific men connected with Captain Hall's expedition would render science an important service if they could notice the trend and bearing of all the glacial scratches they may observe upon denudated surfaces wherever they land. It would be advisable for them, if possible, to break off fragments of such glaciated rocks, and mark with an arrow their bearing. It would be equally important to notice how far the loose materials, pebbles, boulders, etc., differ in their mineralogical character from the surface on which they rest, and to what extent they are themselves polished, rounded, scratched, or furrowed, and also what is the nature of the clay or sand which holds them together. It would be particularly interesting to learn how far there are angular boulders among these loose materials, and what is their position with reference to the compacted drift made up of rounded, polished, and scratched pebbles and boulders. Should an opportunity occur of tracing the loose materials of any locality to some rock *in situ*, at a greater or less distance, and the nature of the materials should leave no doubt of their identity, this would afford an invaluable indication of the direction in which the loose materials have traveled. Any indication relating to the differences of level among such materials would add to the value of the observation. I have purposely avoided all theoretical considerations, and only call attention to the facts which it is most important to ascertain, in order to have a statement as unbiased as possible.

Reports from Captain C. F. Hall to the Secretary of the Navy.

United States Steamship *Polaris*, Brooklyn Navy Yard,
June 16, 1871.

Sir,—I have the honor to report, agreeably to your instructions, that the steamship *Polaris*, under my command, left Washington Navy Yard at 12.30

P.M., Saturday, 10th inst., and arrived at the Brooklyn Navy Yard at 7 o'clock A.M., Wednesday, 14th inst., in sixty-two hours running time, having anchored two nights and laid over another night at Quarantine. The working of the engine was quite equal to my expectations, and with some slight alterations of machinery, suggested by Mr. M'Kean, who will have reported to the Department, will, it is believed, prove entirely satisfactory. The officers and crew have taken hold of their work with energy and exemplary conduct. * * * I hope to be in readiness for sailing within a week or ten days.

I have the honor to be, very respectfully, your obedient servant,

C. F. HALL,
Commanding United States North Polar Expedition.

Hon. GEO. M. ROBESON, Secretary of the Navy, Washington, D. C.

Steamship *Polaris*, Brooklyn Navy Yard, June 27, 1871.

SIR,—I have the honor to report the due receipt of my official instructions, bearing date 9th inst., and waiting my arrival here on the 14th inst., but not heretofore formally acknowledged. I have also received official letter, 24th inst., with printed copies of the official instructions; likewise, by preceding mails, the printed blanks, in six languages, for deposit in copper cylinders, etc. In reply to inquiries, 24th inst., I beg to state that I hope to have the expedition in entire readiness to sail on the 29th inst., having had two or three days' unavoidable detention, and shall spare no exertions to that end. I shall then hope to reach St. Johns, Newfoundland, on or before the 10th of July, and Holsteinborg, Greenland, on or about the 25th of July prox. Before sailing I shall have the honor to report somewhat more in detail. Very respectfully, your obedient servant,

C. F. HALL,
Commanding United States North Polar Expedition.

Hon. GEO. M. ROBESON, Secretary of the Navy, Washington.

United States Steamship *Polaris*, Harbor of New London, Connecticut, July 2, 1871.

SIR,—I have the honor to inform you that * * * * I left New York with the *Polaris* on the 29th of June, at 7 o'clock P.M., and passing through Hell Gate came to this port, where I arrived at 11 A.M. of the 30th.

It was my intention to proceed on my cruise directly from New York, up to the middle of the afternoon of the 29th, but finding that my assistant engineer, Wilson, had really run, that it was difficult to get a reliable engineer with whom I was acquainted to fill his place at New York, and thinking it would be imprudent to go to sea with only one engineer, I resolved to run into this port and procure one, where I was well acquainted, and where I knew I should succeed in getting one to suit me. In this I have been entirely successful. I have employed in Wilson's place Alvin A. Odell, who served during the last war of the rebellion as an acting assistant engineer of the navy, and was honorably discharged.

While at New York, it became necessary to turn into hospital Nathaniel

J. Coffin,* our carpenter. His place is not filled. It is my intention to fill it with a serviceable man at St. Johns, Newfoundland, at the same rate that was to be allowed him.

The following changes in my muster-roll have occurred since I left Washington with the ship:

1. T. L. Berggren, fireman, deserted.
2. William Jessup, seaman, deserted.
3. John Porter, steward, discharged for incapacity.
4. Charles Branett, cook, deserted. In the place of Branett I shipped Joseph Wolf, who also deserted.
5. I have shipped an excellent cook here at New London, by the name of William Jackson, of Arctic experience.
6. William Lindermann, seaman, shipped in place of William Jessup.
7. By consent of Admiral M. Smith, I shipped Frederick Jamka, as additional seaman, on the 27th of June.
8. In the place of Berggren, I shipped at New York John W. Booth, an experienced fireman. * * *

It is my purpose to go to sea to-morrow morning at 3 o'clock. * * * Inclosed is muster-roll of the officers and crew.

Very respectfully, your obedient servant, C. F. HALL,
Commanding North Polar Expedition.

Hon. GEO. M. ROBESON, Secretary of the Navy, Washington, D. C.

[*By Telegraph from New London, Connecticut.*]

July 3, 1871.

To HON. GEORGE M. ROBESON:

SIR,—The *Polaris*, at this hour, 4 o'clock A.M., under way, going out of New London harbor for sea. The company now all of glorious material, and with bright hopes. Letter by mail. C. F. HALL,
Commanding North Polar Expedition.

United States Steamship *Polaris*, St. Johns, Newfoundland, July 19, 1871.

SIR,—I beg to advise you that the ship is now all ready to proceed. For the last two days I have been detained in endeavoring to obtain a carpenter, but have not succeeded, and therefore leave without one. It has taken a full week to repair the steam machinery. * * *

Your obedient servant, C. F. HALL,
Commanding North Polar Expedition.

Hon. GEO. M. ROBESON, Secretary of the Navy, Washington, D. C.

Holsteinborg, Greenland, August 1, 1871,
Lat. 66° 57', long. 53° 53' 45".

SIR,—I have the honor to inform you that, in accordance with my letter dated July 19, 1871, at St. Johns, Newfoundland, the *Polaris* left that port

---

* Nathaniel J. Coffin was sent in the *Congress* to rejoin the *Polaris*.

at 3.30 P.M. of the same day, and proceeded direct to Greenland, in compliance with your orders. On the afternoon of July 27, we dropped anchor in the harbor of Fiscanaes, Greenland, one of the Danish ports, lat. 63° 8' N., where the expedition remained until the morning of July 29, when we took our course along up the coast to this place, arriving here at 10 o'clock A.M., July 31, yesterday.

We were agreeably surprised, on entering Holsteinborg harbor, to find in it the Swedish expedition, consisting of a brig and a small steamer, under the command of Fr. W. Von Otter, which had been to Disco and Upernavik, and is now on its return, proposing to resume the homeward voyage to-day, *via* St. Johns, Newfoundland. The commander kindly offering to forward any dispatches I may have for you, I gladly improve the opportunity.

With the exception of a gale that broke upon us on the evening of the day we left Fiscanaes—a gale that made the sea, with its icebergs all around us, vie in wild grandeur with "Greenland's icy mountains," along which we were coasting—the weather has been exceedingly fine from Washington to Holsteinborg.

I should sooner than this have remarked that the object of stopping at the port of Fiscanaes was to secure, if possible, Hans Christian, the Esquimau dog-driver and hunter of Dr. Kane's expedition up Smith Sound, and later, of Dr. Hayes's expedition, 1860-'61, in the same direction; but there we ascertained that this native was at Upernavik. The Swedish expedition not only confirm the news that Hans is at Upernavik, but add that he is anxiously awaiting the arrival of the American North Polar expedition, that he may join it.

From St. Johns to this place the *Polaris* has been under steam, with the exception of a little over two days, one of which we were becalmed, and waited through the day for a breeze. Fifty tons of coal were consumed from St. Johns to this port. Northerly winds have prevailed, but what has been unfavorable to our sailing has cleared the west coast of Greenland of the ice-pack, leaving us, beyond all doubt, a clear passage to the northward of Baffin Bay.

The Swedish expedition bring highly favorable news from Upernavik: that the season for Arctic navigation is better than known for several years; that no ice, save occasional bergs, is to be found between here and Disco, and between the latter place and Upernavik none has been seen for several weeks.

Governor Elberg, of the Holsteinborg district, residing here, has proffered our expedition all the aid he can. I find in him the same genial soul as in 1860, when I spent seventeen days in this harbor.

Your orders are that, if I do not hear of the transport before reaching Holsteinborg, for me to remain at that (this) port waiting for her and the supplies as long as the object of the expedition will permit a delay for that purpose. After waiting as long as is safe, under all these circumstances, as they may present themselves, I am, if I do not hear of the transport, to proceed to Disco.

In compliance with these instructions, I remain here waiting for the transport, say till Saturday morning, August 5; then, if nothing is seen or heard of her, the anchor of the *Polaris* will be weighed, and our course taken for Goodhavn (Lively), in the island of Disco.

On arriving there my plan will be to make arrangements for sending on a native force immediately, to raise about fifty tons of coal at the Rittenbek mine, situated near the extreme eastern angle of Disco Island. Specimens of this coal, which is bituminous, Baron Von Otter has just exhibited to me. It is found to be of fine quality for steaming purposes, and must be made to answer our necessities in case the transport does not arrive at Goodhavn by the 15th of August, the latest day I can prudently delay in resuming the voyage for the North Pole. I am confident, however, that the transport will be in time for the expedition at the proposed places of rendezvous. Governor Elberg has only fifteen tons of coal (brought here in a Danish ship) on hand, and signifies his willingness to let me have two-thirds of it. He informs me that it is quite doubtful our getting a supply at Disco, otherwise than by mining it, as already foreshadowed.

I expected to get a supply of reindeer furs here, but none are to be had, the reindeer, of late years, having been nearly all killed off in this neighborhood. On Disco, Upernavik, and Smith Sound, we must depend for our winter furs and our sledge-dogs.

Baron Von Otter, of the Swedish expedition, has paid two visits to the *Polaris*, taking deep interest in its object. He has kindly furnished me, for you, an abstract of his work performed since leaving Upernavik. [See table on opposite page.]

The columns of latitude and longitude will show to you the eastern limit of the ice-pack in Davis Strait, a matter in which you will be interested, as it demonstrates the wide iceberg channel for the *Polaris* to navigate through in her upward course. * * *

The whole ship's company have continued in good health and spirits since leaving New York, and all remain sanguine that next year our discovery will be the North Pole. I have the honor to be, sir,

Your obedient servant,   C. F. HALL,
Commanding United States North Polar Expedition.

Hon. GEO. M. ROBESON, Secretary of the Navy.

United States Steamship *Polaris*, Goodhavn, Greenland, August 17, 1871,
Lat. 69° 14' 41'' N., long. 53° 34' W.

SIR,—I have the honor to report the proceedings of the North Polar expedition since sending you my dispatches from Holsteinborg, dated August 1, 1871, which dispatches were forwarded by the hands of Baron Von Otter, commanding the Swedish expedition. Therein I indicated to you my purpose to remain at Holsteinborg till Saturday, August 5, for the transport, and then, if it had not arrived, to weigh the anchor of the *Polaris* and proceed at once to this port. I, however, only waited till the P.M. of August 3, when I took my departure, arriving here at Goodhavn at 2 P.M. of the 4th, just twenty-four hours from the time of leaving Holsteinborg. I

APPENDIX.

DEEP-SEA SOUNDINGS AND SEA TEMPERATURES OBTAINED ON BOARD OF HIS SWEDISH MAJESTY'S STEAM VESSEL *INGEGERA*, BETWEEN THE LATITUDES OF HOLSTEINBORG AND UPERNAVIK, JULY, 1871.

Date.	Latitude where soundings were obtained.	Longitude.	Depth in fathoms.	Temperature Surface. Swedish thermometer. C.	Temperature Bottom. Mill Cas. thermometer. F.	Remarks: Intermediate surface temperature.	
July.							
10	(Disco.) Mouth of N. Fiord,		54	+ 6.5	30	+ 6.5.	
10	3 miles off the said Fiord.		27	+ 4.5	35.5	+ 6.0, 6.0, 6.2, 5.8, 5.5, 6.0, 6.0.	
10	70 22 N.	54 22 W.	146	+ 6.0	27.0	+ 6.0, 5.8, 2.0, 2.0, 3.0, 4.5.	
11	70 35 N.	54 20 W.	161	+ 4.5	28.5	+ 4.5, 5.0, 6.0, 5.0, 7.8, 8.0, 8.5, 4.3, 4.0, 7.0.	
12	70 38 N.	52 00 W.	274	+ 7.0	26.75	+ 7.0, 8.0. (Among glacier-ice.)	
13	70 43 N.	52 08 W.	410	+ 7.0	24.25	+ 7.0, 8.0, 8.2, 10.0.	
13	70 53 N.	52 18 W.	397	+ 8.0	27.0	+ 10.0, 10.0, 10.0, 6.0, 10.0.	
13	71 07 N.	52 42 W.	235	+10.0	26.75	+ 10.0, 8.0, 6.0, 6.8, 9.0, 4.0.	
14	71 27 N.	53 58 W.	122	+ 4.0	27.0	+ 6.0, 9.0, 5.0, 6.0, 4.5, 7.5, 8.2, 8.5, 7.0, 6.0, 5.6, 7.0, 8.0, 7.0, 5.2, 6.0.	
18	72 37 N.	56 52 W.	67	+ 6.8	26.0	+ 6.0, 4.2, 5.3, 5.0, 5.5.	
18	72 32 N.	58 08 W.	116	+ 4.2	26.75	+ 4.2, 3.5, 2.6, 3.4.	
19	72 20 N.	59 39 W.	172	+ 3.4	25.75	+ 1.5, 0.8, 0.5, 1.5.	
19	72 04 N.	59 50 W.	227	+ 1.2	26.75	+ 1.8, 5.0, 4.5, 5.0.	
20	71 10 N.	58 56 W.	199	+ 6.2	25.75	+ 6.0, 6.0, 6.5, 5.6, 5.0., 2.5, 3.8.	
21	70 23 N.	59 01 W.	276	+ 1.6	23.75	+ 1.5, 1.0, 0.2, 0.2, 0.8, 1.5, 0.8, 0.0, 0.5, 1.5.	
22	69 52 N.	58 32 W.	199	+ 1.6	25.0	+ 1.5, 2.4, 0.5, 0.2, 1.0.	
22	69 19 N.	58 18 W.	183	+ 1.2	26.0	+ 1.0, 5.0, 3.0, 4.0, 3.0, 2.0, 2.5, 3.5, 4.0, 0.2, 4.0, 2.0, 0.0, 0.5, 0.8, 0.2, 1.0.	Along the west Ice.
24	68 08 N.	58 47 W.	169	+ 1.4	27.25	+ 1.4, 0.5, 0.5, 1.5, 1.0, 0.0, 1.0, 1.8, 0.8,	
24	67 57 N.	59 07 W.	279	+ 0.8	26.0	+ 1.0, 0.8, 1.0, 3,0, 0.2, 1.2, 0.0, 0.0.	
25	67 25 N.	58 45 W.	930	± 0.0	25.0	0.0 (among heavy ice), 0.5, 0.0.	
25	67 26 N.	58 29 W.	692	+ 0.4	25.5	+ 0.4, 2.0, 2.0, 2.5.	
26	67 43 N.	57 27 W.	128	+ 3.2	31.0	± 3.4, 4.8, 4.2, 0.	
26	67 50 N.	57 04 W.	132	+ 4.0	31.0	+ 4.2, 4.8 5.0.	
26	67 59 N.	56 33 W.	98	+ 4.8	31.75	+ 5.0, 4.8.	
26	68 08 N.	56 03 W.	48	+ 5.1	30.75	+ 5.0, 5.2.	
27	68 26 N.	54 55 W.	264	+ 5.4	31.75	+ 5.0, 5.5.	
27	68 27 N.	54 37 W.	215	+ 4.6	31.75	+ 5.5, 4.6.	
27	68 01 N.	54 47 W.	131	+ 5.9	32.0	+ 4.6, 5.9.	
	To Holsteinborg...		...	......	......	+ 5.9, 4.8.	

Holsteinborg, August 1, 1871.

Fr. W. Von Otter,
Commanding Swedish Greenland Expedition.

soon found that the highest official of North Greenland, Karrup Smith, chief inspector, was not at home, but was on his annual tour to the principal settlements of his district, and was not expected to return for two or three weeks. I lost no time, however, in paying my respects to the inspector's lieutenant, Fr. Lossen, who received me very cordially. As he could converse only in the Danish tongue, the wife of Inspector Smith became our interpreter. On explaining to Lieutenant Lossen the object of the expedition visiting this port, he exhibited some degree of hesitation in rendering to it that aid and co-operation we had a right to expect; but great was my success when the Hon. Mrs. Smith voluntarily took upon herself the task to advocate warmly the necessity of the Danish Government, through all its officials, to aid, in whatever way and to whatever extent might be desired, so great and glorious a country as the United States.

By the advice of the chief inspector's wife, Mrs. Smith, I sent off a boat-party to attempt to find and recall her husband. First Mate H. C. Chester was the officer I detailed for this service. He left here at meridian of August 6 for Jacob Haven, lat. 69° 13' N., long. 51° 00' W., and not finding the inspector there, proceeded northward to Rittenbek, lat. 69° 41' N., long. 51° 12' W., where he was found. The inspector at once responded to my request, and arrived here with his boat, in company of Mr. Chester's, at 2 A.M. of August 11. Mr. Chester performed this service with alacrity and fidelity. The whole distance voyaged was one hundred and seventy-five miles, and performed entirely under oars, except two hours' sailing.

Before the return of Mr. Chester and the arrival of Inspector Smith, the United States steamship *Congress*, under the command of Captain Davenport, came into port, relieving me of a mountain load of anxiety. This eventful day was August 10. I have omitted to state that, before leaving Holsteinborg, I wrote out full particulars of my purposes and plans for information of the commander of the transport on her arrival there, and left the same in charge of Governor Elberg, with instructions that the same should be dispatched by boat to the transport the moment she should appear in the offing, that the delay in making harbor at Holsteinborg might be obviated. Captain Davenport, in his zeal and great good judgment, had decided, however, to crowd on all sail direct from St. Johns, Newfoundland, to Goodhavn (this port). I had made up my mind, and had so stated it in my letter left at Holsteinborg for the commander of the transport, to prolong our stop at this port to August 15, and not later, for the transport; in the mean time to obtain coal, as indicated in my letters to you from Holsteinborg.

On the arrival of the inspector of North Greenland, Karrup Smith, he signified at once his readiness to extend to the North Polar expedition all the aid and co-operation as desired by your letters to him. The large Government store-house he has thrown open for the stores and provisions to be left here on deposit for the future use of the North Polar expedition, and take the same in trust, promising to have all possible care taken of them by his people till called for. In proof of the spirit with which Inspector Smith enters into this expedition, in behalf of the Danish Government he represents, all compensation is refused for the use of said store-house and for the trust referred to.

There are two persons in the employ of the Danish authority at Upernavik, the same being within the district of Inspector Smith, that I consider will be of great service to the expedition I have the honor to command, and therefore, after mature consideration, I have concluded that I shall meet your approval in trying to secure their services. The one is Hans Christian, the dog-driver and hunter of Dr. Kane's expedition of 1853–'55, and the other a Dane, by the name of Jensen, formerly of Dr. Hayes's expedition. Hon. Inspector Smith, although short of help in Government service, assures me that every possible exertion will be made by him, and by those in authority under him, to supply these men to this expedition. * * *

At meridian, the anchor of the *Polaris* will be weighed, when her prow will be turned to the north.

In two days we ought to be, and shall most likely be, in the port of Upernavik; on concluding business there, which will occupy two or three days, the only work remaining to be done will be to push on directly for the North Pole.

The season being so far advanced, my plan and purpose is now, that on arriving at the cape, your orders direct me to make Cape Dudley Digges, to steam directly for Smith Sound, and thence make all possible attempts to find passage on the *west* side of the sound from Cape Isabella, up to Kennedy Channel, and thence on and up to the very pole itself. It is advisable that a deposit of provisions and ammunition should be made from the *Polaris* on some island near Cape Alexander, at the entrance east side of Smith Sound, for our preservation in case she should become ice-wrecked in the desperate battle she is about to engage, and for which she has been so thoroughly prepared by a most thoughtful and liberal Government. This is a most extraordinary open season for Arctic navigation, as reported by all who have already tried it, therefore your honor may rest assured that this expedition will improve the opportunity to its fullest extent.

I close this hasty dispatch by acknowledging the reception of your letter of June 18, inclosing the appointment of Mr. Geo. E. Tyson as assistant navigator, also that of Mr. Odell as assistant engineer of the *Polaris*, which appointments I have delivered to these parties, giving them great satisfaction and encouragement. I have the honor to be, yours, respectfully,

C. F. HALL,
Commanding the United States Polar Expedition.

Hon. GEO. M. ROBESON, Secretary of the Navy, Washington, D. C.

NOTE.—I ought not to close this dispatch without expressing to *the President, to you, and to my God*, my heart's gratitude for the perfect manner in which every thing has been carried out by your Department to insure success in the object had in view by the Congress of the United States. Your obedient servant, C. F. HALL.

Hon. GEO. M. ROBESON, Secretary of the Navy.

United States Steamship *Polaris*, Upernavik, Greenland, August 20, 1871,
North lat. 72° 46'; west long., 56° 02'.

SIR,—I have the honor to report the progress of the United States steamship *Polaris* since August 17, 1871, the date of my last dispatches, which were to be delivered to you by Captain Davenport, commanding the United States steamship *Congress*.

I heaved anchor and left the harbor of Goodhavn at 2 P.M., August 17, after an affectionate farewell from the commander and officers of the *Congress*, who, by your order, have given me all possible assistance in carrying out the final arrangements of the expedition.

At the moment of starting, Hon. Karrup Smith paid me a short visit, bidding farewell, and intrusting to my care a package to the address of Dr. Rudolph, governor of Upernavik, whom he acquainted with the character of the expedition, and instructed in regard to my wants.

Hon. Karrup Smith also informed me he had learned from a Danish vessel at anchor outside the harbor, that Dr. Rudolph was about to leave Upernavik and return to Denmark on a Danish vessel loaded with blubber and skins, accumulated at Upernavik during the preceding year. Knowing my principal object in proceeding to Upernavik, which is that of procuring skins and securing the services of Hans Christian and Jensen, he advised me to make all possible speed during the voyage to Upernavik in order to have an interview with Governor Rudolph, in whose willingness and capability of complying with my wishes he placed the fullest confidence.

Traversing a distance of two hundred and twenty-five miles in thirty-three and a half hours, we arrived at 11.30 P.M., August 18, at Upernavik, where, after experiencing some difficulty in arousing the inhabitants from their deep slumbers, Mr. Elberg, the new governor of the settlement and son of the well-known governor of Holsteinborg, immediately arrived on board the vessel with the hoped-for tidings of Dr. Rudolph's presence at Upernavik. To him I intrusted the letters of Hon. Karrup Smith to Dr. Rudolph, who soon afterward gave me the pleasure of his amiable acquaintance in company with Governor Elberg.

Most readily and cordially Governor Rudolph complied with all my wishes, and agreed to dispatch two kyacks (one-man boats of the natives), with letters to Hans Christian, of Proven, about fifty miles to the southward, and to Jensen, of Tossac, the same distance to the northward, informing them of my desire of their services, and to secure their readiness of leaving their respective residences whenever one of our boats should call on Hans, and the steamship on Jensen.

Governor Rudolph also proposed that Governor Elberg should accompany me on our voyage to Tossac, in order to assist me in securing the services of Jensen, procuring dogs, dog and seal skins, and whatever might be required, and also to enable Jensen, who holds the position of governor of Tossac, to leave his post by transferring the public property and accounts to some other suitable person.

I have also, by the courtesy of Dr. Rudolph, been enabled to procure some dogs, dog and seal skins, as also a small addition of coal.

At noon of August 19, I dispatched a boat and boat's crew, under command of Mr. Chester, the first mate, with orders to proceed to Proven and return with the utmost speed with Hans Christian and family, when I at once shall heave anchor and start for Tossac.

P.S.—The Danish brig, which is to return to Denmark with Governor Rudolph, and also to carry our mail, starts at midnight, and I therefore am compelled to close my report; but I may remark that Governor Rudolph, who has been in this portion of the country for a period of over thirty years, thinks this year to be more favorable for any northern voyage than any year gone or to come.

Mr. Chester has returned with Hans Christian and family, traversing a distance of one hundred miles in a remarkably short time, considering that he had to depend upon oars, the wind being most unfavorable during the entire time. He reached Proven at 11 P.M., August 19, and started on his return at 8 A.M., August 20, arriving here at 8.30 P.M.

My intention to start on the arrival of Mr. Chester I am compelled to abandon, as Governor Elberg, who will kindly assist me in securing the services of Jensen, and procuring an additional supply of furs, can not leave this place before 12 o'clock to-morrow, and I therefore have set this time for our departure. Very respectfully, your obedient servant,

C. F. HALL,
Commanding United States Naval Polar Expedition.

Hon. GEO. M. ROBESON, Secretary of the Navy.

August 21, 12.30 A.M. Just received news from Tossac that Jensen is willing to accompany the expedition.

C. F. HALL.

Instructions to CAPTAIN DAVENPORT.

Navy Department, July 18, 1871.

SIR,—As soon as the United States steamship *Congress*, under your command, is ready, and has received on board the stores for the use of the expedition toward the North Pole, you will proceed with her, with all dispatch, by such route as you may find most expeditious, to the coast of Greenland.

If you can make, without any extraordinary risk, the port of Lively, in the island of Disco, you will proceed to that place and deliver to the steamship *Polaris* the stores you have for the use of the expedition, taking the receipt of Captain Hall for the same, if you find the *Polaris* there, or if she arrives within a reasonable time. Should she not arrive at Lively, however, within a reasonable time after your arrival there, you will land your supplies for her, and see them stored, if possible, under the direction of the authorities, to be delivered to Captain Hall on his arrival.

If you find that you can not safely enter the port of Lively, in Disco, you will make the best of your way to Holsteinborg, and there deliver your stores to the *Polaris*, or leave them for her in like manner as above directed, after waiting for her there for a reasonable time. If you find the *Polaris* at Holsteinborg, and learn that the *Congress* can enter Lively, proceed in company to that port; or if you should at Holsteinborg, or otherwise, learn that she has proceeded to Lively, you will endeavor to reach her there, as it is the earnest wish of the Department that the *Polaris* receive on board her stores at that port, if possible. In case of the happening of any contingency not contemplated by these instructions, the Department must leave the course to be pursued to your judgment, with the understanding and direction that you are sent for the purpose, if practicable, of actually communicating and leaving the supplies with the *Polaris* at Lively, and if not practicable there, then at Holsteinborg; and that it is of the greatest importance that she should receive the supplies, and at a port as near Lively as possible; and that if, for any reason, you should not be able to communicate with her or hear of her before the season makes it necessary for you to leave, you are to leave the supplies for her, as above directed, at Lively, if possible, and if not, at Holsteinborg.

In any event, it may be well, if you can do so without too much delay, to endeavor, on your way to Lively, to communicate with Holsteinborg, and ascertain the whereabouts of the *Polaris*. If she has not arrived there, you can leave word of your departure to Disco, and instruct Captain Hall to follow you without delay.

I inclose, for your information and guidance, a pamphlet which contains the orders issued by the Department to Captain Hall. I also inclose a letter to the Danish authorities at Lively or Holsteinborg, requesting them to render you assistance and facilities in prosecuting your duties under these orders. * * * Very respectfully, etc.,

GEO. M. ROBESON,
Secretary of the Navy.

Captain H. K. DAVENPORT,
Commanding United States Steamship *Congress*, New York.

The Secretary of the Navy to the Danish Authorities.

Navy Department, July 18, 1871.

SIR,—The United States steamship *Congress*, under the command of Captain H. K. Davenport, of the United States Navy, visits the port within your jurisdiction for the purpose of delivering supplies to the expedition toward the North Pole, under the charge of Charles F. Hall.

In the event of the *Congress* not meeting the *Polaris* on her arrival, and of her being compelled to return to the United States before the vessels can communicate, I have given directions that the supplies be stored until Captain Hall arrives in the port to receive them.

I have the honor to request that you will afford to the commander of the *Congress* such assistance and facilities as may be in your power and consistent with your duty toward your Government to enable him to obey the orders from this Department. Accept the assurance that any and all help you may be able to render Captain Davenport will be appreciated and acknowledged by the Government of the United States.

Very respectfully, your obedient servant, GEO. M. ROBESON,
Secretary of the Navy.

His DANISH MAJESTY'S GOVERNOR, at Lively, or Holsteinborg,
Coast of Greenland.

Letters from Captain H. K. DAVENPORT *to the Secretary of the Navy.*

United States Steamship *Congress* (second rate),
St. Johns, Newfoundland, August 2, 1871.

SIR,—I sent you a telegram announcing my arrival at this port on the 1st inst.

I made the passage from New York in five days and eighteen hours, under steam and sail. The weather was light, dense fog prevailing most of the way, so much so that I got observations of the sun but on three occasions. I have filled up with coal, and hope to do as well the remainder of the voyage. * * *

I expect to return to this port about the first week of September; but if I am not "up to time" there need be no cause for anxiety, as the winds may be unfavorable for a quick passage, in addition to detention at Disco.

I was surprised to learn that the *Polaris* left here only six days prior to my arrival. She will not be many days ahead of me in arriving at Disco, unless some unforeseen occurrence impedes my progress.

The first icebergs which we have seen were off this port. There were about fifteen in sight at one time; something of a premonition of what we are to expect. I shall telegraph you immediately on my return.

As the mail leaves here but once a fortnight, the consul will telegraph you when we get off.

Very respectfully, your obedient servant, H. K. DAVENPORT,
Captain, Commanding *Congress*.

Hon. Geo. M. Robeson, Secretary of the Navy, Washington, D. C.

United States Steamship *Congress* (second rate). At sea, August 21, 1871,
Lat. 63° 48' N. long., 58° 3' W.

SIR,—I inclose copy of a correspondence between his Danish Majesty's inspector at Goodhavn (Port Lively) and myself, which speaks for itself.

If what I said meets with the approval of the Department, I trust that the recognition of the civilities extended to the *Congress* and the *Polaris* will not be lost sight of by the Government.

Very respectfully, your obedient servant, H. K. DAVENPORT,
Captain, Commanding *Congress*.

Hon. Geo. M. Robeson, Secretary of the Navy, Washington, D. C.

Goodhavn, August 12, 1871.

DEAR SIR,—I intended this forenoon, accompanied by Mrs. Smith, to pay a visit to you and to Captain Davenport, but feeling myself not quite well, and adding to this the bad weather, I beg you to excuse me for not paying my respects to-day. You would oblige me very much by giving Hon. Captain Davenport my compliments, and pray him to excuse this delay.

At home I am at your service, if you should wish any information that I am able to give. I beg you, sir, to give my thanks to Dr. Bessel for the great care he takes of our sick people.

The store-house of the opposite shore is open to you—when you address to Mr. Lossen—whenever you wish to commence removing the provisions; but it is necessary to place lowest such things that are not damaged by moisture.

I beg you be so kind as to tell me at what time the service of the Sunday will be over, as I hope then to have the honor of visiting the *Polaris* and the *Congress*. I am, dear sir, very respectfully yours,

KARRUP SMITH.

Hon. Captain C. F. HALL, United States Steamship *Polaris*.

Goodhavn, August 13, 1871.

Sir,—It was a great honor and satisfaction to me, by your letter of yesterday, to receive your acknowledgment of our good will concerning the ships of the United States, *Congress* and *Polaris*.

By this I assure that every effort shall be made to keep the deposit for *Polaris* instrusted to our care as safe as our own stores.

I am, sir, with the highest respect, your obedient servant,

KARRUP SMITH,
Inspector in North Greenland.

Hon. Captain DAVENPORT, etc., etc., etc.,
Commanding United States Steamship *Congress*.

United States Steamship *Congress* (second rate),
Goodhavn, Disco, August 7, 1871.

Sir,—On the eve of my departure for the United States, I embrace the occasion to express my warmest acknowledgments for the kindness and courtesy which you have been good enough to extend to myself and those under my command, as well as to Captain Hall and the expedition which he has in charge; but more than all, for the promptness with which you placed the Government store-house at my disposition for the stores of the *Polaris*.

I shall not fail to bring to the attention of the Government these manifestations of good will which have characterized His Danish Majesty's representative in this part of his dominions, and I am sure that I but anticipate the recognition of these civilities by the Secretary of the Navy of the United States, through our minister at Copenhagen.

In bidding you adieu, I beg leave to subscribe myself, with high respect, your obedient servant,

H. K. DAVENPORT,
Captain United States Navy, Commanding *Congress*.

His Danish Majesty's Inspector,
KARRUP SMITH, Esq., etc., etc., etc., Goodhavn.

United States Steamship *Congress* (second rate),
St. Johns, Newfoundland, August 28, 1871.

Sir,—I beg leave to submit the following detailed report of my proceedings since leaving this port on the 3d of August.

I left St. Johns early on the morning of that date and proceeded, under steam or sail, or both, as circumstances warranted, direct for Goodhavn* (Port Lively), in the island of Disco.

Nothing of special interest occurred during the passage. We passed numerous icebergs, but, fortunately, fell in with no "pack-ice," the season, so I am informed, being unusually open.

I arrived at Goodhavn on the 10th of August, seven days from St. Johns, and fifteen from New York, including two days' stoppage here, coaling ship—less than thirteen running days.

On approaching the harbor, a boat was observed coming out, on board of which was Captain Hall, of the *Polaris*. He came on board, and accom-

---

* Goodhavn is the proper name of the port. The Danes do not call it *Port Lively*.

panied me in to the anchorage. His delight at seeing the *Congress* was unspeakable; but it fell short of his astonishment at the rapidity of my passage, and the dispatch of the Department in forwarding the supplies for his expedition, particularly as he himself had arrived at Goodhavn but six days before me. * * *

As soon as I anchored, I requested Captain Hall to call upon the *Inspector of North Greenland*—this is the proper title of the chief functionary of the Danish Government in that part of the world, the title of *governor* referring to the local magistrate of the settlement—and inform him that I would call the next day to pay my respects to him, etc., etc.

Accordingly, the next morning, the 11th, at 11 o'clock, accompanied by Captain Hall, I landed, in full uniform. As soon as I touched the shore, I was saluted by a battery of six-pounders, and was met by the governor, or local magistrate. The inspector received me at his door, with every mark of official and personal respect and consideration. Upon entering his house we were presented to his wife, who was our interpreter throughout our visit, although her husband, to a limited degree, understood and spoke English. After being seated and the usual phrases of common politeness had been interchanged, I presented him with your letter, and took occasion to make a few remarks touching the object of my visit; the interest of our Government and people in the expedition of Captain Hall; and assured him that any attention, civility, or courtesy, extended to Captain Hall or his associates, by the Danish authorities throughout Greenland, would be heartily appreciated by the President of the United States, and would be duly acknowledged by our Government. I did not say much, but I endeavored to "speak to the point"—Hall's expedition.

Inspector Smith, after expressing his pleasure at seeing me, replied that he would do all in his power to further the views of our Government; that he would assist Captain Hall in every possible way in his power, and that he was sure the authorities throughout Greenland would, with pleasure, give him a helping hand; and added that the Government store-house was at my immediate disposal, for the stores which I had on board for the *Polaris*. * * *

My visit was agreeable in every respect, and I left with the assurance on the part of the inspector that every thing which we wanted would be granted, in so far as their means would admit, with promptness as well as pleasure.

On the 12th—the day after the above—I hauled the *Polaris* alongside of the *Congress* and put on board of her all the coal, provisions, and stores she could carry. After disposing of her, I landed the residue, putting the provisions and stores in the Government store-house; and built a "crib" outside, into which I stored the coal.

On the 17th, the *Polaris* being ready for sea, I went on board, made a few remarks to the officers and crew, at the request of Captain Hall, and bade them farewell. She got under way at 2 o'clock P.M., and, amidst the cheers of my ship's company, took her departure for the Polar regions. The day was beautifully bright, and the temperature that of a May morn in the latitude of Washington.

On the 19th of August, at 7 o'clock A.M., I got under way, and left Goodhavn for this port. Throughout my stay in that hospitable though hyperborean region, every thing was done by the authorities which courtesy or politeness could indicate. * * *

Inspector Smith kindly volunteered and accompanied a party from the ship to visit a glacier, some nine miles from the anchorage. We had to go four miles in a boat, and proceeded thence on foot, five miles, over the worst imaginable route, through bogs, over hills, mountains, and down rocky valleys, enough to appall one who had never seen such a country; and all this, too, with a knowledge of what he would have to encounter, having been over the route before. He led the way, as our guide, with a degree of modesty and perseverance which won not only our admiration and astonishment, but also our warmest regard, respect, and esteem.

On the 17th I had the honor of entertaining the inspector and wife at dinner in my cabin, and tried as best I could to reciprocate, in my feeble way, the civilities of which I had been the recipient. They had previously afforded me the pleasure of their company to a luncheon, upon which a number of the officers of the ship were good enough to assist me in entertaining them.

The Department will, I trust, overlook the prolixity of this letter, when it is remembered that it is a "detailed report," which I hope may not be without interest. * * *

I am happy to say that all were in good spirits, and perfect harmony prevailed on board of the *Polaris* when she left Goodhavn; and I may add that the presence of the *Congress*—there may be something in her name—had a charmingly beneficial influence upon the entire ship's company of that little craft, in addition to the impression which she made upon the inhabitants of Goodhavn. She is the largest ship which has ever entered that port, either man-of-war or merchantman, and the largest they had ever seen, and her dimensions filled with admiration and amazement all who visited her, from the chief functionary to the most humble Esquimau.

When I left Goodhavn, I had but three and a half days' coal on board, and for the first twenty-four hours after leaving carried both steam and sail to drive me through the narrow part of Davis Strait, where I stopped the engines, put out fires, and made the rest of my way under sail alone, with variable winds, calms, and gales, until I got within two hundred miles of St. Johns, when I lighted fires and ran into port, arriving here at 3 o'clock P.M.

In conclusion, I beg leave to say that I have carried out the views and instructions of the Department, with an honest zeal, to the best of my ability and understanding, and with all the dispatch of which I was capable.

Very respectfully, your obedient servant, H. K. DAVENPORT,
Captain, Commanding *Congress.*

Hon. GEO. M. ROBESON, Secretary of the Navy, Washington, D. C.

NOTE.—I omitted to mention, in its proper place, that the salute fired in my honor, on landing, was duly returned, and also that the inspector was received with proper military honors when he visited the ship, and a salute of fifteen guns was fired on his leaving. H. K. D.

APPENDIX.

Prayer at the North Pole.

Written for the use of the North Polar Expedition, by Rev. Dr. Newman, of Washington, to be used only on reaching the Pole.

Great God of the universe! our hearts are full of joy and gladness for all Thy marvelous goodness unto us. We have seen Thy wonders upon the deep, and amidst the everlasting hills of ice, and now we behold the glory of Thy power in this place so long secluded from the gaze of civilized man. Unto Thee, who stretchest out the north over the empty place, and hangest the earth upon nothing; who hath compassed the waters with bounds until day and night come to an end; we give Thee thanks for what our eyes now behold, and for what our hearts now feel.

Glory be to God on high, and on earth peace, good will toward men! We praise Thee; we bless Thee; we worship Thee; we glorify Thee; we give Thee thanks for Thy great glory, O Lord God, our heavenly King! God the Father Almighty! Praise Him all ye His works. Praise Him sun, moon, and stars of light. Praise Him ye heaven of heavens, and ye waters that be above the heavens. Praise the Lord from the earth, ye dragons and all deeps, fire and hail, snow and vapor, stormy winds fulfilling His word; praise Him frost and cold, snow and ice, day and night, summer and winter, seas and floods. Praise Him all ye rulers and peoples of the earth. Let every thing that hath breath praise the Lord. Glory be to the Father, and to the Son, and to the Holy Ghost, as it was in the beginning, is now, and ever shall be, world without end.

In Thy name, O Lord, we consecrate this portion of our globe to liberty, education, and religion, and may future generations reap the advantage of our discoveries. Bless the nation that sent us forth; bless the President of our great republic; bless all the people of our favored land, whose national banner we now wave over this distant country.

And now may the God of our fathers guide and direct our returning footsteps to those who wait to greet us with joy in the homes and land we love. May no evil befall us; no sin stain our souls; no error lead us astray from Thee and duty. Hear us for the sake of Him who hath taught us to pray: Our Father who art in Heaven, hallowed be Thy name; Thy kingdom come; Thy will be done on earth as it is in heaven; give us this day our daily bread; forgive us our trespasses as we forgive them who trespass against us; lead us not into temptation, but deliver us from evil; for Thine is the kingdom, the power, and the glory forever. Amen!

Extract of Letter from J. CARSON BREVOORT, *Esq., to the Editor.*

"Brooklyn, December 20, 1873.

* * * "The *Polaris* expedition was an official one, and I had nothing to do with it beyond lending to Captain Hall a few books, and consulting with him about his proposed line of search. I do not know whether the 'Blue-books' were saved or not. * * * Kane had a lot of my books with him in the *Advance*, which he abandoned in Rensselaer Harbor; Hayes had some, which have been returned. Hall had a copy of 'Luke Fox's Voyage of 1635,' and it bears an indorsement as follows:

"'This book belongs to my friend, J. Carson Brevoort.

"'To-morrow, March 31, myself and native party, consisting of 13 souls, start on my sledge-journey to King William Land.

"'C. F. HALL,
"'29th (Snow-house) Enc't, near Fort Hope, Repulse Bay, Lat. 66° 32′ N., long. 86° 56′ W.

"'Friday, March 30, 1866.'

"I value this highly, as you may suppose. Hall brought me a pair of walrus-teeth, from King William Land, and a musk-ox skin and horns. His object, as you perhaps know, on this trip was to discover the records of the Franklin expedition of 1845 by passing a summer on King William Land. But his plans were frustrated by the hostility of the natives, and the timidity of some of his own party.

"Soon after his return he announced to me his determination to induce the Government to send out a Polar expedition, and he pushed the scheme, like another Columbus, untiringly and patiently. He was the only real and genuine leader of the *Polaris* expedition. Had all the others been, like himself, enthusiastic, and impressed with the like ambitious purpose, the *Polaris* would have wintered near the pole itself, and have *come out near Spitzbergen or Behring Strait*. He was a prudent, far-seeing leader withal, counting on having to meet all possible obstacles from natural causes, but had left out of his calculations the opposition he might meet from those less zealous than himself. I warned him on this last point, and cautioned him about his companions. If you have read the story of Columbus and of the Pizarros, or of Sebastian Cabot and Captain Sperts, you will understand what he must have felt when forced to turn away from the threshold of the Polar basin.

"Hall never put his full thoughts on paper, and his letters are more of a business than of a scientific nature. He preferred talking over his plans with those who understood him. His nature was gentle, kind, and patient. His companions, Ebierbing and Tookilooto (Joe and Hannah) now at Groton, near New London, can bear testimony to his uniform gentleness and sympathetic, unselfish disposition. His own published book is full of outpourings of this description, and further testifies to his deep and fervent Christian convictions.

"I could write much more about Hall, whom I admired, respected, and loved, both as a leader and as a man, but I must close.

"Any thing that I have here said you may use; but I feel that it is not much to the purpose. Yours respectfully,

"J. CARSON BREVOORT."

To E. VALE BLAKE.

Carbondale, Penn., December 22, 1873.

Your letter was received quite recently, having laid a long time in the Secretary's office, and this will excuse my not answering it before.

In answer to your question respecting the "lookout kept up for those separated from the ship on the eve of the 15th of October, 1872," I will state the facts as far as they came under my observation. As soon as it was light enough to see, Mr. Chester was sent to the mast-head with a glass to look for our comrades. He reported that he saw on a piece of ice what appeared like barrels and boxes of provisions, but could not see any thing that looked like men. I believe that Henry Hobby also went to the mast-head with like success. We were not at all surprised when we heard that they could not be seen. * * * During the winter I greatly regretted that I did not go up to the mast-head myself, but I never had an idea that I would have seen them. * * * We might have interested the natives in their behalf, however. Give my regards to Captain Tyson, whom I hope soon to see. Respectfully yours,

R. W. D. BRYAN.

OFFICIAL EXAMINATION OF THE POLARIS SURVIVORS.

Washington, October 11, 1873,
On Board United States Steamship *Tallapoosa*.

Present, Hon. Geo. M. Robeson, Secretary of the Navy; Commodore Reynolds, and Captain Howgate, of the Signal Service. The principal queries, addressed to all the parties examined, related—1st, to the possibility of getting farther north than 82° 16'; 2d, to the circumstances of Captain Hall's death; 3d, to the abandonment of the *Polaris;* 4th, to the kind and amount of effort made by the party on shore to get sight of their late companions left adrift on the ice-floe; and, 5th, as to special facts of interest, scientific or otherwise, observed by the witnesses.

The first examined was Captain S. O. Buddington; he testified to the effect that he did not think the *Polaris* could have made any farther northing than she did; that he "did not see any chance to get north" of lat. 82° 16'; also, that it was impossible to keep the *Polaris* afloat after the storm on the 15th of October, 1872. The journal of events which he presented to the Board of Inquiry was written for him by Mr. Mauch. "No formal survey of the ship was held" to decide on her sea-worthiness after they got ashore; that the "original log-books were buried at Life-boat Cove," and only copies brought home. In regard to Captain Hall's death, he thought it resulted from natural causes. As for the ice-floe party, he said that Mr. Chester was up aloft "nearly the whole time, from the time we started in until we got ashore," and that he sent Mr. Hobby up; that if the ice-floe party had been seen, he could not have reached them. He "did not think there was any refraction" on October 16. In regard to the charges which had been made against him by members of the ice-floe party, he admitted nearly every thing except as to his unwillingness to go north, which, in his opinion was impossible; he admits, however, saying to Noah Hays, in regard to some carpenter's shavings, "They will do for the devilish fools on the sledge-journey." This he says Captain Hall, to whom it referred, overheard. He also admitted the occurrence of a difficulty with Captain Hall at Disco; also breaking open a locker, which brought on himself a reprimand from Captain Hall; also that Captain Hall had written a letter addressed to him containing strictures upon his conduct, which letter Captain Hall subsequently burned; also admitted the general charge of drinking, and the specific charge of taking surreptitiously Dr. Bessel's alcohol, and of an altercation with the doctor about it; did not think he was at any time incapacitated for duty in consequence. Speaking of Captain Tyson, he said that the latter was in the habit " of complaining bitterly about the management generally," but they had no trouble to speak of. Captain Buddington stated distinctly that " he never left the ship" to travel, or otherwise; did not discover that any tidal wave came from the Pacific; thought he saw land to the north-east of Repulse Harbor; saw land to the west, above Cape Union; thinks the North Pole might be reached by the route taken by the *Polaris;* did not agree very well with Dr. Bessel. He also stated that some wheat, accidentally spilled on the ground, took root on shore of Polaris Bay, and grew to the height of two or three inches.

Mr. H. C. CHESTER, first mate, stated that at the highest point the *Polaris* reached he could see a dense water-cloud to the north; that when the consultation of officers was held (August 30, 1871), near lat. 82° 16' N., the man aloft sung out that there was a lead close to the east shore; thinks the vessel could have gone on; that both he and Tyson gave that as their opinion; thought "if some one else had been sailing-master the ship would have gone farther north. If he had had command, he would have tried it." On his sledge-journey with Captain Hall, in the early part of October, 1871, they staid two days at Cape Brevoort; also stated that at that point the sledge-journey ceased, but that he and Captain Hall walked northward for eight and a half hours to the highlands at Repulse Harbor [Captain Hall does not mention this in his last dispatch to the Secretary of the Navy, dated at Cape Brevoort]; that from these heights he saw a cape on the north-west coast, extending sixty miles north, and also land trending off to the east of Robeson Channel; it was a clear day; "a dense water-cloud that extended round in a sort of semicircle, some parts lighter than others," was seen to the north. Captain Hall's health was "first-rate" while on this journey; thought Hall's death natural. After separation of the party, " did not know of any one being at the mast-head to look out for their late companions but himself," but thought one other man went up when he was not there. In the course of the day, October 16, thought he was up altogether an hour and a half; was up and down the mast-head every ten or fifteen minutes, until they neared the land. Saw a piece of floe with provisions on it, but no men. He copied the log-books, and re-copied his first copy into a smaller form, because the large books were too heavy to take in the boat. Thinks Captain Buddington "a good whaling-captain, but that he has not enthusiasm for the North Pole;" he at times " depreciated Captain Hall, using improper language among the seamen on the main deck; he did not speak respectfully of the commander or the expedition," being sober at the time. "His" (Captain Buddington's) "idea was, that the enterprise was all d—d nonsense; thought the scientific work was all nonsense, too; he regarded the whole thing as foolishness. Mr. Chester could not decide in his own mind which way the tide came from —whether from the north or south.

Mr. MORTON, second mate, thought it was dangerous to attempt to get north after coming to anchor in Polaris Bay. Watched with Captain Hall during his sickness; confirms the statement concerning Captain Hall's suspicions of foul play; thinks he was delirious at the time; says Dr. Bessel was kind to him; thinks that after the storm of November 27, 1871, the *Polaris* might have been prevented from resting on the spur of the berg which so wrenched and strained her. After Hall's death, resigned his charge of the provisions; "found it would be an unpleasant situation;" says Captain Buddington "had not firmness enough to send an *order* to Chester when the latter was at Newman Bay, but he sent a *request*." At the time of the separation of the party, thought the ice was safer than the ship. Saturday, June 1, 1873, was the day agreed upon for leaving Life-boat Cove in the boats, but they were prevented by a gale, which continued over Sunday. They started the 3d, and reached Northumberland Island at midnight of that day; "thought some people were not very sorry at Hall's death;" heard Captain Buddington "mutter disrespectfully about him," and use a "good many careless expressions;" thinks the lookout on the 16th of October might have amounted altogether "to an hour during the day;" thinks the tide at Robeson Channel comes from the south, and that the climate is milder at Polaris Bay than at Kane's winter-quarters; less snow at the higher latitude. He "found grass in patches as high as your ankle."

Washington, October 16, 1873.

By invitation, Surgeon-general Barnes, of the United States Army, and Surgeon-general Beal, of the United States Navy, were present to listen to such statements of Dr. Bessel's as related to the sickness and death of Captain Hall.

Dr. EMIL BESSEL: At Disco, at the time of the difficulty between Captain Hall and Meyers, "I told him" (Captain Hall) "I preferred to go on shore myself if Mr. Meyers was dismissed." On the day we reached lat. 82° 16', Captain Buddington showed me a dark cloud hanging quite low over the horizon at a pretty good distance to the north, ahead of us; could get some glimpse of land north. I placed it in lat. 84° 40' N. "At the consultation among the officers, Messrs. Chester, Tyson, and Morton suggested going ahead;" he (Dr. Bessel) did the same, only suggested going over to the west coast. Captain Buddington said he "did not see any chance to go farther; Captain Hall was very anxious to go on." Captain Tyson and one of the men from the mast-head reported they "saw plenty of open water;" it could not be seen from the deck. Of Hall's sickness he says:

Before I saw him, there had been vomiting; the cabin was very warm; it was about fifteen feet long and eight wide; there were eight berths in it; seven people slept there, including the captain; they all slept there during his sickness. [One of these was the colored cook.] His pulse was very irregular—from sixty to eighty—he all at once became comatose. I applied mustard poultices to his legs and breast, put blisters on his neck, and cold-water applications to his head; in twenty-five minutes he recovered consciousness; then found a condition of hemiplegia; the left side of the face, and the left arm and side, were paralyzed, also the muscles of the tongue, the point of which deflected to the left; gave him castor-oil, and three or four drops of croton-oil; he slept during that night; complained next day of difficulty of swallowing, and numbness of the tongue; part of the time he could not speak distinctly; gave more castor and croton oil, and he recovered pretty well from the paralysis.

Oct. 26. He was restless, with but little appetite; ate, I think, some preserved food, peaches or pine-apple; complained of chilliness; the temperature of his body varied from 83° to 111°; at this time his mind was unaffected; gave him hypodermic (under the skin) injection of one and a half grains of quinine, to see the effect; better in the evening; took some arrow-root and soup.

Oct. 27. Appetite improved, but still numbness of the tongue.

Oct. 28. First symptoms of mind wandering; thought some one was going to shoot him; he accused every body; thought he saw blue gas coming out of persons' mouths; would not take clean stockings from Chester; thought they were poisoned; appeared to trust Joe and Hannah most; made one and another taste the food, even that which came out of the sealed cans opened in his cabin; he continued thus till Saturday, November 4; he would not let me see him from October 29 till November 4; gave him no treatment during that time; he took pills and medicines of his own; he asked me for pills, and to satisfy him I made some of bread, and Hannah gave them to him; partial paralysis of tongue continued; in showing his tongue it was always deflected toward the left.

Nov. 5. I bathed his feet with warm mustard water; tried to do so on the 6th; he thought I was going to poison him with the bath.

Nov. 7. 1 A.M. He jumped out of bed, asked for Captain Buddington and Hannah; Mr. Hays called me from the observatory; when I came, he asked for water, and drank some.; I found the pupil of his left eye dilated, that of the right contracted; he went back to bed; said he felt worse, and spoke with more difficulty; he then became comatose; you could hear gurgling, or râle, in his throat; tried with a pin if sensation remained; there was some on the right side, but none on the left; finally there were reflectory, or spasmodic, motions of the muscles of the left side, and occasionally on the right; this was November 8; he died at 3.25 A.M. of that day. Thinks the tides at Thank God Harbor came from the North Pacific; thinks Captain Tyson could not have traveled overland in sledges, because there was not snow enough; the ground was too bare. [See testimony of the carpenter, Mr. Coffin, as to wheels.] During his stay at Newman Bay (with the

Tyson boat-party) there was not water enough to float the boat; the ice kept pouring down the whole time through Robeson Channel in small pieces, hummocks, and at last heavy fields. Robeson Channel was not frozen during the whole winter, except a few days in March. The ice poured down from north to south, with two exceptions. On one occasion it went north for thirty minutes, and on another for fifteen minutes; he thinks, from geological features, that Greenland has been split off from the west land at some period; thinks that the observations made by the *Polaris* party prove the insularity of Greenland. [No surveys were made far enough to the north-east to prove that.] All the geological specimens, the skins, the skeletons of musk-cattle, and most of the photographic apparatus, were lost, with many instruments of different kinds; saved some instruments, and part of the records of the scientific work performed; the wide-bore thermometers, supplied by the United States Signal Corps, indicated correctly temperature to $-40°$; the Cassella (English), narrow-bore, would stop at $-35°$. Could not account for the fact that the ice-floe party were not to be seen, though the latter could distinguish the smoke-stack of the *Polaris*. In the spring of 1873 attempted to go overland to Thank God Harbor; did not succeed, because neither natives nor seamen would accompany him; went to "Brother John Glacier," and staid four days, making observations on rate of progress, limits of *nevé*, etc. After being picked up by the *Ravenscraig* on June 23, remained on board of that vessel, which was beset in the ice, until July 4; then Captain Allen crossed to the westward near Lancaster Sound; on the 7th of July was transferred, with others of the party, to the Scotch whaler *Arctic*, Captain Adams; on board of this vessel was Captain Markham, R.N. The *Arctic* sailed down Prince Regent Inlet to Fury Beach; landed there, and examined the remains of the wreck of the British ship *Fury*, lost by Captain Parry in 1824; found a lot of the canned provisions supposed to have belonged to that vessel; tested them, and found them still good; also saw two English muskets, marked 1850. [These and the provisions were probably left by the *Prince Albert*, Captain Penny, who was at Fury Beach in 1851.] He, in company with Captain Markham, also went on a boat excursion to the south side of Creswell Bay; saw there thirty deserted huts made of the skulls of the Greenland whale; saw ninety-six skulls. During the appearance of the aurora borealis, in no instance could the least amount of electricity be detected; the record or journal he kept was miscarried in England, and he had not yet recovered it. He had kept no "sick-list" during the entire cruise, because no one was sick.

EMIL SCHUMAN, chief engineer: Heard Captain Hall thank Dr. Bessel for his kindness to him during his sickness; also heard him (Captain Hall) say to Captain Buddington, five or six days before he died, that, "in case he died, he (Captain Buddington) should go to the North Pole, and not come back till he had reached it;" and Captain Buddington had to promise that he would. After Captain Hall died, "Captain Buddington took charge of the papers," and read them, "and we all read them." I could not read Captain Hall's handwriting; I tried to do so. There was no disorder in the ship; there may have been in the forecastle, but I did not know it; I am so constituted that I would not have heard any, if it were to take place; I would go away; thought Captain Buddington did all he could; went up aloft once to look for the lost party; the machinery of the *Polaris* worked well.

H. HOBBY, seaman: During Captain Hall's illness, Mauch, the captain's clerk, came into the cabin and told the chief engineer and myself "that there had been some poisoning round there," not meaning that Captain Hall had taken it, but "that the smell was in the cabin." Captain Buddington told me Captain Hall was dead, and said, "We are all right now." I said, "How do you mean by that?" He says, "You sha'n't be starved to death now." I told him "I never believed I would." At Disco we heard that Captain Buddington, Dr. Bessel, and Mr. Schuman were going to leave; some of the men said they were going to leave too. Captain Buddington told us in Washington, at the Navy Yard, "that in regard to all matters of eating we had to come to him. From Disco on, the rations were shortened. We spoke about it to Captain Buddington. We never got the thing made better till Captain Hall found it out himself. After he found it out, we then had abundance. About 12 o'clock (of October 16) went up to the topmast and looked for the party on the ice; could see nothing of them. In the spring Dr. Bessel wanted me to go to the North Pole with him on a sledge-journey, with fifty pounds of pork and sixty pounds of bread. I thought that a very foolish idea. He promised me two hundred dollars if I would go higher with him than Parry had been. After the *Polaris* was beached, we (some of the seamen) made fast only a single hawser; we were not told to do this, but we took the responsibility. If I had had any thing to say, I would have secured her properly at that time. On meeting the *Ravenscraig*, I and some of the men and Mr. Chester would have preferred to remain with the boats, feeling sure we could get to Disco with them. Captain Buddington and some others wanted to go aboard the vessel. When the ship went on Providence Berg in November, 1871, Captain Buddington said it was the safest place we could have. All of us said the contrary. I had never seen a ship setting on the ground the whole winter, and this was the same; she soon commenced to keel over. When she had set about a fortnight, then the captain thought it would not be a good plan to leave her there; if he had come to this conclusion before, she could have been gotten off in about an hour's time; we could have sawed her off; there were only two or three inches of ice on the port side; that is, where she got her break in the stern. I went up twice to look for the separated party; there was no one looking from the mast-head about 4 P.M. [The time when the ice-floe party saw her tied up.] I think we could have gone farther north after we got into winter-quarters. The reason we did not go was because Captain Buddington said "it was not safe to go farther north;" this was spoken in the presence of every one. When the consultation was had, they all said they wanted to go north, with the exception of Captain Buddington. There was open water to the northward at that time. I know there were a couple of officers who were greatly relieved by his (Captain Hall's) death; the doctor was one of them; I think Captain Buddington was also; I never heard them say so; I could see it by their works; Mr. Meyers said "that now the officers would have something to say."

H. SIEMAN, seaman: The thing I did not like was there not being any sledge-journeys. I did not see Captain Hall during his sickness. I asked Captain Buddington for permission, but never had the privilege. I asked Dr. Bessel about Captain Hall, and he told me that he would not get over his sickness; this was before he got so very sick the second time. In my opinion, the expedition died with Captain Hall; was on shore with a telescope some miles behind Cape Brevoort. I think I saw land northward on the east and west sides. I heard a glacier discharge below our winter-quarters, where the place called Southern Fiord is.

A. A. ODELL, second engineer: We had a very good crew; every thing went on peaceably; do not think the *Polaris* was exactly of the right build for a ship to go north, but she was very strong. The machinery was in good condition. It was as much as we could do to get into shore at Lifeboat Cove.

N. J. COFFIN, carpenter: At Disco, when Captain Davenport came on board the *Polaris* and read the object of the voyage, Captain Hall stated in the cabin before all of us that he had been insulted by Dr. Bessel. This was the time he read off the duties of every man. When Captain Hall came back from his sledge-journey, *I had orders to make some wheels. I made three of them; then, at his death, I was ordered to discontinue them.* Captain Hall encountered a great deal of bare ground, and he wanted to go over that, when he could not use the sleds on account of there being no snow. He was calculating upon another journey right off. I saw him twice while he was sick. I had a piece of furniture to fix, and made that an excuse to go into the cabin; and Mr. Morton asked me once to go in and open a keg of tamarinds. I asked if it would do him any harm if I were to call in; they thought it not advisable to disturb him. I asked Hans (who was with him on his sledge-journey) what he thought of his sickness. Hans said that he traveled hard on the journey; and while they were building houses (snow-huts) Captain Hall did not do any work in the cold, and that did not do him any good. I heard Mr. Mauch talking with Hays; he (Mauch) was something of a chemist; he was telling Hays that the alcohol they burned had tartar-emetic in it, and that the fumes of it acted as a poison when burned. He said he thought that hurt Captain Hall. I asked him what he was saying, and he told me the same thing. He told me he thought it had a great deal of effect on Captain Hall's health.

I do not think there was a piece of the stem torn out below the six-foot water-mark. There was not, to my knowledge, and I made several surveys of the vessel. In regard to the ice-floe party, I had an idea—whether it was only imagination or not I do not know—but I thought I saw a large number of men on the piece of ice that was nearly like a berg, and a number sufficiently great to indicate that it was our party. I saw no provisions, or any thing else; they were near enough for me to take in the whole outline of them; they were on a piece of ice that was floating—moving with the current very rapidly; the time I thought I saw these men on the ice was just before dark. During the day we had looked, but did not see them. [This was no doubt a reality, and the time corresponds with that when all of the ice-floe party were collected together looking at the *Polaris* tied up.] Thought it was perhaps a mirage. I reported this fact to Mr. Chester, I think. I went in bathing at Polaris Bay; did not feel uncomfortable until about an hour after; a storm came up, and it became quite chilly before I got on board.

N. HAYS, rated as seaman, but employed as coal-passer in the fire-room: I was below a good deal, and did not know as much of what was going on as those did who were on deck. I presume no one knew less than I did. I had no chance to observe any thing. Seven hours I was on duty, to five off, while the vessel was under steam. I did not go on deck at all to do any duty; I do not remember any thing very distinctly; I kept a sort of a journal; I saw Captain Hall when he came back (from his sledge-journey); I had it in mind up to a little while ago that he said he had been unwell two or three days; but I found, on inquiry among the rest, that he had told them no such thing, and therefore I must be mistaken about that. After the separation, when morning came, I do not know that we looked for our comrades right away; I thought of them all the time, but our attention was drawn to the shore; I was that morning at the wheel; there was a marine-glass lying there; I took that, and scanned the horizon two or three times to the southward, to Littleton Island, and to the shore, but I could see nothing of our comrades; others looked also. Mr. Chester was at the mast-head once or twice, I believe, but he was on deck most of the time; no one was at the mast-head continually; Mr. Chester said he saw something on the ice; the general opinion seemed to be that it was "black-ice" (*i. e.*, a cavity in the ice which at a long distance looks black); I should think that was fifteen miles off; we could not have got to them. Dr. Bessel said he would make a confidant of me in regard to his enterprise of going up north. He was gone about a day and a half. He told us he had crossed the channel (Smith Sound), and had been a little over a degree above the position of the house (at Thank God Harbor), but I don't think it was possible by any means. He told Captain Buddington he was going inland to examine a glacier. The discipline was good under Hall; afterward I think the men did what they were ordered from principle, and not from what they considered necessity. One day I was over at the observatory with Dr. Bessel; I was there a good part of the time then, in the winter. He appeared to be very light-hearted, and said "that it" (Hall's death) "was the best thing that could happen for the expedition." I think those were the words he used. The next day he was laughing when he mentioned it. I was much hurt at the time, and told him "I wished he would select somebody else as an auditor if he had any such a thing to say." I never passed a day in the Arctic regions but what at some time of the day, in the shade, I observed salt-water ice making, though at the same time the sun was pouring down incessantly for twenty-four hours in the day, and the thermometer at that time, 40° to +50°.

W. F. Campbell, fireman: The coffee that Captain Hall drank, on return from his sledge-journey, was made purposely for him, though several others had some at the same time. I saw him several times during his sickness, but did not speak to him. I heard that some of the men asked Dr. Bessel what he thought, and the doctor told them that " he would never get over it;" this was when he was first sick. Think he died a natural death. After the separation, saw some provisions on the ice, but no human beings. I went up to the mast-head about ten o'clock in the morning. I had a cat on board that we took from Washington; it staid with us both winters in the ship, and finally ran away from us at Hakluyt Island, as we came down in our boats. The Esquimaux at Life-boat Cove had never seen a cat before, and were much interested in it. They have a name for it in the Esquimau language, though they have not the animal.

Washington, December 24, 1873.

At 12 M. Hon. Geo. M. Robeson, Secretary of the Navy; Admiral Reynolds, Professor Spencer F. Baird, and Captain Howgate assembled at the Navy Department, for the purpose of taking the statements of the last three of the survivors rescued from the steamship *Polaris*, who arrived in New York November 6, 1873.

R. W. D. Bryan, astronomer and chaplain: On reaching our highest latitude, August 30, 1871, I believe the consultation was called, because Captain Buddington had told Captain Hall that they had gone as far as they could. A good part of the time it was foggy, and it was snowing; for a short time we had very clear weather; then I could see the land on the east side, which seemed to end in a point; I saw also the land on the west side; I did not go up aloft; far ahead we saw what the sailors call a "water sky;" right around the vessel there was quite a space between the different floes, so that I was personally very much provoked that they did not go up farther. Afterward I learned that a correct judgment could not be formed by looking from the deck. I suppose even now they could have gone on for perhaps half a mile, but I am very well satisfied they could not have gone any farther.

I never supposed that Captain Hall was so sick that he would die until he did die, although Dr. Bessel used to say that "if he had another attack he would die;" but then I did not believe it. The doctor at one time wanted to administer a dose of quinine, and the captain would not take it. The doctor came to me and wanted me to persuade Captain Hall to take it. I did so, and I saw him prepare the medicine; he had little white crystals, and he heated them in a little glass bowl; heated the water apparently to dissolve the crystals. That is all I know about any medicine. It was given in the form of an injection under the skin in his leg. The night before he died, as he went to bed, he appeared very rational indeed; I remember this very distinctly. The doctor was putting him to bed, and tucking his clothes around him, when the captain said, "Doctor, you have been very kind to me, and I am obliged to you." I noticed that particularly, because it was a little different from what he had been saying to the doctor. Mr. Morton told me that all the evidence that he had that he was dead was a cessation of breathing. Just before he died, he had heard his regular breathing, and then all of a sudden it ceased, and then commenced again; it ceased twice, and then altogether. The separation of the ice-floe party was entirely accidental, unless some person maliciously cut the rope, which I have no idea was the case. The morning after, Mr. Chester came down from the mast-head and reported that he could see a piece of ice, with provisions, but not any signs of the men; that satisfied us all, because we had an idea that the wind drifted us away from them, and the current took them down. We had no idea at all that any one could see them. It just satisfied us at once that they were too far off to be seen. That is the reason no one else went up to look.

Q. "Didn't any body else go up to the mast-head?"

A. No, sir, not that I recollect of; and I think I recollect pretty accurately, because I remember I reproached myself all winter that I did not do it. Henry Hobby might have gone up when I was not looking, but I do not remember any body else but Mr. Chester. I was transferred from the *Ravenscraig* to the *Intrepid*, with Mr. Mauch and Mr. Booth; then, when Captain Walker, of the *Erick*, was ready to sail, he offered to take us, and we went in her. Had heard after Captain Hall's death that both Bessel and Buddington had expressed "relief," as though they had been under some kind of restraint which was not pleasant, and they were glad it was over. There was no difficulty between Captains Buddington and Tyson in regard to the business aboard the ship. After some lengthened conversation, perhaps, there would be a want of some little cordiality; but after a short time they would be very friendly; think some excuse ought to be made for Buddington's appearing occasionally under the influence of liquor, "because so little affected him." Others took the liquor when they could get it; I frequently saw a person with a key belonging to a closet in the cabin where the doctor had stored liquor open the door to get some of it. It was Mr. Schuman, the engineer; he made a key to that door. I do not remember any other one. I believe the doctor medicated several of his cans of alcohol—put in some tartar-emetic. I put all my records out on the ice, containing astronomical, magnetic, and other observations; they are all lost. I had one little plant that Dr. Bessel did not have in his collection. No records or instruments were left on board the *Polaris*. Some of these were put in the cairn at Life-boat Cove; I think I can answer that there is not much there. We did not leave any thing valuable. The log that has been spoken of that was preserved was the log that Mr. Chester wrote. He found he had made a mistake in the first one; that he had left out a day in it, or something of that kind; and so, instead of correcting the mistake, he started a new one, and copied the whole thing up to that date; then he kept on writing the log. He had two large books; then he condensed these large books into a log-book that he brought back here, and those two large books (his first copy) were buried, when we left the house, with the instruments. The old log, that had been copied twice, was left knocking around the house.

J. B. MAUCH, shipped as seaman: Acted part of the time as captain's clerk. I kept Captain Buddington's journal, and Captain Hall's; that is, I kept my own, and Captain Hall copied his from mine. He did not write much. The record which he made at Cape Brevoort I copied. He dictated to me from the original. I put that among his records. In his writing-desk was the original paper from which he dictated to me. [This desk, with the paper referred to, was saved on the ice-floe; an extract from it is given in the fac-simile writing on page 128 of this volume. The whole document is in the Appendix.] Do not know whether any of Captain Hall's papers were destroyed or not. At Life-boat Cove we had as many as one hundred and one natives with us at different times. The old log-book was buried. Mr. Chester wrote a new one. I think you can get the corrections, barometrical and thermometrical, from the observations in my records.

J. W. BOOTH, fireman: Only one little accident happened to the engine; the blow-off pipe gave out at St. Johns, and one of the nuts of the reversing-link came off. When at our highest latitude, we had from one hundred and ten to one hundred and twenty tons of coal left. In coming up Kennedy Channel and Smith Sound, the propeller was making about sixty-five turns. After going into winter-quarters, the machinery was all taken apart to preserve it, and to keep the pipes from bursting; and it was put in order, so that it might be put together again in the spring. The engine was working better when we beached her than it ever did before. The apparatus which was put on board to burn blubber was never used after it was tried in Brooklyn. I helped to put it together for them.

INDEX.

ADMIRALTY, British, 62.
Advance, the U. S. steamship, 57.
Agassiz on glaciers, 450.
All ice and icebergs, 266.
Allen, Captain, 405.
Allowance per diem on floe, 229; reduced, 289.
Ambition, a pure, 74.
American expeditions, 34, 64.
American flag raised, 150.
Annual reports of discoveries by Captain Hall, 121.
Ansel Gibbs, the whaler, 97.
Antelope, the bark, 97.
Apoplexy, cases of, 162.
Appropriation for North Polar Expedition, 100.
Arctic adventurers, 20.
Arctic chronology, 413–421.
Arctic, the whaler, 407.
Augustina, 263.
Aurora borealis, 172, 226, 254, 361.
Austin, Captain, 55.

BABY born on *Polaris*, 192; named Charlie Polaris, 192; on the ice-floe, 263.
Back, Lieutenant (Sir George), 48.
Baffin, William, 32.
Baird, Professor Spencer F., 2, 433.
Bait, cheese for, 363.
Baiting porpoises, 358.
"Ballecners," 85.
Barrow, Sir John, 44.
Battery of ice-blocks, 321.
Battle between bergs, 325.
Bay, Newman, 180.
Bay, Polaris, 148.
Bay, Roberts, 336.
"Bear" and "Spike," 214.
Bears, habits of, 290, 309; liver poisons crew, 298; shot by Joe, 177, 323; shot by Captain Tyson, 309; shot by Lieutenant H. C. White, 370.
Beechey, Captain, 42; Beechey Island, 57.
Belcher, Sir Edward, 60, 149.
Bellot, René, 61.
Beset in the ice, 98.
Bessel, Dr. Emil, 129; storm-bound, 166;

lost in the darkness, 169; snow-blind, 187.
"Billy's" adventure, 378.
Birds, Arctic, 87, 392; blown out to sea, 384; in the rigging, 385.
Bladder-nose seal, 289.
"Blauket," the seal's, 235.
Blow-holes of seal, 217; of narwhals, 282.
Blubber, value of, for lamps, 210; as food, 281.
Boat-journeys, 185–188.
Boat-steerer, 92.
Boats on *Polaris*, 103.
Books, none on ice-floe, 257.
Booth, Sir Felix, 46.
Boothia Peninsula, 50.
Braine, Commander, 342; Bradford, his choice of bergs, 302.
Brevoort, Cape, 161; J. Carson, 104, 467.
British Naval Board, obtuseness of, 61.
Bryan, R. W. D., astronomer to *Polaris* expedition, 131; letter of, 469; testimony of, 478.
Buddington, Captain James, 342; Sidney O., 131, 406, 407.
Burial at sea, 77.

CABOT, John and Sebastian, 26.
Canaries in Greenland, 381.
Canary-bird rations, 290.
Cannibalism, rumors of, 60; feared, 230.
Cape Constitution, 177.
Cape Mercy, 95, 97.
Chamisso Island, 43.
Charts, defective, 148.
Children imperiled, 201; in snow-hut on floe, 263; in the boat, 310; at St. Johns, 339; at Brooklyn Navy Yard, 350.
Chipman, Assistant Ice-pilot, 358.
Chivalry, modern, 73.
Christmas on *Polaris*, 171; on the ice-floe, 232.
"Cleaning house" on the floe, 274.
Cleanliness impossible, 303.
Clothing in Arctic regions, native fur, 83;

sheep-skin good substitute, 396; civilized, 146, 252, 280; inventory of Captain Tyson's, 276; luxury of clean, 277.
Coal near Disco, 421; at Ivgitut, South Greenland, 379, 381.
Cold, excessive, 78; no barrier to travel, 411; effect on crews' courage, 264.
Collinson, Captain, 61.
Commagere, Frank Y., 347; his yarn, 357.
Compasses, variation of, 44; deviation corrected, 356.
Conclusions arrived at, 411, 412.
Congress, U. S. steamship, 144.
Congressional action, 100.
Constellations, beauty of the Arctic, 247.
Cook, Captain, 34.
Crozier, Captain, 122.
Cumberland Gulf, 95; Captain Tyson "at home" there, 375.
Curious record, 373.
Cylinders, copper, thrown overboard, 148, 149; buried on shore, 161, 187.

DALY, Judge Charles P., letter to, 424.
Danes in Greenland, 83, 371.
Danish officials, 135, 137.
Dannet, Captain, 51.
Dante's ice hell, 278.
Davenport, Captain, 146.
Davis, John, 29.
De Haven, Lieutenant, 57, 85.
De Long, Lieutenant, 343.
Dealy Island, 61.
Dease's journey overland, and by boat, 48.
Death of Captain C. F. Hall, 162, 472.
Deer-meat plenty, 119; deer seen, 377.
"Devil's Thumb," 83.
Dirt in snow-huts, 280, 297; unavoidable on floe, 303.
Discipline, lack of, 221.
Disco, description of, 83, 225; hopes centred on, 242.
Discoveries of Captain C. F. Hall, 424; discoveries, real and supposed, of others, 22–24.
Doctor's clerk, 361.
Dogs, Esquimaux, 154, 177, 215, 396.
Dovekies appear, 282.
Dramatic series of events, 407.
Drift of the ice-floe with Captain Tyson and party, 197–331; De Haven's, 84; M'Clintock's, 63; Parry's, 45; British ship *Resolute*, 95; crew of the *Hansa*, 419.
Drift-wood, at Newman Bay, 187.
Drink bear's blood, for want of water, 323.
Drying-nets of Esquimaux, 211.
Duck Islands, 84; ducks, 87.
Dundee, news from, 407.

EASTER-SUNDAY, 317.
Eating frozen seal entrails, 237, 258; every part but bones and gall, 270; his "jacket," 282; seal-skin, hair on, 256; scrap blubber from lamp, 260.
Ebierbing, "Joe," the Esquimau, 221.
Economizing paper, 226, 251.
Eider-ducks, 88.
Elberg, Governor, 137.
Electric clouds, 168.
Era, schooner, voyages in, 98; log, records from, 98.
Erebus and *Terror*, 51; last seen by white men, 51; fate of, 119.
Esquimaux, first impressions of, 80; huts, 209; tents and dress, 83, 401, 409; nursery usage, 339; half-breed belles, 372; music and dancing, 312; boats, 143, 270, 316; sledges, 287, 349, 396; dogs, 154, 177; lamps, 210; unthrifty, 213; tribes of west coast of Davis Strait neglected by Christian world, 377; civilized vices, 377; difference of disposition in tribes, 121.
Exercising in a space three feet square, 278.
Expeditions, pre-Columbian, 25; mercantile, 25; early English, 27; to the eastward, 27; due north, 35, 36; overland, 33, 41, 50; modern, 44; early American, 34; modern American, 64; *Polaris*, 100; first and second Grinnell, 56, 64; via Behring Strait, 34, 413; to the northwest, 413; via Spitzbergen coast, 34, 45; scientific, 44, 419; relief, 53–63; sledge, 155, 176, 180; boat, 41, 45, 186.
Explorers, rival, 25.

FAILURES and successes, 71.
Fauna, Arctic, 392.
Fiords of Greenland, 177, 178, 381.
Fire in coal-bunker, 368; fire "training," 357.
Firing signals, 327.
Fiscanaes, 143.
Flag presentation, 126; raising, 150.
Floe separates from *Polaris*, 197; list of persons on the floe, 202; size of, 201, 207, 312; floe breaks up, 312–324; becomes a "pack," 302.
Fog, pest of northern regions, 374; a black, 385.
Forlorn Hope, the, 63.
Fox, Captain Luke, 104, 415; Channel, 31, 415; Arctic, 155, 181, 229; voyage of the, 63.
Franklin, Sir John, 44; last expedition of, 51; date of death, 64; name stricken from British Navy List, 417; relics of, 57, 64, 120; Lady, 56, 60, 63.
Frobisher's gold, 27; relics collected by C. F. Hall, 120; "strait" resolved into "bay," 117.

INDEX. 483

Frolic, U. S. steamship, brings floe waifs to Washington, 341.
Frozen mercury, 254.
Fury, British steamship, consort of *Hecla*, 40; wreck of, 43; beach, 421.

GALES, violent, 168, 174, 197, 320, 382.
Game plenty, 392; scarce on floe, 230-232; at extreme north, 181, 392.
Geographical Society, 124; Captain Hall addresses the, 126; reports to, 414.
George Henry, the whaler, 92.
Georgiana, the brig, 96.
German efficiency, 306.
German element on *Polaris*, 134; on floe, 231; expeditions, 418, 419.
Gilbert, Sir Humphrey, 28.
Glaciers, how to observe, 450; in Greenland, 266; extinct, 187, 391; extent of the Humboldt, 266; discharge of, 475.
Glorious spectacle, a, 55.
Goodhavn, harbor of, 83, 135.
"Gradgrinds," ancient, 69.
Grant, President, 343.
Grave of Captain Hall, 190; re-arranged, 192.
Greenland, early settlement of, 24; rugged coast, 362; interior plateaux of, 394; *mer-de-glace*, 266; inhabitants, 80, 371; officials, 135, 137; "captains," 38.
Greer, Commander James A., 347; his success, 354.
Grinnell Land, 22; first expedition, 56; second, 60.
Ground-tackle lost, 193.

HALL, Captain Charles Francis, 64; early life and traits, 113; Arctic expeditions of, 116, 118; lives with the Esquimaux, 125; letters of, 102, 105, 111, 132; his last public address, 127; idiosyncrasy, 137; geographical reports, 121; his early discoveries, 119; at New London, Conn., 99; persevering efforts, 125; appointed to command U. S. steamship *Polaris*, 101; his sledge-journey, 155; his premonitions, 127; his illness and death, 162, 472; *fac-simile* of his writing, 128.
Hall Land named, 166.
Hannah, wife of Ebierbing, 221; learns white manners, 303; afraid of being eaten, 230.
Hans, native Esquimau, 136; mistaken for a bear, 218; kills and eats two dogs, 213; his family hut on the floe, 263; at St. Johns, 336; reminiscences of, 248, 274; Mrs. Hans's summer arrangements in Brooklyn Navy Yard, 350; returns to Greenland on *Tigress*, 350; compelled to wash, 362.

Harbor, Thank God, latitude of, 153.
Harpoon metamorphosed, 275.
Hartstene, Lieutenant, 63.
Hawk, interviewed aloft, 385.
Hayes, Dr. I. I., 24, 67.
Henry's, Professor, instructions to Scientific Corps, 431.
Hillgard, Professor, 433.
Hobson, Lieutenant, 64.
Holsteinborg, 80.
Hope abandoned, 208; the preserver of man, 261.
Hudson, Henry, 31; Bay Company, 48, 50.
Hunger, pains of, 232.
Hunting bears, 177; musk-cattle, 180; seal, 79; for "specimens," 378.
Hurricane, Arctic, 175, 278.

ICE, groaning, 195; fresh-water ice melted to drink, 195, 279, 402; salt-water, to season soup, 279; pash, 204, 322; young, 270; frightful noises of breaking, 296.
Icebergs, how formed, 266; their different forms, 267; various histories, infinite variety, 267; excitement on nearing, 363; narrow escape from, 385; a beautiful sight, 363; naïve comments on, 359.
Igloo, or native hut, 209.
Igloolik Island, 426.
Illusions, Arctic, 24.
Inglefield, Captain, 61.
Insects, Arctic, 392.
Instructions, official, to Captain Hall, 108; to Scientific Corps, 109.
Intrepid, the whaler, 407.
Isabel, voyage of, 60, 61.
Isabella, the, 48.
Ivgitut, South Greenland, 379.
Ivory, article of commerce, 273; royal chair of Denmark made of, 273.

JANSEN, Governor, 146; sharp on a trade. 139.
"Joe" and Hannah, 348; his hunting qualities, 219, 249; his valuation of seal, 284; his opinion of sailors, 278; shoots a bear, 177, 323; his shrewdness on *Tigress*, 351.

KANE, Dr. E. K., 59, 60, 34.
Kellet, Captain, 53.
Kennedy, Captain, 23, 24, 61; Channel, 148.
King, Dr., 48.
Kingituk, 146.
Kryolite, 379, 281.
Kyack, 270, 286, 287, 314.

LABRADOR, coast of, 269, 295.

Lake discovered by Captain Hall, 161, 425; Great Bear, 48; Musk Ox, 48.
Lamont, James, F.R.S., 419.
Lamps, native, 210; improvised on floe, 211; heating power of, 293, 294.
Latitude, highest attained by *Polaris*, 150; of Thank God Harbor, 153; of ice-floe party when rescued, 382.
Lemmings, 181.
Life-boat Cove, 400.
Littleton Island, 365, 399.
Livers of bear and oogjook poisonous, 298.
Log-book records, 98, 119; mutilations of, 354.
Lupton, Colonel James, 2; Cape, 152.

M'CLELLAN, the bark, 77.
M'Clintock, Captain, 58, 59, 63.
M'Clure, Captain, 61.
Mackenzie River, 42.
Magazine, guarding the, 357.
Magnetic pole discovered, 47; effect on compasses, 44.
Magnetism, terrestrial, 410, 436.
Meek, Professor, 433.
Melting ice for drink, 279.
Melville Bay, 79, 80, 343; ice-pack in, 389; Island, 54.
Mental coercion, providential, 62.
Mer-de-glace, the feeder of Humboldt Glacier, 266.
Mercury frozen, 254.
Meteorology, 438.
Meyers, Fred., 130; surveys Newman Bay, 180; separated from party, 313; frozen, 314; a picture of famine, 318; grotesque misery of, 319.
Middle ice, 79.
Milk, how procured on the ice-floe, 310.
Mineralogy, 349, 378, 379.
Mistakes of explorer, 22.
Mock moons, 175; mock suns, 176.
Modern facilities for Arctic research, 72.
Mollemokes, tricks on, 88.
Monticello, the bark, 118.
Musk-cattle, habits of, 181; Labrador species, 154; skins used for bedding, 201; peculiar strategy of, 181; found far north, 392.

NARWHALS, description of, 272, 273; good to carry off shot, 277, 281.
National Academy of Sciences, 428.
Newcomb, Professor, 433.
Newman Bay, 161, 181; Dr. at Disco, 144; writes three prayers for *Polaris* expedition, 145.
New-year's-day on the floe, 237.
New whaling-ground, 427.
Night, Arctic, 220, 247; a fearful, 321.
Nipped in ice, danger of, 168.
Nomenclature, Arctic, 21.

North Pole, orders to reach, 30, 34; invitation to visit, 105; probable geographical features of, 388; how to recognize the spot, 127; Captain Tyson's plan to get there, 393.
Northumberland Island, 352, 362.
North-west passage, 28; made by Captain M'Clure over ice, 61; by Captain Collinson in his ship, 61; probably made by Franklin, 119.
Nothing to read on the floe, 257.

OBSERVATIONS, proof of genuineness, 109; faulty, 229, 243, 247.
Oil-boiler, 194, 479.
Oogjook, or monster seal, 291; thirty gallons of oil taken from, 291; liver poisons the crew, 298.
Oomiak, or woman's boat, 104, 316.
One full meal, 220.
One short meal a day, reduced to, 289.
"Open Polar sea" *non est*, 148.
Orray Taft, the whaler, 97.
Osborne, Lieutenant Sherrard, 57, 58.
Overhanging bergs, 316.

PAINS of hunger, 222.
Paraselene, 175.
Parhelia, 176.
Parker, Captain, 91.
Parry, Captain, 37, 42, 45.
Peabody, George, 60.
Pemmican, how composed, 215; made into "tea" or "soup," 233.
Pendulum experiments, 70.
Penny, Captain, 55, 56.
Periwinkle, U. S. steamship, 101.
Pim, Lieutenant, 61.
Plateaux of Hall Land and Greenland, 394.
Plover, British steamer, 53, 56.
Polar Sea discussed, 148, 388.
"Polaris, Charlie," Hans's baby, 192, 263, 323.
Polaris, U. S. steamship commissioned, 101; description of, 102-104; officers and crew, list of, 112; reaches highest latitude, 150; drifts southward, 193, is abandoned, 400; presented to Esquimaux chief, 401; founders, 367; a twice-copied log brought home, 478.
Pole, magnetic, 47; geographical, 127.
Pond Bay renamed, 22.
Porpoises, angling for; how taken, 358.
Port Foulk as base of supplies, 394.
"Pounding-day" on the floe, 274.
Prayers used by the Polaris expedition, 145, 160; composed by Dr. Newman, 159.
Profit and loss of Arctic exploration, 67.
Providence Berg named by Captain Hall, 151; it splits in two, 168.

INDEX. 485

Provisions on the ice-floe, 202; how divided, 212.
Puney, Joe and Hannah's adopted daughter, 211, 262.

QUAYLE, Captain, 85; John, 92, 93.
Questions, curious, 332.

RAE, Dr. John, 50.
Rations weighed out by the ounce, 212; effect of insufficient, 222.
Ravenscraig, the whaler, 405.
Records, orders for keeping, 109.
Reindeer, 173, 404.
Reliance on God alone, 269.
Relics of Frobisher's expedition, 117; of Franklin's, 120; of Captain C. F. Hall, 159; of U. S. steamship *Polaris*, 355.
Refraction, curious effects of, 24; causes mistakes, 24.
Relief parties, 53-63, 64, 119.
Rensselaer Harbor, ice encumbered, 395.
Reporter of *New York Herald* on *Tigress*, 347.
Rescue, the tender to *Advance*, 57; lost, 97; rescue parties (see Relief) of ice-floe party, 331; of *Polaris* survivors, 405.
Resolute, British ship, finding of the, 93; drift of, 95.
Resolution Island, 78.
Repulse Bay, 97; harbor, 150.
Robeson, Hon. George M., Secretary United States Navy, 2; Channel, 149.
Roquette, M. De la, 37.
Ross, Sir John, 37, 47; James C., 47, 54.
Rudolph, Governor, 137.

SAILORS' tricks, 87; parlor, 50.
Saturday-night usage, 369.
Scarcity of game on floe, 275.
Scenery, Arctic, 153.
Schoonmaker, commander U. S. steamship *Frolic*, 341; report of, 341.
Scientific instructions, 431; notes, 410; devotees, 71.
Scotch hospitality, 406; whalers, 371, 377.
Sea-drenched, 315.
Seals, different species, 236; meat heat-giving, 290; blubber of, 210; how caught, 217, 230; blood-soup, 250; saving the blood on floe, 272; a pet, 89; small Greenland, 286; divided *a la Esquimaux*, 235; bladder-nose, 78, 289; size of a rare specimen, 195; on the ice-floe eat the whole but gall and bones, 270; six hundred killed by the crew of the *Tigress*, 334.
Separation of floe from *Polaris*, 198; locality of, 398.

Sheddon, Robert, 53.
Shot used for weights, 212.
Signal set on the floe, 204; fire, of blubber, 326.
Silence of Arctic night, 170.
Simpson and Dease's journey, 46.
"Sick-list," none kept on *Polaris*, 473; all hands after the rescue, 335; crew on the floe from eating oogjook-liver, 298.
Sixty hours of storm-turmoil, 301.
Sledges, native, 396; improved, 349; of unframed skins, 287; excursion of Captain Hall, 155; of Dr. Bessel, 176; of Captain Tyson, 180.
Smith Sound, currents of, 388; pack-ice of, 395; little ice in winter in, 389; Inspector Karrup, 135.
Smithsonian Institution, 413.
Snow, red, analyzed, 37; snow-blindness, 189; huts, how built, 209; Mr. W. P., 56.
Snowed under, 202; in, 264.
Solemn entry in journal in view of death, 259.
Soup, ice-floe recipes for, 223, 224, 310.
Southern fiord, 177.
Sphinx, the northern, 19.
St. Johns, port of, 133; trouble at, 134.
Stars, beauty of, 247.
Stealing food, 250, 318.
Steam first used on Arctic waters, 46.
Steamship sighted from floe, 204, 326.
"Stone-fever," a bad case of, 378.
Succi's chronic cry, 263.
Suicide, reflections on, 294.
Summer tents of natives, 125, 401.
Sumner, Senator, 100; head-land, 423.
Sun, Arctic summer, 153, 187, 188; disappears, 157; re-appears, 175, 246; mock, 176.
Sunday on *Polaris*, 142, 166; on *Tigress*, 360.
"Symmes's Hole," 148.

"TABLE-TURNING" without spirits, 379.
"Tea," pemmican, 233.
Terror, British steamship (see *Erebus*), 51.
Tessuisnk (or Tossac), harbor of, 138.
Thanksgiving-day on *Polaris*, 169; on the floe, 233.
Thermometer useless; mercury frozen, 254.
Tides, 173, 410.
Tigress, the, rescues ice-floe party, 335; bought by United States Government, 344; sails in search of *Polaris* and survivors of expedition, 350; reaches Littleton Island, 353; a stormy voyage—returns to Brooklyn Navy Yard—pleasant remembrance of officers, 387; list of officers and crew, 347.

INDEX.

Tobacco, praise of, 283, 293; a present of, 293; out of, 332.
Tobias, the little boy, sick on the floe, 257, 263; recovered, 302.
Too-koo-lito (see Hannah).
Tossac (see Tessuisak).
Tough eating, 235.
True Love, the, an old English whaler, 91.
"Turning in" on the floe, 326.
Tyson, Captain George E., 76; early life and first whaling voyage, 77; as boat-steerer on the *George Henry*; sights the British ship *Resolute*, 92; goes over the ice to her with three companions, 93; reports his treasure-trove to his captain, 95; sails as master of brig *Georgiana*, 96; takes the first whale ever caught in Repulse Bay, 97; subsequent voyages of, 98; meets Captain C. F. Hall in Arctic regions, 97; supplies him with a boat, 98; joins *Polaris* expedition, 99; makes sledge-journey north, 180; goes on boat expedition to Newman Bay, 186; proposes pedestrian trip, 188; separated from *Polaris*, 198; his fortunes on the ice-floe, 198–328; drifts one hundred and ninety-six days, 331; sails in searching steamship *Tigress*, 356; visits Buddington's deserted camp, 366; journal of cruise, 356–386.

UNEXPLORED area, 73.
United States Arctic expeditions, 125.
Upernavik, 138, 371; a good time there, 372.
Useless cruising, 383.
Uses of Arctic exploration, 68.

VELOCITY of Arctic winds, 176.
Victory, the first steamship in Arctic seas, 46.
Visited by land-birds, 319, 384.
Visitor, an unmannerly, 250.

WALK to Captain Hall's grave, 175; a ten-miles' walk on the floes, 296.
Walker, Cape, 60, 62.
Walrus, hard to kill, 147; the sealer, 340.
Water, drinking—how obtained on floe, 279.
Waves, force of, 382.
Weakness, evidences of, 213, 286.
Wellington Channel, 59.
Whale-meat "drugs," 90.
Whales, different species, 85; "right" and white, size of, 86; prolonged struggle with a, 86.
White, Lieutenant, United States Navy, 370; Captain E. W., letter of, 423.
Winds, Arctic, 167, 440.
Winter-quarters, of Barentz, 29, 420; in Cumberland Gulf, 89; preparing for, at Thank God Harbor, 158; Buddington's, 402; Dr. Hayes's, 394; Dr. Kane's, 389; at Niountelik, 98.
Without water to drink, 313.
Woolen clothing in Arctic climate, 146.
Wrangel, Baron, his open Polar sea, 33.

YORK, Cape, 364.
Young ice, 261, 271.

ZERO, forty degrees below, 255, 473.
Zoology, Arctic, 390, 412.

THE END.

VALUABLE & INTERESTING WORKS
FOR PUBLIC AND PRIVATE LIBRARIES,
Published by HARPER & BROTHERS, New York.

☞ *For a full List of Books suitable for Libraries, see* HARPER & BROTHERS' TRADE-LIST *and* CATALOGUE, *which may be had gratuitously on application to the Publishers personally, or by letter enclosing Six Cents in Postage Stamps.*

☞ HARPER & BROTHERS *will send any of the following works by mail, postage prepaid, to any part of the United States, on receipt of the price.*

FLAMMARION'S ATMOSPHERE. The Atmosphere. Translated from the French of CAMILLE FLAMMARION. Edited by JAMES GLAISHER, F.R.S., Superintendent of the Magnetical and Meteorological Department of the Royal Observatory at Greenwich. With 10 Chromo-Lithographs and 86 Woodcuts. 8vo, Cloth, $6 00.

HUDSON'S HISTORY OF JOURNALISM. Journalism in the United States, from 1690 to 1872. By FREDERICK HUDSON. Crown 8vo, Cloth, $5 00.

PIKE'S SUB-TROPICAL RAMBLES. Sub-Tropical Rambles in the Land of the Aphanapteryx. By NICOLAS PIKE, U. S. Consul, Port Louis, Mauritius. Profusely Illustrated from the Author's own Sketches; containing also Maps and Valuable Meteorological Charts. Crown 8vo, Cloth, $3 50.

TYERMAN'S OXFORD METHODISTS. The Oxford Methodists: Memoirs of the Rev. Messrs. Clayton, Ingham, Gambold, Hervey, and Broughton, with Biographical Notices of others. By the Rev. L. TYERMAN, Author of "Life and Times of the Rev. John Wesley," &c. Crown 8vo, Cloth, $2 50.

TRISTRAM'S THE LAND OF MOAB. The Result of Travels and Discoveries on the East Side of the Dead Sea and the Jordan. By H. B. TRISTRAM, M.A., LL.D., F.R.S., Master of the Greatham Hospital, and Hon. Canon of Durham. With a Chapter on the Persian Palace of Mashita, by JAS. FERGUSON, F.R.S. With Map and Illustrations. Crown 8vo, Cloth, $2 50.

SANTO DOMINGO, Past and Present; with a Glance at Hayti. By SAMUEL HAZARD. Maps and Illustrations. Crown 8vo, Cloth, $3 50.

SMILES'S HUGUENOTS AFTER THE REVOCATION. The Huguenots in France after the Revocation of the Edict of Nantes; with a Visit to the Country of the Vaudois. By SAMUEL SMILES, Author of "The Huguenots: their Settlements, Churches, and Industries in England and Ireland," "Self-Help," "Character," "Life of the Stephensons," &c. Crown 8vo, Cloth, $2 00.

HERVEY'S CHRISTIAN RHETORIC. A System of Christian Rhetoric, for the Use of Preachers and Other Speakers. By GEORGE WINFRED HERVEY, M.A., Author of "Rhetoric of Conversation," &c. 8vo, Cloth, $3 50.

EVANGELICAL ALLIANCE CONFERENCE, 1873. History, Essays, Orations, and Other Documents of the Sixth General Conference of the Evangelical Alliance, held in New York, Oct. 2–12, 1873. Edited by Rev. PHILIP SCHAFF, D.D., and Rev. S. IRENÆUS PRIME, D.D. With Portraits of Rev. Messrs. Prouier, Carrasco, and Cook, recently deceased. 8vo, Cloth, nearly 800 pages, $6 00.

PRIME'S I GO A-FISHING. I Go a-Fishing. By W. C. PRIME. Crown 8vo, Cloth, $2 50.

ANNUAL RECORD OF SCIENCE AND INDUSTRY FOR 1873. Edited by Prof. SPENCER F. BAIRD, of the Smithsonian Institution, with the Assistance of Eminent Men of Science. 12mo, over 800 pp., Cloth, $2 00. (Uniform with the *Annual Record of Science and Industry for 1871 and 1872.* 12mo, Cloth, $2 00.)

VINCENT'S LAND OF THE WHITE ELEPHANT. The Land of the White Elephant: Sights and Scenes in Southeastern Asia. A Personal Narrative of Travel and Adventure in Farther India, embracing the Countries of Burma, Siam, Cambodia, and Cochin-China (1871–2). By FRANK VINCENT, Jr. Magnificently Illustrated with Map, Plans, and numerous Woodcuts. Crown 8vo, Cloth, $3 50.

MOTLEY'S LIFE AND DEATH OF JOHN OF BARNEVELD. Life and Death of John of Barneveld, Advocate of Holland. With a View of the Primary Causes and Movements of "The Thirty Years' War." By JOHN LOTHROP MOTLEY, D.C.L., Author of "The Rise of the Dutch Republic," "History of the United Netherlands," &c. With Illustrations. In Two Volumes. 8vo, Cloth, $7 00.

TYNG ON A CHRISTIAN PASTOR. The Office and Duty of a Christian Pastor. By STEPHEN H. TYNG, D.D., Rector of St. George's Church in the City of New York. Published by the request of the Students and Faculty of the School of Theology in the Boston University. 12mo, Cloth, $1 25.

PLUMER'S PASTORAL THEOLOGY. Hints and Helps in Pastoral Theology. By WILLIAM S. PLUMER, D.D., LL.D. 12mo, Cloth, $2 00.

POETS OF THE NINETEENTH CENTURY. The Poets of the Nineteenth Century. Selected and Edited by the Rev. ROBERT ARIS WILLMOTT. With English and American Additions, arranged by EVERT A. DUYCKINCK, Editor of "Cyclopædia of American Literature." Comprising Selections from the Greatest Authors of the Age. Superbly Illustrated with 141 Engravings from Designs by the most Eminent Artists. In elegant small 4to form, printed on Superfine Tinted Paper, richly bound in extra Cloth, Beveled, Gilt Edges, $5 00; Half Calf, $5 50; Full Turkey Morocco, $9 00.

THE REVISION OF THE ENGLISH VERSION OF THE NEW TESTAMENT. With an Introduction by the Rev. P. SCHAFF, D.D. 618 pp., Crown 8vo, Cloth, $3 00.

This work embraces in one volume:

I. ON A FRESH REVISION OF THE ENGLISH NEW TESTAMENT. By J. B. LIGHTFOOT, D.D., Canon of St. Paul's, and Hulsean Professor of Divinity, Cambridge. Second Edition, Revised. 196 pp.

II. ON THE AUTHORIZED VERSION OF THE NEW TESTAMENT in Connection with some Recent Proposals for its Revision. By RICHARD CHENEVIX TRENCH, D.D., Archbishop of Dublin. 194 pp.

III. CONSIDERATIONS ON THE REVISION OF THE ENGLISH VERSION OF THE NEW TESTAMENT. By J. C. ELLICOTT, D.D., Bishop of Gloucester and Bristol. 178 pp.

NORDHOFF'S CALIFORNIA. California: For Health, Pleasure, and Residence. A Book for Travelers and Settlers. Illustrated. 8vo, Paper, $2 00; Cloth, $2 50.

MOTLEY'S DUTCH REPUBLIC. The Rise of the Dutch Republic. By JOHN LOTHROP MOTLEY, LL.D., D.C.L. With a Portrait of William of Orange. 3 vols., 8vo, Cloth, $10 50.

MOTLEY'S UNITED NETHERLAND'S. History of the United Netherlands: from the Death of William the Silent to the Twelve Years' Truce—1609. With a full View of the English-Dutch Struggle against Spain, and of the Origin and Destruction of the Spanish Armada. By JOHN LOTHROP MOTLEY, LL.D., D.C.L. Portraits. 4 vols., 8vo, Cloth, $14 00.

NAPOLEON'S LIFE OF CÆSAR. The History of Julius Cæsar. By His late Imperial Majesty NAPOLEON III. Two Volumes ready. Library Edition, 8vo, Cloth, $3 50 per vol.

Maps to Vols. I. and II. sold separately. Price $1 50 each, NET.

HAYDN'S DICTIONARY OF DATES, relating to all Ages and Nations. For Universal Reference. Edited by BENJAMIN VINCENT, Assistant Secretary and Keeper of the Library of the Royal Institution of Great Britain; and Revised for the Use of American Readers. 8vo, Cloth, $5 00; Sheep, $6 00.

MACGREGOR'S ROB ROY ON THE JORDAN. The Rob Roy on the Jordan, Nile, Red Sea, and Gennesareth, &c. A Canoe Cruise in Palestine and Egypt, and the Waters of Damascus. By J. MACGREGOR, M.A. With Maps and Illustrations. Crown 8vo, Cloth, $2 50.

WALLACE'S MALAY ARCHIPELAGO. The Malay Archipelago: the Land of the Orang-Utan and the Bird of Paradise. A Narrative of Travel, 1854-1862. With Studies of Man and Nature. By ALFRED RUSSEL WALLACE. With Ten Maps and Fifty-one Elegant Illustrations. Crown 8vo, Cloth, $2 50.

WHYMPER'S ALASKA. Travel and Adventure in the Territory of Alaska, formerly Russian America—now Ceded to the United States—and in various other parts of the North Pacific. By FREDERICK WHYMPER. With Map and Illustrations. Crown 8vo, Cloth, $2 50.

ORTON'S ANDES AND THE AMAZON. The Andes and the Amazon; or, Across the Continent of South America. By JAMES ORTON, M.A., Professor of Natural History in Vassar College, Poughkeepsie, N. Y., and Corresponding Member of the Academy of Natural Sciences, Philadelphia. With a New Map of Equatorial America and numerous Illustrations. Crown 8vo, Cloth, $2 00.

WINCHELL'S SKETCHES OF CREATION. Sketches of Creation: a Popular View of some of the Grand Conclusions of the Sciences in reference to the History of Matter and of Life. Together with a Statement of the Intimations of Science respecting the Primordial Condition and the Ultimate Destiny of the Earth and the Solar System. By ALEXANDER WINCHELL, LL.D., Professor of Geology, Zoology, and Botany in the University of Michigan, and Director of the State Geological Survey. With Illustrations. 12mo, Cloth, $2 00.

WHITE'S MASSACRE OF ST. BARTHOLOMEW. The Massacre of St. Bartholomew: Preceded by a History of the Religious Wars in the Reign of Charles IX. By HENRY WHITE, M.A. With Illustrations. 8vo, Cloth, $1 75.

RECLUS'S THE EARTH. The Earth: a Descriptive History of the Phenomena and Life of the Globe. By ÉLISÉE RECLUS. Translated by the late B. B. Woodward, and Edited by Henry Woodward. With 234 Maps and Illustrations, and 23 Page Maps printed in Colors. 8vo, Cloth, $5 00.

RECLUS'S OCEAN. The Ocean, Atmosphere, and Life. Being the Second Series of a Descriptive History of the Life of the Globe. By ÉLISÉE RECLUS. Profusely Illustrated with 250 Maps or Figures, and 27 Maps printed in Colors. 8vo, Cloth, $6 00.

Harper & Brothers' Valuable and Interesting Works. 3

LOSSING'S FIELD-BOOK OF THE REVOLUTION. Pictorial Field-Book of the Revolution; or, Illustrations, by Pen and Pencil, of the History, Biography, Scenery, Relics, and Traditions of the War for Independence. By BENSON J. LOSSING. 2 vols., 8vo, Cloth, $14 00; Sheep, $15 00; Half Calf, $18 00; Full Turkey Morocco, $22 00.

LOSSING'S FIELD-BOOK OF THE WAR OF 1812. Pictorial Field-Book of the War of 1812; or, Illustrations, by Pen and Pencil, of the History, Biography, Scenery, Relics, and Traditions of the Last War for American Independence. By BENSON J. LOSSING. With several hundred Engravings on Wood, by Lossing and Barritt, chiefly from Original Sketches by the Author. 1088 pages, 8vo, Cloth, $7 00; Sheep, $9 50; Half Calf, $10 00.

ALFORD'S GREEK TESTAMENT. The Greek Testament: with a critically revised Text; a Digest of Various Readings; Marginal References to Verbal and Idiomatic Usage; Prolegomena; and a Critical and Exegetical Commentary. For the Use of Theological Students and Ministers. By HENRY ALFORD, D.D., Dean of Canterbury. Vol. I., containing the Four Gospels. 944 pages, 8vo, Cloth, $6 00; Sheep, $6 50.

ABBOTT'S FREDERICK THE GREAT. The History of Frederick the Second, called Frederick the Great. By JOHN S. C. ABBOTT. Elegantly Illustrated. 8vo, Cloth, $5 00.

ABBOTT'S HISTORY OF THE FRENCH REVOLUTION. The French Revolution of 1789, as viewed in the Light of Republican Institutions. By JOHN S. C. ABBOTT. With 100 Engravings. 8vo, Cloth, $5 00.

ABBOTT'S NAPOLEON BONAPARTE. The History of Napoleon Bonaparte. By JOHN S. C. ABBOTT. With Maps, Woodcuts, and Portraits on Steel. 2 vols., 8vo, Cloth, $10 00.

ABBOTT'S NAPOLEON AT ST. HELENA; or, Interesting Anecdotes and Remarkable Conversations of the Emperor during the Five and a Half Years of his Captivity. Collected from the Memorials of Las Casas, O'Meara, Montholon, Antommarchi, and others. By JOHN S. C. ABBOTT. With Illustrations. 8vo, Cloth, $5 00.

ADDISON'S COMPLETE WORKS. The Works of Joseph Addison, embracing the whole of the "Spectator." Complete in 3 vols., 8vo, Cloth, $6 00.

ALCOCK'S JAPAN. The Capital of the Tycoon: a Narrative of a Three Years' Residence in Japan. By Sir RUTHERFORD ALCOCK, K.C.B., Her Majesty's Envoy Extraordinary and Minister Plenipotentiary in Japan. With Maps and Engravings. 2 vols., 12mo, Cloth, $3 50.

ALISON'S HISTORY OF EUROPE. FIRST SERIES: From the Commencement of the French Revolution, in 1789, to the Restoration of the Bourbons, in 1815. [In addition to the Notes on Chapter LXXVI., which correct the errors of the original work concerning the United States, a copious Analytical Index has been appended to this American edition.] SECOND SERIES: From the Fall of Napoleon, in 1815, to the Accession of Louis Napoleon, in 1852. 8 vols., 8vo, Cloth, $16 00.

BALDWIN'S PRE-HISTORIC NATIONS. Pre-Historic Nations; or, Inquiries concerning some of the Great Peoples and Civilizations of Antiquity, and their Probable Relation to a still Older Civilization of the Ethiopians or Cushites of Arabia. By JOHN D. BALDWIN, Member of the American Oriental Society. 12mo, Cloth, $1 75.

BARTH'S NORTH AND CENTRAL AFRICA. Travels and Discoveries in North and Central Africa: being a Journal of an Expedition undertaken under the Auspices of H. B. M.'s Government, in the Years 1849-1855. By HENRY BARTH, Ph.D., D.C.L. Illustrated. 3 vols., 8vo, Cloth, $12 00.

HENRY WARD BEECHER'S SERMONS. Sermons by HENRY WARD BEECHER, Plymouth Church, Brooklyn. Selected from Published and Unpublished Discourses, and Revised by their Author. With Steel Portrait. Complete in 2 vols., 8vo, Cloth, $5 00.

LYMAN BEECHER'S AUTOBIOGRAPHY, &c. Autobiography, Correspondence, &c., of Lyman Beecher, D.D. Edited by his Son, CHARLES BEECHER. With Three Steel Portraits, and Engravings on Wood. In 2 vols., 12mo, Cloth, $5 00.

BOSWELL'S JOHNSON. The Life of Samuel Johnson, LL.D. Including a Journey to the Hebrides. By JAMES BOSWELL, Esq. A New Edition, with numerous Additions and Notes. By JOHN WILSON CROKER, LL.D., F.R.S. Portrait of Boswell. 2 vols., 8vo, Cloth, $4 00.

SARA COLERIDGE'S MEMOIR AND LETTERS. Memoir and Letters of Sara Coleridge. Edited by her Daughter. With Two Portraits on Steel. Crown 8vo, Cloth, $2 50.

SHAKSPEARE. The Dramatic Works of William Shakspeare, with the Corrections and Illustrations of Dr. JOHNSON G. STEEVENS, and others. Revised by ISAAC REED. Engravings. 6 vols., Royal 12mo, Cloth, $9 00.

4 Harper & Brothers' Valuable and Interesting Works.

DRAPER'S CIVIL WAR. History of the American Civil War. By JOHN W. DRAPER, M.D., LL.D., Professor of Chemistry and Physiology in the University of New York. In Three Vols. 8vo, Cloth, $3 50 per vol.

DRAPER'S INTELLECTUAL DEVELOPMENT OF EUROPE. A History of the Intellectual Development of Europe. By JOHN W. DRAPER, M.D., LL.D., Professor of Chemistry and Physiology in the University of New York. 8vo, Cloth, $5 00.

DRAPER'S AMERICAN CIVIL POLICY. Thoughts on the Future Civil Policy of America. By JOHN W. DRAPER, M.D., LL.D., Professor of Chemistry and Physiology in the University of New York. Crown 8vo, Cloth, $2 50.

DU CHAILLU'S AFRICA. Explorations and Adventures in Equatorial Africa with Accounts of the Manners and Customs of the People, and of the Chase of the Gorilla, the Crocodile, Leopard, Elephant, Hippopotamus, and other Animals. By PAUL B. DU CHAILLU. Numerous Illustrations. 8vo, Cloth, $5 00.

BELLOWS'S OLD WORLD. The Old World in its New Face: Impressions of Europe in 1867-1868. By HENRY W. BELLOWS. 2 vols., 12mo, Cloth, $3 50.

BRODHEAD'S HISTORY OF NEW YORK. History of the State of New York. By JOHN ROMEYN BRODHEAD. 1609-1691. 2 vols. 8vo, Cloth, $3 00 per vol.

BROUGHAM'S AUTOBIOGRAPHY. Life and Times of HENRY, LORD BROUGHAM. Written by Himself. In Three Volumes. 12mo, Cloth, $2 00 per vol.

BULWER'S PROSE WORKS. Miscellaneous Prose Works of Edward Bulwer, Lord Lytton. 2 vols., 12mo, Cloth, $3 50.

BULWER'S HORACE. The Odes and Epodes of Horace. A Metrical Translation into English. With Introduction and Commentaries. By LORD LYTTON. With Latin Text from the Editions of Orelli, Macleane, and Yonge. 12mo, Cloth, $1 75.

BULWER'S KING ARTHUR. A Poem. By EARL LYTTON. New Edition. 12mo, Cloth, $1 75.

BURNS'S LIFE AND WORKS. The Life and Works of Robert Burns. Edited by ROBERT CHAMBERS. 4 vols., 12mo, Cloth, $6 00.

REINDEER, DOGS, AND SNOW-SHOES. A Journal of Siberian Travel and Explorations made in the Years 1865-'67. By RICHARD J. BUSH, late of the Russo-American Telegraph Expedition. Illustrated. Crown 8vo, Cloth, $3 00.

CARLYLE'S FREDERICK THE GREAT. History of Friedrich II., called Frederick the Great. By THOMAS CARLYLE. Portraits, Maps, Plans, &c. 6 vols., 12mo, Cloth, $12 00.

CARLYLE'S FRENCH REVOLUTION. History of the French Revolution. Newly Revised by the Author, with Index, &c. 2 vols., 12mo, Cloth, $3 50.

CARLYLE'S OLIVER CROMWELL. Letters and Speeches of Oliver Cromwell. With Elucidations and Connecting Narrative. 2 vols., 12mo, Cloth, $3 50.

CHALMERS'S POSTHUMOUS WORKS. The Posthumous Works of Dr. Chalmers. Edited by his Son-in-Law, Rev. WILLIAM HANNA, LL.D. Complete in 9 vols., 12mo, Cloth, $13 50.

COLERIDGE'S COMPLETE WORKS. The Complete Works of Samuel Taylor Coleridge. With an Introductory Essay upon his Philosophical and Theological Opinions. Edited by Professor SHEDD. Complete in Seven Vols. With a fine Portrait. Small 8vo, Cloth, $10 50.

DOOLITTLE'S CHINA. Social Life of the Chinese: with some Account of their Religious, Governmental, Educational, and Business Customs and Opinions. With special but not exclusive Reference to Fuhchau. By Rev. JUSTUS DOOLITTLE, Fourteen Years Member of the Fuhchau Mission of the American Board. Illustrated with more than 150 characteristic Engravings on Wood. 2 vols., 12mo, Cloth, $5 00.

GIBBON'S ROME. History of the Decline and Fall of the Roman Empire. By EDWARD GIBBON. With Notes by Rev. H. H. MILMAN and M. GUIZOT. A new cheap Edition. To which is added a complete Index of the whole Work, and a Portrait of the Author. 6 vols., 12mo, Cloth, $9 00.

HAZEN'S SCHOOL AND ARMY IN GERMANY AND FRANCE. The School and the Army in Germany and France, with a Diary of Siege Life at Versailles. By Brevet Major-General W. B. HAZEN. U.S.A., Colonel Sixth Infantry. Crown 8vo, Cloth, $2 50.

TYERMAN'S WESLEY. The Life and Times of the Rev. John Wesley, M.A., Founder of the Methodists. By the Rev. LUKE TYERMAN, Author of "The Life of Rev. Samuel Wesley." Portraits. 3 vols., Crown 8vo, Cloth, $7 50.

VÁMBÉRY'S CENTRAL ASIA. Travels in Central Asia. Being the Account of a Journey from Teheran across the Turkoman Desert, on the Eastern Shore of the Caspian, to Khiva, Bokhara, and Samarcand, performed in the Year 1863. By ARMINIUS VÁMBÉRY, Member of the Hungarian Academy of Pesth, by whom he was sent on this Scientific Mission. With Map and Woodcuts. 8vo, Cloth, $4 50.

THOMSON'S LAND AND THE BOOK. The Land and the Book; or, Biblical Illustrations drawn from the Manners and Customs, the Scenes and the Scenery of the Holy Land. By W. M. THOMSON, D.D., Twenty-five Years a Missionary of the A. B. C. F. M. in Syria and Palestine. With two elaborate Maps of Palestine, an accurate Plan of Jerusalem, and several hundred Engravings, representing the Scenery, Topography, and Productions of the Holy Land, and the Costumes, Manners, and Habits of the People. 2 large 12mo vols., Cloth, $5 00.

DAVIS'S CARTHAGE. Carthage and her Remains: being an Account of the Excavations and Researches on the Site of the Phœnician Metropolis in Africa and other adjacent Places. Conducted under the Auspices of Her Majesty's Government. By Dr. DAVIS, F.R.G.S. Profusely Illustrated with Maps, Woodcuts, Chromo-Lithographs, &c. 8vo, Cloth, $4 00.

EDGEWORTH'S (MISS) NOVELS. With Engravings. 10 vols., 12mo, Cloth, $15 00.

GROTE'S HISTORY OF GREECE. 12 vols., 12mo, Cloth, $18 00.

HELPS'S SPANISH CONQUEST. The Spanish Conquest in America, and its Relation to the History of Slavery and to the Government of Colonies. By ARTHUR HELPS. 4 vols., 12mo, Cloth, $6 00.

HALE'S (MRS.) WOMAN'S RECORD. Woman's Record; or, Biographical Sketches of all Distinguished Women, from the Creation to the Present Time. Arranged in Four Eras, with Selections from Female Writers of each Era. By Mrs. SARAH JOSEPHA HALE. Illustrated with more than 200 Portraits. 8vo, Cloth, $5 00.

HALL'S ARCTIC RESEARCHES. Arctic Researches and Life among the Esquimaux: being the Narrative of an Expedition in Search of Sir John Franklin, in the Years 1860, 1861, and 1862. By CHARLES FRANCIS HALL. With Maps and 100 Illustrations. The Illustrations are from Original Drawings by Charles Parsons, Henry L. Stephens, Solomon Eytinge, W. S. L. Jewett, and Granville Perkins, after Sketches by Captain Hall. 8vo, Cloth, $5 00.

HALLAM'S CONSTITUTIONAL HISTORY OF ENGLAND, from the Accession of Henry VII. to the Death of George II. 8vo, Cloth, $2 00.

HALLAM'S LITERATURE. Introduction to the Literature of Europe during the Fifteenth, Sixteenth, and Seventeenth Centuries. By HENRY HALLAM. 2 vols., 8vo, Cloth, $4 00.

HALLAM'S MIDDLE AGES. State of Europe during the Middle Ages. By HENRY HALLAM. 8vo, Cloth, $2 00.

HILDRETH'S HISTORY OF THE UNITED STATES. FIRST SERIES: From the First Settlement of the Country to the Adoption of the Federal Constitution. SECOND SERIES: From the Adoption of the Federal Constitution to the End of the Sixteenth Congress. 6 vols., 8vo, Cloth, $18 00.

HUME'S HISTORY OF ENGLAND. History of England, from the Invasion of Julius Cæsar to the Abdication of James II., 1688. By DAVID HUME. A new Edition, with the Author's last Corrections and Improvements. To which is Prefixed a short Account of his Life, written by Himself. With a Portrait of the Author. 6 vols., 12mo, Cloth, $9 00.

JAY'S WORKS. Complete Works of Rev. William Jay: comprising his Sermons, Family Discourses, Morning and Evening Exercises for every Day in the Year, Family Prayers, &c. Author's enlarged Edition, revised. 3 vols., 8vo, Cloth, $6 00.

JEFFERSON'S DOMESTIC LIFE. The Domestic Life of Thomas Jefferson: compiled from Family Letters and Reminiscences by his Great-Granddaughter, SARAH N. RANDOLPH. With Illustrations. Crown 8vo, Illuminated Cloth, Beveled Edges, $2 50.

JOHNSON'S COMPLETE WORKS. The Works of Samuel Johnson, LL.D. With an Essay on his Life and Genius, by ARTHUR MURPHY, Esq. Portrait of Johnson. 2 vols., 8vo, Cloth, $4 00.

KINGLAKE'S CRIMEAN WAR. The Invasion of the Crimea, and an Account of its Progress down to the Death of Lord Raglan. By ALEXANDER WILLIAM KINGLAKE. With Maps and Plans. Two Vols. ready. 12mo, Cloth, $2 00 per vol.

KINGSLEY'S WEST INDIES. At Last: A Christmas in the West Indies. By CHARLES KINGSLEY. Illustrated. 12mo, Cloth, $1 50.

SPEKE'S AFRICA. Journal of the Discovery of the Source of the Nile. By Captain JOHN HANNING SPEKE, Captain H.M. Indian Army, Fellow and Gold Medalist of the Royal Geographical Society, Hon. Corresponding Member and Gold Medalist of the French Geographical Society, &c. With Maps and Portraits and numerous Illustrations, chiefly from Drawings by Captain GRANT. 8vo, Cloth, uniform with Livingstone, Barth, Burton, &c., $4 00.

STRICKLAND'S (MISS) QUEENS OF SCOTLAND. Lives of the Queens of Scotland and English Princesses connected with the Regal Succession of Great Britain. By AGNES STRICKLAND. 8 vols., 12mo, Cloth, $12 00.

KRUMMACHER'S DAVID, KING OF ISRAEL. David, the King of Israel: a Portrait drawn from Bible History and the Book of Psalms. By FREDERICK WILLIAM KRUMMACHER, D.D., Author of "Elijah the Tishbite," &c. Translated under the express Sanction of the Author by the Rev. M. G. EASTON, M.A. With a Letter from Dr. Krummacher to his American Readers, and a Portrait. 12mo, Cloth, $1 75.

LAMB'S COMPLETE WORKS. The Works of Charles Lamb. Comprising his Letters, Poems, Essays of Elia, Essays upon Shakspeare, Hogarth, &c., and a Sketch of his Life, with the Final Memorials, by T. NOON TALFOURD. Portrait. 2 vols., 12mo, Cloth, $3 00.

LIVINGSTONE'S SOUTH AFRICA. Missionary Travels and Researches in South Africa; including a Sketch of Sixteen Years' Residence in the Interior of Africa, and a Journey from the Cape of Good Hope to Loando on the West Coast; thence across the Continent, down the River Zambesi, to the Eastern Ocean. By DAVID LIVINGSTONE, LL.D., D.C.L. With Portrait, Maps by Arrowsmith, and numerous Illustrations. 8vo, Cloth, $4 50.

LIVINGSTONES' ZAMBESI. Narrative of an Expedition to the Zambesi and its Tributaries, and of the Discovery of the Lakes Shirwa and Nyassa. 1858-1864. By DAVID and CHARLES LIVINGSTONE. With Map and Illustrations. 8vo, Cloth, $5 00.

M'CLINTOCK & STRONG'S CYCLOPÆDIA. Cyclopædia of Biblical, Theological, and Ecclesiastical Literature. Prepared by the Rev. JOHN M'CLINTOCK, D.D., and JAMES STRONG, S.T.D. 5 vols. now ready. Royal 8vo. Price per vol., Cloth, $5 00; Sheep, $6 00; Half Morocco, $8 00.

MARCY'S ARMY LIFE ON THE BORDER. Thirty Years of Army Life on the Border. Comprising Descriptions of the Indian Nomads of the Plains; Explorations of New Territory; a Trip across the Rocky Mountains in the Winter; Descriptions of the Habits of Different Animals found in the West, and the Methods of Hunting them; with Incidents in the Life of Different Frontier Men, &c., &c. By Brevet Brigadier-General R. B. MARCY, U.S.A., Author of "The Prairie Traveller." With numerous Illustrations. 8vo, Cloth, Beveled Edges, $3 00.

MACAULAY'S HISTORY OF ENGLAND. The History of England from the Accession of James II. By THOMAS BABINGTON MACAULAY. With an Original Portrait of the Author. 5 vols., 8vo, Cloth, $10 00; 12mo, Cloth, $7 50.

MOSHEIM'S ECCLESIASTICAL HISTORY, Ancient and Modern; in which the Rise, Progress, and Variation of Church Power are considered in their Connection with the State of Learning and Philosophy, and the Political History of Europe during that Period. Translated, with Notes, &c., by A. MACLAINE, D.D. A new Edition, continued to 1826, by C. COOTE, LL.D. 2 vols., 8vo, Cloth, $4 00.

NEVIUS'S CHINA. China and the Chinese: a General Description of the Country and its Inhabitants; its Civilization and Form of Government; its Religious and Social Institutions; its Intercourse with other Nations; and its Present Condition and Prospects. By the Rev. JOHN L. NEVIUS, Ten Years a Missionary in China. With a Map and Illustrations. 12mo, Cloth, $1 75.

THE DESERT OF THE EXODUS. Journeys on Foot in the Wilderness of the Forty Years' Wanderings; undertaken in connection with the Ordnance Survey of Sinai and the Palestine Exploration Fund. By E. H. PALMER, M.A., Lord Almoner's Professor of Arabic, and Fellow of St. John's College, Cambridge. With Maps and numerous Illustrations from Photographs and Drawings taken on the spot by the Sinai Survey Expedition and C. F. Tyrwhitt Drake. Crown 8vo, Cloth, $3 00.

OLIPHANT'S CHINA AND JAPAN. Narrative of the Earl of Elgin's Mission to China and Japan, in the Years 1857, '58, '59. By LAURENCE OLIPHANT, Private Secretary to Lord Elgin. Illustrations. 8vo, Cloth, $3 50.

OLIPHANT'S (MRS.) LIFE OF EDWARD IRVING. The Life of Edward Irving, Minister of the National Scotch Church, London. Illustrated by his Journals and Correspondence. By Mrs. OLIPHANT. Portrait. 8vo, Cloth, $3 50.

RAWLINSON'S MANUAL OF ANCIENT HISTORY. A Manual of Ancient History, from the Earliest Times to the Fall of the Western Empire. Comprising the History of Chaldæa, Assyria, Media, Babylonia, Lydia, Phœnicia, Syria, Judæa, Egypt, Carthage, Persia, Greece, Macedonia, Parthia, and Rome. By GEORGE RAWLINSON, M.A., Camden Professor of Ancient History in the University of Oxford. 12mo, Cloth, $2 50.

SMILES'S LIFE OF THE STEPHENSONS. The Life of George Stephenson, and of his Son, Robert Stephenson; comprising, also, a History of the Invention and Introduction of the Railway Locomotive. By SAMUEL SMILES, Author of "Self-Help," &c. With Steel Portraits and numerous Illustrations. 8vo, Cloth, $3 00.

SMILES'S HISTORY OF THE HUGUENOTS. The Huguenots: their Settlements, Churches, and Industries in England and Ireland. By SAMUEL SMILES. With an Appendix relating to the Huguenots in America. Crown 8vo, Cloth, $2 00.

www.ingramcontent.com/pod-product-compliance
Lightning Source LLC
Chambersburg PA
CBHW020833020526
44114CB00040B/600